Birds, Scythes and Combines

Birds, Scythes and Combines provides a historical perspective to changes in farmland bird populations in Britain over the past 250 years. Despite the scale of change in habitats and agricultural methods in the late eighteenth and nineteenth centuries due to enclosure and the spread of high farming, early avifaunas show that farmland birds were little affected. Specialised species of fen and marsh were lost, often as much due to persecution as to habitat destruction, but farmland birds benefited from the appearance of extensive new resources, which aided their adaptation to the altered habitats created by the new farming methods. In addition, many old permanent grass habitats were little altered, leaving a major reservoir of important habitats unchanged. In contrast, modern farming methods, with changes in grassland management, in herbicide use and in harvesting methods particularly, have led to a collapse in the diversity of farmland and to a consequent steep decline in the population size of a high proportion of the farmland birds we see today.

MICHAEL SHRUBB is well placed to write this book, as a retired farmer with a lifelong interest in birds (particularly British farmland birds), which he has been studying since the late 1950s. He also has extensive knowledge of birds in Europe. He has served on the Councils and Committees of both the Royal Society for the Protection of Birds and the British Trust for Ornithology and organised the national Lapwing Survey for the latter in 1987. He has held a number of posts in county and regional ornithological societies, including being Recorder for Sussex from 1966 to 1972 and Editor of *Welsh Birds* from 1991 to 2001. He is author of *Birds of Sussex: Their Present Status* (1979) and *The Kestrel* (1993) and co-author (with R. J. O'Connor) of *Farming and Birds* (1986). He has published a number of papers in the ornithological literature on farming and birds, on Lapwings and on Kestrels.

Birds, Scythes and Combines

A history of birds and agricultural change

MICHAEL SHRUBB

CAMBRIDGE
UNIVERSITY PRESS

CAMBRIDGE
UNIVERSITY PRESS

University Printing House, Cambridge CB2 8BS, United Kingdom

Cambridge University Press is part of the University of Cambridge.

It furthers the University's mission by disseminating knowledge in the pursuit of education, learning and research at the highest international levels of excellence.

www.cambridge.org
Information on this title: www.cambridge.org/9780521814638

© Michael Shrubb 2003

This publication is in copyright. Subject to statutory exception and to the provisions of relevant collective licensing agreements, no reproduction of any part may take place without the written permission of Cambridge University Press.

First published 2003
First paperback edition 2012

A catalogue record for this publication is available from the British Library

Library of Congress Cataloguing in Publication data
Shrubb, Michael.
Birds, scythes, and combines : a history of birds and agricultural change /
by Michael Shrubb.
p. cm.
Includes bibliographical references (p.).
ISBN 0 521 81463 4
1. Agriculture – Great Britain – History. 2. Agricultural ecology – Great Britain – History.
3. Bird populations – Great Britain – History. 4. Birds – Habitat – Great Britain – History.
5. Birds, Protection of – Great Britain – History. I. Title.
S455 .S584 2003
577.5'5'0941 – dc21   2002074065

ISBN 978-0-521-81463-8 Hardback

Cambridge University Press has no responsibility for the persistence or accuracy of URLs for external or third-party internet websites referred to in this publication, and does not guarantee that any content on such websites is, or will remain, accurate or appropriate.

Contents

	Introduction	vii
	Acknowledgements	xi
1	The agricultural background	1
2	The farmland birds	18
3	Arable farming systems: high farming and before	43
4	Enclosure	66
5	Some thoughts on hedges	114
6	Drainage	127
7	Weeds, weeding and pesticides	156
8	Arable farming systems: after 1945	183
9	Grassland and stock	200
10	Winter food resources	252
11	Labour, machines and buildings	278
12	Exploitation and persecution	306
13	Conclusions	322
	Appendix 1 Estimating areas of important traditional feeding sites for seed-eating birds in British farmland	329
	Appendix 2 Scientific names of birds	333

Bibliography of avifaunas 336
References 341
Index 363

Introduction

This book aims to provide some historical perspective to continuing change in agriculture and its effects on bird populations. The area I have covered is the whole of Britain except the Isle of Man. I have differentiated between countries where important differences arose, for example Parliamentary Enclosure in England and Wales but not Scotland. Otherwise my observations should be read in a British context. If I have quoted English examples most frequently it is because I know England best.

I sense the need for such a perspective. The late twentieth century has seen major changes in the ecology of farmland of which a striking manifestation has been declining populations of many common farmland birds. This did not happen in the nineteenth century, despite the scale of habitat change. This is relevant to understanding many of the changes which have occurred in the modern era and ornithologists now studying these problems will increasingly lack direct knowledge of the impact of farming changes in the nineteenth and the first half of the twentieth centuries on bird populations and habitats, as these recede in time. I doubt, too, if modern bird-watchers have any real conception of what we have lost from our farmlands, even in my time. As a youngster starting birding in the late 1940s, I found the family farm swarming in winter with plovers, snipe, wildfowl, finches and buntings and larks. Three pairs of Kestrels bred, two to three pairs of Barn Owls, seven pairs of Little Owls, 12 pairs of Redshank, whilst breeding Turtle Doves, Lapwings, Grey Partridges, Corn Buntings etc. were abundant. Little of this remains today. I still find it chastening and difficult to comprehend how lifeless the farm has been rendered by modern farming, compared to how I first remember it. The collapse in numbers started in the mid 1970s. It is possible to argue convincingly that right up to 1970 the majority of farmland birds exhibited a strong trend towards long-term population stability. Even the impact of organochlorine seed dressings and insecticides in the 1950s did not undermine that, for populations readily recovered once these substances were removed. That situation has altered sharply in the last quarter of the twentieth century, because a small group of changes, unique to the modern era, have combined to cause a collapse in the diversity of modern farmland.

I have been observing the interactions between farming and birds since the late 1950s, mainly on our family farm, Oakhurst, extending to some 600 acres (243 ha) on the Selsey Peninsula in West Sussex, but latterly also in an area of pastoral farming in north Breconshire. When I first started helping on the family holding as a schoolboy, it was a true mixed farm, managed on a straightforward high-farming system (see Chapter 3). We grew wheat, barley, oats, clover, turnips, swedes, mangolds, sugar beet, often linseed and occasionally beans. The holding consists of two farms, so two lambing flocks were kept and yearling lambs were also bought in annually, to graze the large area of roots. A large area of old damp permanent pasture was summer-grazed by beef cattle, finished in yards during the winter. Power was provided by a mixture of horses and tractors. The corn was all cut with the binder, ricked and progressively threshed through the late autumn and winter. We bought our first sprayer in 1952 and stopped using horses about the same time. As with many farmers' sons of my generation, therefore, I grew up with the farming system developed in the late eighteenth and early nineteenth centuries, which persisted through the Victorian era and until after 1945.

Progressively over the next 40 years the system we used was changed and simplified, until today Oakhurst has no livestock and grows only winter cereals and peas. Much of the grass is in permanent set-aside; the remainder was ploughed or sold away. I therefore worked through the entire gamut of the modern agricultural revolution. This gave me an unrivalled opportunity to observe the impact of a major revolution in farm management on birds at first hand and my study area extended also over five neighbouring farms, where I had unfettered access. Some of the resulting observations have been published in various journals and reports, more were incorporated in *Farming and Birds*, which Raymond O'Connor and I wrote in the 1980s, and some are newly presented here.

I have taken the discussion back to 1750, although useful ornithological material gets increasingly difficult to find beyond the early nineteenth century. In assessing their effects on birds, therefore, one tends to be limited largely to examining farming changes that have occurred since 1800, when adequate information on the status and distribution of birds starts to emerge. I have concentrated in this account upon what Thirsk (1997) termed mainstream agriculture, the production of basic commodities such as grain and meat. The alternative crops Thirsk described were limited in area, designed to supplement income from mainstream activities rather than replace them. However, as she shows, their impact on the subsequent thrust of mainstream agriculture has been of great significance. Methods and systems of cropping and livestock management developed in times of recession, often on the margins of farming, have been widely adopted in mainstream agriculture with the return of prosperity in both the eighteenth and twentieth centuries. This process is likely to be repeated.

Besides contemporary journals a major source has been the great series of regional and county avifaunas published in the second half of the nineteenth century, many of which included much valuable historic data and information about major habitat changes throughout the nineteenth century. Equally valuable have been the series of national avifaunas published during the century, particularly Montagu, McGillivray and Yarrell. The tradition of the county avifaunas has continued to the present and has provided a further mine of information, as have county bird reports. The modern literature on farmland birds is voluminous. I have made a particular analysis of about 45 species, which are especially typical of agricultural habitats, to act as a suite of indicator species – harriers and other ground-feeding raptors, Hobby, bustards, gamebirds and Corncrake, waders, pigeons and doves, Barn and Little Owls, larks, wagtails, chats, corvids, sparrows, finches and buntings, Red-backed Shrike and Black-headed Gull. In Chapter 2 birds have been categorised as field or ground species etc. on the basis of my systematic counts in Sussex.

One important caveat about the sources must be made. Nineteenth-century ornithologists usually lacked the type of census data we take for granted today. Their assessments of status change rested to a much greater extent on knowledge of distribution and the opinions of local observers and correspondents. However the idea that nineteenth-century ornithologists overlooked significant changes in population, particularly declines, because these could have occurred without overall distribution changing, rests largely on modern methods of recording and mapping distribution by the 10-km square, with which this can certainly happen. But nineteenth-century ornithologists did not use this method. They depended on intimate knowledge of local areas, both of themselves and their correspondents, as I still do today. I do not say that Curlews or Skylarks are declining in my Welsh area because they have gone from certain

10-km squares but because I find fewer or none in various sites where they were numerous 10 years ago. The judgements of nineteenth-century ornithologists rested upon the sum of many such observations. Nor can one read men like Stevenson, Yarrell, the Ticehursts, Harvie-Brown, Bolam, Nelson, Lilford, Forrest and others without accepting what a profound understanding of their areas and birdlife they had. One handicap to recording they did have was the lack of good optical equipment. Some species were inadequately recorded as a result. Savi's Warbler was only found in Britain (by ornithologists) in 1840; it was extinct within 20 years. The harriers were not properly separated until the 1850s at least, Willow Tit wasn't found before 1900 and I believe that the Cirl Bunting was similarly overlooked, with ornithologists working its distribution out once it was realised how widespread it was, rather than recording its colonisation and spread. For the bulk of common farmland birds these problems did not arise.

There is an extensive agricultural literature for the period. We are fortunate in Britain in having a long series of official statistics, the June Census Statistics gathered annually by the Agricultural Departments since 1866, to provide basic information on the extent of crops and numbers of stock. They are not perfect and it is often difficult to follow changes through long periods because categories are changed or crops added or dropped and so on. I know of no reason, however, for supposing that the basic patterns they show are unacceptable. Crop areas are given in the units of measurement in which they will be found in the sources I have used, with metric equivalents where necessary.

I have drawn largely upon the long series of county accounts of farming published in the *Journal of the Royal Agricultural Society of England* during the middle years of the nineteenth century and the similar series for Scotland in the *Transactions of the Highland and Agricultural Society*. The same journals published many detailed papers on the development and use of important machinery, on new methods, on drainage, on hedge management and so on. I have made much use of standard histories, particularly Ernle, who drew extensively and quoted copiously from the eighteenth and early nineteenth-century Reports to the Board of Agriculture, a source I have also consulted. Cobbett, despite his turgid political commentaries, is also a useful source of information and often included observations about birds. Finally major sources of information have been the journals *Agricultural History Review* and *Economic History Review*. A full list of references is included.

Acknowledgements

I am grateful to many people for their help with this book. Dr. John Evans, formerly of the Agricultural Development and Advisory Service (ADAS), Roy Leverton and Martin Peers have each read substantial parts of the manuscript and Dr. Jeremy Greenwood, Director of the British Trust for Ornithology (BTO) has read the whole. I am grateful to all of them for their many helpful suggestions, criticisms and corrections, which greatly improved the manuscript. Errors and omissions are, of course, my own.

I owe a considerable debt of gratitude to the Librarian at the BTO, Carole Showell, who has been indefatigable in finding and obtaining for me numerous increasingly obscure nineteenth-century references. Her help, and that of her predecessor Philip Jackson, has been essential to an author lurking in Welsh hills. The Librarian at the Royal Society for the Protection of Birds (RSPB) also obtained some important references for me.

The BTO gave permission to use previously unpublished data on field boundaries, gathered as an adjunct of the 1987 Lapwing Survey, and to reproduce the Woodpigeon data of Figure 10.2 and the maps of Figure 10.3. Many friends on the BTO's staff helped greatly with fruitful discussions, access to their own unpublished work and by steering me towards references I had overlooked. Cambridge University Press gave permission to reproduce the drawing of Figure 4.5 and the Royal Agricultural Society of England allowed the use of the drawings of Figure 5.1. The bulk of the photographs have been drawn from the splendid archive held at the Museum of Rural Life at Reading University. I am very grateful to the Archivist of the collection, Caroline Benson, for her care and help in assembling them, and to John Evans for steering me towards this source in the first place.

1
The agricultural background

I define farmland simply as land that is used for agriculture. Most open country in Britain is so used to some extent and therefore qualifies. The habitat elements comprising farmland are divided into 'improved farmland', comprising the area identified in the June Census of Agriculture as 'crops and grass' – tillage crops, plus ley, plus agriculturally improved permanent grass – and 'semi-natural' habitats – hedges and ditches, woodland and copses on farm holdings and unimproved (rough) grazing and moorland. Such semi-natural habitats are integral to farmland; rough grazings are an essential part of upland sheep-farming systems, for example, and small woodlands were once essential features of lowland farm economies (Rackham 1986) and may remain important for field sports. Although semi-natural habitats are often the most significant for specialised species, birds make extensive and important use of improved farmland, particularly in exploiting food resources.

I have started this account somewhat arbitrarily at 1750. Modern agriculture can be fairly confidently dated from around this date. High farming (defined on p. 6), beginning around the mid eighteenth century, was the first modern farming system, in the sense that land was then increasingly adapted to suit the farming system rather than land use adapted to the nature of the land (Chapter 3). The impetus for change was provided by the Industrial Revolution and a rapid increase in population, which nearly doubled over the century (Trevelyan 1944). This rise in population, coupled with the emergence of increasingly large urban manufacturing communities divorced from the land, created a need for large-scale commercial farming, which high farming supplied. Ernle (1922) noted that, by the second half of the eighteenth century, trade in agricultural produce was rapidly becoming wholesale rather than retail. This is a good indication of a switch to a commercial agricultural economy rather than a peasant one.

Britain has a varied geology in a comparatively small area, with much variation in soil types and fertility even at a local level. For agriculture soils are placed in five grades of excellence, grade 1 being the best. This grading system considers both inherent fertility and practical farming factors such as good natural drainage and free working characteristics. Thus light loams may be graded higher than heavier clays for the latter reasons, although inherent fertility may favour the clays. Soils also vary in acidity although acidity may have more impact on natural vegetation than farming, because the farmer can vary it by applying lime. Liming was one of the crucial steps to improving agriculture in the more acid soils of northern and western Britain (Smout 2000). In southern England chalk dug from the Downs was often used instead. This underlines an important point – the extent to which agriculture has altered soil characteristics by applying fertilising agents. A classic example is provided by the Holkham estate in Norfolk, an area once primarily sand and heath. The increase in fertility wrought by Coke through rotation farming and extensive marling (digging out the subsoil, especially chalky clays, from pits and trenches and spreading and mixing it into the topsoil to improve fertility and soil structure) is measured by the ten-fold rise of his rent roll in about 20 years. Looking at the area today it is difficult to imagine its eighteenth-century past. Such patterns were widely repeated in similar soils and probably affected historic plant distributions. Seeds migrate with operations such as marling and chalking, which may also suppress local plants. Wilkinson (1861) noted that chalking the sand and gravel soils of the Stratfield Saye estate in Hampshire led to the loss of wild chamomile there and on our family farm dressing fields with chalk often resulted in infestations of charlock (*Sinapsis*).

Climate also has an important impact on farming. Thus Manley (1952) suggested that the agricultural depression between 1816 and 1840 was partly caused by a succession of wet and difficult summers and poor harvests. In the same way the great agricultural depression of the late nineteenth century was partly triggered by the difficulties of managing complex rotations in the very wet years of the late 1870s. Similar difficult weather conditions for agriculture were frequent in the second half of the eighteenth century as well. Ernle (1922) discussed in detail the poor state of much arable farming in England then, especially in the open-field parishes, from contemporary sources. These ascribed the difficulties to communal management in open fields being set at the standard of the worst farmer involved. But nutrients may have been deficient, particularly phosphates (see Newman & Harvey 1997 for a discussion of this problem in the mediaeval period), as they were in grassland (Chapter 9). Smout (2000) suggests that nutrient defiency was a serious contemporary problem in Scotland. The physical problems of adequately cultivating the heavy soils typical of the open-field parishes in poor climatic conditions may also often have been beyond the capacity of eighteenth-century farmers. Chew (1953) observed a similar situation on heavy clayland farms in east Leicestershire in the 1930s. Much early high farming was developed on light, warm and free-draining soils, easier to cultivate than clays in difficult climatic conditions. In view of the need to increase food production the drive to enclose the heaths and commons, large areas of which embraced such soils and were best suited to high farming at this period, had a convincing basis in genuine necessity. It did not require the spurious sounding arguments with which contemporary authors so often surrounded it.

Periods of agricultural change have always changed the appearance of the countryside and often the social structure of rural communities. Like any economic activity, farming is subject to cycles of prosperity and recession. Prosperity promotes surplus, surplus depresses prices, particularly for basic farm products such as grain, and triggers recession. That, in turn, also encourages the search for diversity in crops, methods or land uses, a pattern equally discernible in the sixteenth and seventeenth centuries and today (Thirsk 1985, 1997). Prosperity engendered by high prices for a rather narrow range of farm products encourages the conversion of non-farming land to agriculture (Rackham 1986), whereas recession causes abandonment.

Three broad trends are discernible in the prosperity of farming between the mid eighteenth century and today, with marked long-term changes from prosperity to recession after about 1875 and back to prosperity from 1947. From about 1987 there was what may be seen as a transition period, when politicians have sought to reduce agricultural production whilst maintaining rural prosperity, leading to schemes such as 'set-aside' and the encouragement of industrial crops, such as linseed. These have proved quite beneficial to birds but short-lived. By the end of the 1990s that pattern appeared to be changing yet again. The sections that follow examine these periods in outline and set the scene for the detailed examination of changes in the following chapters.

Figure 1.1. John Worlidge's drill of the 1660s. Such early machines lacked an efficient means of regulating seed rates, a problem unsolved until 1839. (Courtesy of Rural History Centre, University of Reading.)

1750–1875

The period between 1750 and 1875 was broadly one of agricultural prosperity, although there was serious recession between 1816 and 1840. The whole period saw important technical innovations, major habitat changes and eventually the remarkably widespread adoption of a single arable system, high farming. The most significant technical innovations were the development of the seed-drill (Figures 1.1–1.3), threshing drum and winnowing machine and the mechanisation of the hay and corn harvests (Chapter 11). Of habitat changes, much the most significant was the Parliamentary enclosure of commons and the scale of drainage (Chapters 4 and 6). Generally, however, change was not rapidly implemented in the way we regard as typical today. Parliamentary enclosures, for example, were spread over more than a century. Communications were slower and the era lacked the central advisory organisations and colleges that today quickly and widely disseminate new methods. Nevertheless the Board of Agriculture did much to promote farming development in the late eighteenth and early nineteenth centuries, as did landowners and agricultural entrepreneurs such as Coke at Holkham, and the scale of change in this period perhaps influenced our landscape more than anything since.

Many of the methods used by high farming were originally developed in the Low Countries and were recognised and developed in Britain, for example by Lord Townshend at Raynham, Norfolk, before 1750, particularly the value of leys and the importance of drill cultivation (sowing crops in evenly spaced rows to facilitate hoeing) and careful weeding in improving

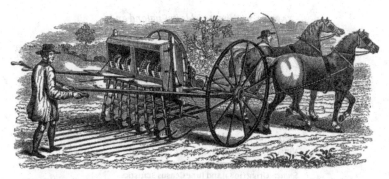

HENSMAN & SON'S
PATENT EIGHT ROW PAIR HORSE STEERAGE CORN & TURNIP DRILL,
TO WHICH
A PRIZE WAS AWARDED by the ROYAL AGRICULTURAL SOCIETY at YORK, 1848.

This Drill is adapted for all sorts of Corn and Seeds; THE HOPPER IS SELF-ACTING, and will deposit as regular up and down hill as on the flat; the axles slide; the Drill can be steered to the greatest possible nicety; and greatly facilitates the use of the Horse Hoe between the rows.

Figure 1.2. A mid-nineteenth-century drill. 'The sower with his seedlip has almost vanished from southern England...the beautiful system of horse-hoeing depends too, of course, entirely on the use of the drill' (Pusey 1851). (Courtesy of Rural History Centre, University of Reading.)

Figure 1.3. A Smythe Suffolk drill, in use in the 1950s. The most successful nineteenth-century design, used for over a century with little modification. Simple and virtually indestructible. (Courtesy of Rural History Centre, University of Reading.)

TABLE 1.1. *Wheat yields in England and Wales between 1700 and 1875*

Year	Average yield (tonnes/ha)	% increase
1700	1.08	
1750	1.15	6.5
1800	1.42	23.5
1850	1.89	33.1
1875	2.02	6.9

Source: Grigg (1989) and June Census Statistics.

yields (Lane 1980, Thirsk 1997). These were important principles of high farming. Earlier agricultural systems and rural economies made much use of the grazing and material resources available in marshes, fens and heaths (Chapter 4). Modern systems, starting with high farming, have instead tended to adapt soil and physical conditions, by techniques such as drainage and enclosure, to suit the farming. Thus there was a much greater variety of habitat in older agricultural systems, whereas almost a defining characteristic of modern systems is increasing uniformity (Chapter 8).

High farming is the term used throughout this book to cover the closely integrated rotations of mixed arable and stock farming of the late eighteenth, nineteenth and early twentieth centuries. Its classic rotation was the four-course Norfolk system, comprising roots, oats or barley undersown, ley, wheat. Its aim was to maintain and improve cereal yields by steadily building up the fertility of the soil. Although the Norfolk rotation has the greatest historical prominence, rotations were widely modified to suit local conditions. During the nineteenth century the basic English high-farming rotation settled around a system of three years cereals, one year roots and one year ley in arable districts (Orwin & Whetham 1964). In gazing districts two-year leys were usually preferred, giving a six-course system. However, the Victorian literature showed that a wide range of variations around this theme was practised. Wilkinson (1861), writing of Hampshire, also showed that a high level of management skill was involved and that rotations were often varied to meet particular weather conditions. Catch crops (inserted between the main courses of a rotation) were widely used. In Scotland, where good cultivable land was comparatively scarce, two- or three-year leys were always preferred and largely took the place of permanent grass.

The success of high farming is illustrated by the increase in wheat yields from 1700 (Table 1.1) and that success led to the system being very

TABLE 1.2. *The basic composition of arable rotations[a] in Britain in 1875*

	England	Wales[b]	Scotland
Arable area (×1000 ha)	5433	450	1376
Percentage in cereals[c]	49	49	40
Percentage in ley	19	35	41
Percentage in fodder roots[d]	15	8	13
Percentage in fallow[e]	4	2	2
Total percentage	87	94	96

[a] Excludes permanent crops such as orchards, now included in tillage.
[b] Includes Monmouthshire.
[c] Areas of wheat, barley and oats only.
[d] Comprises turnips, swedes, mangolds, fodder cabbage and rape and kohl rabi.
[e] Most fallow at 1 June would have been planted to roots later.
Source: June Census Statistics.

widely applied. In Scotland wheat was a much less important crop but oat yields showed a similar pattern (Smout 2000). By 1875 few counties had less than 40% of improved farmland in the standard arable rotations (Figure 1.4). Arable rotations were limited in the range of crops grown (Table 1.2), largely from the reluctance of landlords and their agents to allow crops other than cereals to be grown unless fed to stock, held to be essential to maintain fertility. Most leases also restricted the sale of hay and straw from farms, also obliging stock to be kept. This rigidity eventually undermined the system.

An important feature of the agricultural changes of this period was that they particularly concerned arable farming. There was major loss of semi-natural grassland with the enclosure of commons but the area we would classify today as (improved) permanent grass, which has always included such habitats as permanent pastures, flood meadows, water meadows and hay meadows, changed very much less, providing a core of stability in farmland habitats. Revolution in grassland management is a twentieth-century change (Chapter 9).

1875–1947

In the late 1870s high farming came under severe pressure. The decade included some extremely wet years, culminating in 1879 when neither hay nor grain harvest were gathered by many farmers and severe outbreaks of disease killed large numbers of sheep (*c.* 2 million in East Anglia alone). Bad crops cost as much to grow as good. To the economic difficulties of

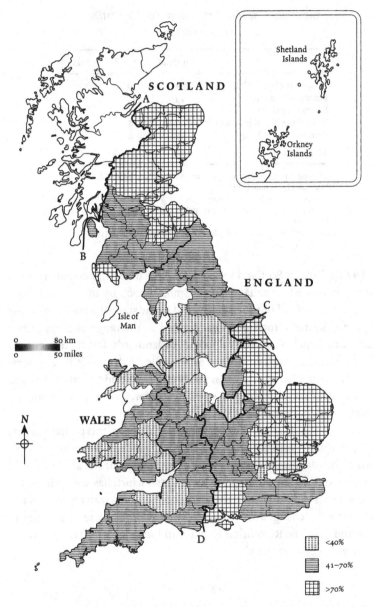

Figure 1.4. Arable land (tillage plus ley) as a percentage of improved farmland in Britain in 1875. The area northwest of the line A–B was excluded, being 90% rough grazings; 70% of improved farmland there was in arable rotations. The area southeast of line C–D comprised the nineteenth-century classification 'arable counties'. Derbyshire, Westmorland and Meirionnydd had less than 30% in arable rotations and are left blank. (Data from June Census Statistics.)

poor crops and dying stock were added the difficulties of keeping waterlogged fields clear of weeds by the constant cultivations required, which affected succeeding crops, particularly on the clay soils where poor natural drainage exacerbated the problems.

Ultimately, however, high farming was undermined by the repeal of the Corn Laws in 1846. These measures controlled cereal prices in favour of the home producer and they particularly affected the wheat market. High farming was developed and spread in an era of protected markets and it proved uneconomic when protection was removed. The impact of repeal was delayed by political circumstances. The Crimean War (1854–6) increased demand in the home market (Orwin & Whetham 1964), and the American Civil War (1861–5) slowed development of the wheat-growing areas of the American West, which later became Britain's main source of bread wheats. Furthermore shipping only developed sufficient capacity to enable bulk imports of grain to replace, rather than supplement, home production at the latter end of the nineteenth century. From the same period the burgeoning rail network and then motor road transport also developed enough bulk capacity to allow flour mills to be sited at ports to take the greatest economic advantage of imported grain. Such port mills accounted for 62% of all flour ground in Britain by 1912 and 76% by 1931 (Smith 1949).

Wheat became uneconomic to grow, with market prices remaining below break-even point (£8.70p per tonne) virtually throughout the period from 1880 to 1940, except during and just after the 1914–18 War; the lowest price recorded was £4.76p in 1934. Other prices followed the same trends (Grigg 1989). Nevertheless wheat prices were most significant, as the economics of high farming depended upon them to an unhealthy extent. To raise wheat yields a high proportion of the rotation's output was ploughed back into the system. In an era of low wheat prices that proportion became uneconomic. Arable farming also suffered from a crucial lack of mechanical power to raise productivity in an era of low prices (Chapter 11).

The most marked feature of the long decline in farming from the mid 1870s to the 1939–45 War was the steady loss of tillage and expansion of grassland of some sort as a proportion of total farmland (Chapter 9). The main impact was felt in the arable counties (Figure 1.4) and arable districts outside these, for example in Wiltshire. Livestock-based enterprises in the north and west benefited from cheaper grain, which reduced feed costs. They also benefited from a more entrepreneurial attitude, supplying local markets particularly with livestock products, and flexible enough to

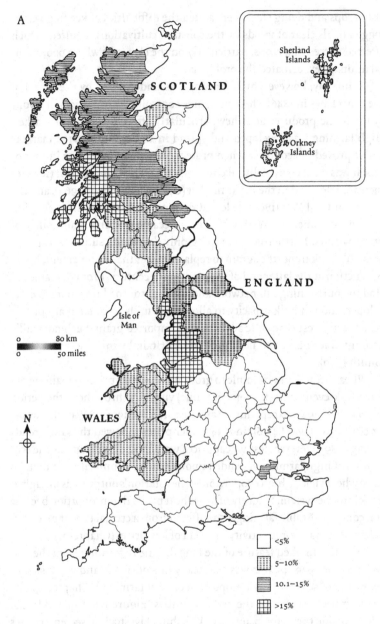

Figure 1.5. Potato and vegetable crops as a percentage of tillage in Britain in (A) 1875 and (B) 1930. (Data from June Census Statistics.)

B

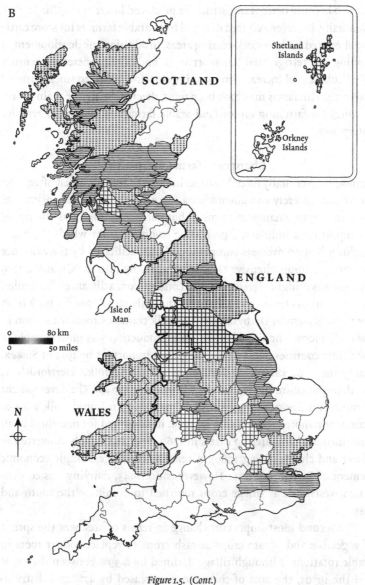

Figure 1.5. (Cont.)

react to changes in those markets, a flexibility which landlords in these areas understood and encouraged (Fletcher 1961). Scotland was markedly less affected than England for much the same reasons (wheat, the primary product of high farming, comprised only 2.4% of crops and grass in 1875) but also because better farm structure produced lower costs (Chapter 11). Eventually, however, reduced demand from arable farmers for store cattle and sheep fed back to stock-rearing areas. In addition the development of canning and refrigerated transport at the end of the nineteenth century led to substantial increases in meat imports and added to stock-farmers' problems. Hill farms may have been most affected, being the main source of stores for fattening on lowland arable farms and lacking alternative enterprises.

Changes in farm enterprises

Economics eventually forced changes in British farming, albeit often after farms were severely run down after a succession of failed tenancies. The most important change in terms of both area and value was the rise of the liquid milk industry, a perishable commodity for which the home producer had no overseas competitor. It was facilitated by the existence by 1870 of a comprehensive railway network. By 1910 the Great Western Railway was bringing fresh milk into London from a distance of 130 miles. Uttley (1931) has left a vivid account of the daily race to get the milk from her father's farm in Derbyshire to the milk train at Cromford station in the early 1900s. The rise of liquid milk production was most notable in the arable counties, where the change was important by 1913 in Sussex, Hampshire, Berkshire, Oxfordshire, Essex, Suffolk, Hertfordshire, Northamptonshire and Nottinghamshire (Smith 1949). The development of motor transport after 1918 assisted dairying for liquid milk sales to penetrate remoter regions and, by 1949, it accounted for one-third of all farm income, compared to *c.* 12% in 1860, and that largely converted to butter and cheese (Grigg 1989); it remains the largest single economic element of British farming. However, since 1945, dairying has concentrated westwards as tillage crops returned to dominate the south and east.

The second most important change in terms of area was the spread of vegetable and potato crops as cash crops to replace fodder roots in arable rotations. Although tillage declined by *c.* 39% between the 1870s and the 1930s, the area of these crops increased by 47% and distribution changed markedly (Figure 1.5). Similar increases occurred in the area of orchards and soft fruit, particularly in Kent, Sussex, Cambridgeshire, Herefordshire, Worcestershire, Angus, Perthshire, Fife and Lanarkshire. A new industry started in the early twentieth century, when sugar-beet growing in eastern England was encouraged by the State from 1926, as a way of preserving the arable farming industry there; by the 1930s 278 000 acres (112 955 ha) were being grown, mainly in East Anglia and the East Midlands, with other important areas in Shropshire and east Yorkshire. It was never an important crop in Scotland, with a maximum of 6500 ha in 1960.

Other changes also emerged in the late nineteenth and early twentieth centuries which, although comparatively insignificant in terms of area, were of considerable long-term consequence in farming's development. The practice of ensiling grass instead of making hay was introduced in 1871, although it was not until the 1960s that a way of ensuring uniform quality was finally worked out (Seddon 1989). A few farmers applied the research of workers such as J.B. Lawes and J.H. Gilbert at Rothamsted in the mid nineteenth century and developed successful enterprises based on continuous corn-growing, using manufactured fertilisers, and such all-tillage farms spread slowly during the 1930s. Other farmers developed enterprises selling hay for the urban horse market, particularly on the southern chalk and in central Scotland, growing timothy, taking two or three cuts annually and boosting yields with nitrogen fertilisers. Such businesses were, however, vulnerable to the decline of horse transport after 1918 and the loss of this market exacerbated the second phase of the agricultural recession in the 1920s and 1930s. The first steps towards chemical weed control were under way by 1911 and by the late 1930s up to 250 000 acres (100 000 ha) were being sprayed by herbicides annually (Murton 1971, Salisbury 1961).

1945–87

After 1945 farming expanded within a markedly changed political framework based on extensive State support. This was embodied in the 1947 Agriculture Act, which set up a system of price guarantees for crops and livestock, either as direct deficiency payments to make up prices to an agreed level, or in the form of acreage payments to support the growing of a crop. Price levels were set in an annual Price Review. The system combined the basic political aims of encouraging home food production and reducing food prices for industrial workers. As with most other aspects of the post-war recovery in farming, the principles of this policy were set well before the war. During 1914–18 the Corn Production Act of 1917 had introduced the idea of guaranteed prices for cereals, although it was repealed when world cereal prices slumped in 1921. The continuing and deepening depression of the 1920s and early 1930s, however, led the Government to reintroduce direct subsidies under the Wheat Act of 1931. State support had already encouraged the introduction of sugar beet as a major crop in the eastern counties in 1926 and Marketing Acts in 1931 and 1933 set up the Milk and Potato Marketing Boards, which controlled and guaranteed prices and production of milk or acted more as a compulsory co-operative for potatoes. Other marketing boards eventually covered wool, hops and eggs. All are now defunct. The British Sugar Corporation was set up in 1936 to exercise a monopoly in sugar-beet production; effectively it was another marketing board (Donaldson *et al.* 1969).

The 1947 Act also set up a system of grant aid for many capital investments on the farm, a principle which continued under European Economic Community (EEC) arrangements (see below). Grants were available for operations such as land drainage, ploughing old grassland, fencing, liming and erecting new buildings, a range similar to that eligible for State loans in the nineteenth century (Chapters 6 and 11). Undoubtedly the thinking behind this was to enable farmers to repair quickly the dilapidations caused by the long recession in farming. Whilst this seems a reasonable decision for 1947, in the context of the ruins to which Europe had been reduced by the 1939–45 War, by the 1960s there was no obvious reason why such grants should have continued. If price support was supposed to ensure profitable farming, capital grants should have been superfluous. They became actively pernicious for such capital grants grossly distort the process of making decisions about capital investment, which come to reflect, not the likely financial consequences of that investment, but what the State will pay. In the uplands the system of price support by headage payments has had similar effects. Decisions on stocking rates reflect not what the land will bear but State payments (Chapter 9).

Another major change in the framework in which farming has operated during the twentieth century has been the decline of estate tenancies in favour of owner-occupied farms. The percentage of holdings occupied by freeholders had fallen to just under 10% in England and Wales by the 1870s but then rose to 72% by 1987 (agricultural statistics). The pattern in Scotland was similar, with the proportion of freeholdings rising to 63% in the same period. These figures represent historical peaks in the proportion of freeholders and the proportion of land they hold. At the same time the number of holdings has fallen and average size increased. The main reason for this change was the declining prosperity of farming as an investment for the greatest change occurred during the second, and most severe, phase of the agricultural recession during the 1920s, when the proportion of farmland held by owner-occupiers rose from 10–12% to 40%. Many estates took advantage of the short spell of comparative prosperity during the period 1914–21 to reduce their exposure to farming.

However, State subsidies did not cause the modern agricultural revolution. The outlines of this appeared before 1939, many before 1914, in a classic demonstration of the principle so ably argued by Thirsk (1997). Undoubtedly farming would have developed along its present lines in any case and the main contribution of subsidies has been in the speed and scale of change. The modern recovery of agriculture can be broadly split into two phases. The period up to about 1965–70 was typified by mechanisation and the reorganisation of field systems. Pastoral farming was concentrating in the west and tillage in the east. Arable farming was largely based on a variant of classic rotation farming, the three-year ley system, three years of cereals, usually barley, alternating with three years of grass ley supporting a dairy herd. The major change in rotations was the decline of fodder roots, replaced by cereal feeds and more intensive management of grass for livestock. The true agricultural revolution of the modern period dated from the late 1960s and was largely based on developments in herbicides (Chapter 7). Other important changes occurred in pesticide usage or stemmed from changes in landownership.

When Britain joined the EEC in 1972 the system set up under the 1947 Act changed to the European system of intervention, a State-run market of last resort designed to maintain a floor beneath which prices were not allowed to fall. More recently State support throughout Europe has moved to a system closer to that set up in Britain under the 1947 Act. However, despite differences in the mechanisms of support, the basic principle that farming should be a State-subsidised industry has not varied. Even policies aimed at reducing production depend largely on State payments to farmers not to grow things. In Britain the scale of support and its impact increased with entry to the EEC, perhaps particularly in upland areas, which then achieved Less Favoured Area status and substantial additional European Union (EU) funds, paid largely as social subsidies. This has led to large increases in sheep numbers in the uplands, with serious effects on farmland birds. In the lowlands access to the European Common Agricultural Policy (CAP) also increased subsidies for cereal-growing but a growing surplus of dairy products resulted in increasing controls on production. The growing of oil seeds, particularly oil-seed rape, was promoted as break crops in cereal rotations. The broad thrust of such changes was to promote the cultivation of cash crops in arable farming at the expense of mixed systems involving stock.

The modern agricultural revolution has equally affected pastoral farming, where there is increasing concentration on producing livestock products from grass. As long ago as 1898 R.H. Eliot, of Clifton Park, Roxburghshire, propounded the principle that the most efficient and economic feed for livestock was properly managed grass and his ideas are widely accepted today, largely through the championship of George Stapledon (Seddon 1989). Thus the most important change in grassland management since 1945 has been large-scale reseeding of pastures. This is underlined by the modern change in the way improved grassland is recorded in the agricultural statistics (Chapter 9). In Scotland both pasture and ley are now simply recorded as improved grass, reflecting similar changes; improved permanent grass has long been less important than ley in Scottish farming. Within grass farmland marked concentrations of enterprises have occurred. The pattern is most clear in dairying where, since 1968, the number of dairy cows has fallen by 18% but the number of holdings with dairy cows by 67%, involving an increase in herd size of 150%. Two-thirds of all dairying in England and Wales is now concentrated in 12 counties, Cumbria, Clwyd, Cheshire, Cornwall, Derbyshire, Devon, Dorset, Dyfed, Lancashire, North Yorkshire, Staffordshire and Somerset.

Post 1987

Just as the 1870s marked the final peak of nineteenth-century prosperity, so 1987 appears to have been that of the post-war era. In the late 1990s a major recession was again emerging. This has some links to the 1870s, in particular in a collapse in cereal prices, now to the levels obtaining in the mid 1970s. But the modern recession's impact has been most severe in pastoral systems in the west, a marked difference. Recession apart, the livestock industry is also in considerable disarray following two animal epidemics, bovine spongiform encephalopathy (BSE) and foot-and-mouth disease. These have, however, provided an opportunity to sensibly restructure the industry, particularly looking at very high stocking densities. Whether that will be taken is not yet clear.

The abandonment of land by farming has been quite widely postulated (*Country Life* 2000). But land is an asset and history shows that use will always be made of it, leading to further habitat changes; abandonment is not necessarily beneficial to farmland birds (Chapter 9). As in the nineteenth century there is also a marked tendency for industrial and commercial wealth to acquire landed property for amenity reasons (Chapter 13).

An important change in direction may well be the large-scale adoption of agri-environment schemes (Chapter 13). It seems certain, in any case, that farming will become subject to increasing bureaucratic control but past experience suggests that not too much environmental optimism should be placed on that. The proposal reported recently in the press, for example, that farmers should certify that their farms meet certain environmental standards before qualifying for subsidy looks well on paper. I question the practicability, effectiveness (how anodyne would standards be?) or even legality, if undertaken unilaterally, of this notion, which exudes a strong smell of fudge. Such a scheme, if enforced, will certainly have a profound effect on arable farm management. We should remember, however, that legislation of this type has a nasty habit of producing unwelcome and unexpected results. Subsidy is not the only option available to farmers. Technology may provide a different route to survival and profitability. Farmland is an important ecosystem in its own right, supporting a distinctive range of birds, and we should not lose sight of that in an enthusiasm for scarifying the industry and replacing it with something else.

2

The farmland birds

Early history

British farmland has been largely made from the wildwood which dominated the landscape after the last glaciation. Open habitats were then largely confined to saltmarshes and coastal dunes, some areas of upland moorland and grassland and flooded valley bottoms. Wildwood was not continuous, however. Although extensive plains habitats were absent, glades and clearings were kept open by storm damage and by large herbivores. The British avifauna, therefore, was once primarily a forest one. Waders and larks would have occurred in the limited open habitats and waders and wildfowl in forest marshes, which probably also attracted *Acrocephalus* warblers and Reed Buntings. Yellowhammers would also have occupied the limited open habitats and also woodland glades and edges, whilst forest is a natural habitat of many finches now more usually regarded as farmland birds. Species such as harriers, Kestrels, larks, Rooks and Starlings must have been local or rare, limited by the scarcity of open grassland for feeding. Vera (2000), however, has recently expressed the view that wood pasture and parkland created by large herbivores formed a much greater part of the original natural habitats of lowland western Europe. If this hypothesis proves valid then open-country species would almost certainly been more widespread than the traditional view above allows.

Land clearance for agriculture began in the Neolithic age (around 4000 BC). Rackham (1990) noted that the first areas to lose wildwood permanently were the East Anglian Brecks, much of the chalk country, the Somerset Levels and the coastal Lake District. Rackham suggested that by 500 BC half of England was no longer wildwood and by the Roman period that 'it was almost as agricultural as it is today'. Over a long period of time, therefore, the loss of wildwood to cultivation and grazed range lands such as moorland, heath and downland led to a fundamental change in the basic character of the avifauna. The most important habitat change was the creation and spread of grassland and heathland habitats, which allowed colonisation or spread by a wide range of steppe and open-country species – harriers and Kestrel, bustards and gamebirds such as partridges and quail, Corncrake, waders, particularly perhaps plovers and Stone Curlew, some owls, larks and pipits, Yellow Wagtail, chats and Corn Bunting. Colonisation of these new habitats would also have encouraged these species to penetrate cultivated land and such birds, together with those of extensive open marshland, still comprise 42% of the farmland avifauna by species (Figure 2.1). Although now comprising only 15% of the farmland avifauna by number, they were once numerous or abundant (Chapter 4).

Neolithic Man did not only clear wildwood for farming, he also converted it to managed woodland, mainly coppice, a process which continued in parallel with clearance for agriculture. Williamson (quoted by Burton 1995) suggested that this process provided a convenient bridge between forest and farm. However, the age structure of coppice is quite different from that of mature natural forest, lacking habitats for birds requiring old timber, which are rather scarce in farmland. Furthermore species such as pigeons, thrushes, finches, buntings and some corvids, which inhabit forest but which are largely ground-feeding, would have had little difficulty in adapting to farmland without such a stepping-stone if trees or bushes remained. They would have benefited from the increased feeding opportunities appearing in farmland. For some of these species exploitation of these resources in winter perhaps provided an important bridge (Chapter 3). Nevertheless Williamson's suggestion probably holds good for the more strictly woodland species now inhabiting farmland, such as Wren, many warblers and tits, which make little use of farmland resources.

Broadly, therefore, it can be said that the conversion from wildwood to agricultural land had three general consequences. It led to colonisation and expansion by plains and open-marshland species, categorised here as ground-nesting/wetland birds. It probably led to increases of species already part of the forest avifauna which now occupy the category of field species and it led to adaptation by some true woodland species to the fragments of woodland persisting in farmland. Despite changes in constituent

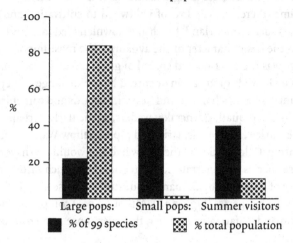

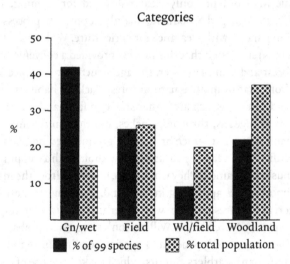

Figure 2.1. Summary of the basic structure of breeding bird populations of farmland. Large populations are defined as >1000000 birds, small populations as <100000. Gn/wet, ground-nesting or wetland species; Wd/field, woodland/field species; for further details of Categories see Table 2.1. 'Summer visitors' here includes both summer visitors and summer migrants as defined on p. 24. Swallow, as an aerial species, is excluded from Categories.

species caused by climatic variations, the basic form of the farmland avifauna was set at this early period and has probably varied rather little since. This is shown by Yapp's (1981) analysis of the birds illustrated incidentally in the illuminations of mediaeval manuscripts. Looking only at the English manuscripts he illustrates there are 56 recognisable species, of which only Spoonbill, White Stork and Common Crane do not remain familiar birds of our countryside, although they still occur. The 10 most frequently illustrated birds are owls, Goldfinch, Crane, Jay, Bullfinch, Swallow, Magpie, Sparrowhawk, Blackbird and White Stork, and crows. These birds are often quite well observed and were clearly familiar to the artists. Altogether in English and European manuscripts Yapp listed 123 species of wild birds, again dominated by species which remain familiar and common today. They are also dominated by birds of field and marsh, only 13% being strictly woodland species. Many of the species illustrated were perhaps quarry species. Besides hawking, these manuscripts illustrate trapping song birds and netting partridges, occupations which continued into the nineteenth century. A similar pattern of continuity is provided by the ordinances proscribing various species as pests since Tudor times. Crows, Rooks, sparrows, Kites, Buzzards, Jays and Bullfinches, for example, have clearly been regarded as troublesome in crops, orchards and poultry yards for centuries.

Farmland bird populations today

Breeding birds

Most bird species probably occur in farmland from time to time; I recorded over 250 species on Oakhurst. A much smaller number, however, can be regarded as regular occupants. Table 2.1 sets out those species I have treated as farmland birds in this analysis. It includes both the breeding and wintering species. I have defined the breeding birds of farmland as: (1) the ground-nesting species of fields, moorland and marshes; (2) species which feed in fields, either entirely or to a significant extent; (3) the common raptors and owls of the open countryside; and (4) the common hedgerow birds included in the farmland index of the Common Birds Census (CBC). This definition embraces a total of 99 species and $c.$ 91 657 000 birds. It also includes some rarer species, e.g. Hobby, Corncrake, Woodlark and Cirl Bunting, which were once much more common in farmland than today. Although other species, particularly water birds such as Coot, breed in farmland occasionally or perhaps regularly in small numbers, I believe that this definition covers all the principal species that should be considered as farmland birds.

As O'Connor & Shrubb (1986) remarked, the birds breeding in farmland can be divided into two broad categories, a core of very numerous and widespread species forming the bulk of the population and a much larger

TABLE 2.1. *The birds of farmland. The breeding population estimates shown are for total birds not pairs, unless otherwise shown*

Species	Category[a]	Status[b]	Breeding population (birds)[c]	Winter population (birds)
Mute Swan	G'n/wet	Res.	25750	27750
Bewick's Swan	Ground	WV		7200
Whooper Swan	Ground	WV		5600
Bean Goose	Ground	WV		450
Pink-footed Goose	Ground	WV		192000
White-fronted Goose	Ground	WV		20000
Greylag	G'n/wet	Res.(FI)	13100	
Greylag	Ground	WV		114000
Canada Goose	G'n/wet	Res.(FI)	46700	61000
Brent Goose (dark-breasted)	Ground	WV		103300
Wigeon	Ground	WV		250000–300000
Teal	G'n/wet	Res.	3000–5200	136000
Mallard	G'n/wet	Res.	200000–260000	500000
Pintail	Ground	WV		27800
Garganey	G'n/wet	SM	30–250	
Shoveler	G'n/wet	SV	2000–3000	
Red Kite	Field	Res.	396	500
Marsh Harrier	G'n/wet	SM	160 (f)	
Hen Harrier	G'n/wet	SV, WV	1170	750–1000
Montagu's Harrier	G'n/wet	SM	16	
Sparrowhawk	Woodland	Res.	64000	170000
Buzzard	Field	Res.	24000–34000	36000–51000
Kestrel	Field	Res., WV	70000–80000	>100000
Merlin	G'n/wet	SV, WV	2600	2000–3000
Hobby	Woodland	SM	3500	
Red Grouse	G'n/wet	Res.	500000	500000
Black Grouse	G'n/wet	Res.	5000–8100 (m)	?40000
Red-legged Partridge	G'n/wet	Res.(FI)	180000–500000	500000
Grey Partridge	G'n/wet	Res.	300000	700000
Quail	G'n/wet	SM	200–600	
Pheasant	G'n/wet	Res.(FI)	1.55 million (f)	8 million
Corncrake	G'n/wet	SM	589 (m)	
Moorhen	G'n/wet	Res.	480000	800000
Oystercatcher	G'n/wet	SV	66000–86000	
Stone Curlew	G'n/wet	SM	<400	
Golden Plover	G'n/wet	SV, WV	45200	250000
Lapwing	G'n/wet	Res., WV	265000	1.5–2 million
Ruff	Ground	WV		700
Dunlin	G'n/wet	SV	18000	
Jack Snipe	Ground	WV		10000–100000
Snipe	G'n/wet	Res., WV	110000	>100000
Curlew	G'n/wet	SV,? WV	66000–76000	5000–7000
Redshank	G'n/wet	SV	61000–67000	
Black-headed Gull	G'n/wet	SV, WV	334000	<1.9 million
Common Gull	Ground	WV	136000	<900000
Stockdove	Field	Res.	336000	350000–400000

TABLE 2.1. *(Cont.)*

Species	Category[a]	Status[b]	Breeding population (birds)[c]	Winter population (birds)
Woodpigeon	Field	Res.	3.9 million	up to 10 million
Collared Dove	Field	Res.	360 000	up to 1 million
Turtle Dove	Field	SM	60 000	
Cuckoo	Woodland	SM	26 000–52 000	
Barn Owl	Field	Res.	6960–7934	11 000–23 000
Little Owl	Field	Res.	10 000–18 000	16 000–33 000
Tawny Owl	Wd/field	Res.	>40 000	150 000
Long-eared Owl	Field	Res., WV	2200–7200	10 000–35 000
Short-eared Owl	G'n/wet	SV, WV	2000–7000	5000–50 000
Wryneck	Wd/field	SM	400–extinct	
Green Woodpecker	Wd/field	Res.	14 400	70 000
Great Spotted Woodpecker	Woodland	Res.	26 400	?c. 50 000
Woodlark	G'n/wet	Res.	30	?200
Skylark	G'n/wet	Res., WV	>2.0 million	25 million
Swallow	Aerial	SM	1.14 million	
Meadow Pipit	G'n/wet	Res., WV	3.8 million	1.5 million
Yellow Wagtail	G'n/wet	SM	100 000	
Pied Wagtail	Field	Res.	600 000	0.75–2 million
Wren	Woodland	Res.[d]	10.2 million	8–14 million
Dunnock	Woodland	Res.[d]	3.4 million	c. 8 million
Robin	Woodland	Res.[d]	5.8 million	c. 7 million
Whinchat	G'n/wet	SM	28 000–56 000	
Stonechat	Field	Res., SV	17 000–44 000	36 000–72 000
Wheatear	G'n/wet	SM	>110 000	
Ring Ouzel	G'n/wet	SM	11 000–22 000	
Blackbird	Wd/field	Res., WV	7.04 million	11–16 million
Fieldfare	Wd/field	WV		750 000
Song Thrush	Wd/field	Res., WV	1.36 million	2.5–4.5 million
Redwing	Wd/field	WV		750 000
Mistle Thrush	Wd/field	Res.	363 400	320 000–650 000
Grasshopper Warbler	G'n/wet	SM	21 000	
Sedge Warbler	G'n/wet	SM	500 000	
Reed Warbler	G'n/wet	SM	80 000–160 000	
Lesser Whitethroat	Woodland	SM	60 000	
Whitethroat	Woodland	SM	1.21 million	
Garden Warbler	Woodland	SM	192 000	
Blackcap	Woodland	SM	660 000	
Chiffchaff	Woodland	SM	780 000	
Willow Warbler	Woodland	SM	3.17 million	
Goldcrest	Woodland	Res., WV	414 000	c. 1.75 million
Spotted Flycatcher	Woodland	SM	168 000	
Long-tailed Tit	Woodland	Res.	281 400	250 000–300 000
Coal Tit	Woodland	Res.	390 000	780 000
Blue Tit	Woodland	Res.[d]	4.5 million	7.08 million
Great Tit	Woodland	Res.[d]	2.1 million	2.1 million
Treecreeper	Woodland	Res.	136 000	340 000
Red-backed Shrike	Wd/field	SM	350–extinct	
Jay	Woodland	Res.	166 400	300 000

(cont.)

TABLE 2.1. (Cont.)

Species	Category[a]	Status[b]	Breeding population (birds)[c]	Winter population (birds)
Magpie	Field	Res.	1 million	0.8–1.7 million
Chough	Field	Res.	680	1500–1600
Jackdaw	Field	Res., WV	679 000	1.5–2.5 million
Rook	Field	Res., WV	2.28 million	3–4 million
Crow	Field	Res.	1.94 million	3–3.5 million
Raven	Field	Res.	14 000	20 000–30 000
Starling	Field	Res., WV	2.2 million	?10 million
House Sparrow	Field	Res.	5.2–9.2 million	5.6–8.4 million
Tree Sparrow	Field	Res.	220 000	375 000
Chaffinch	Wd/field	Res., WV	8.7 million	?23 million
Brambling	Wd/field	WV		up to 1.8 million
Greenfinch	Wd/field	Res.	1.06 million	4–5 million
Goldfinch	Field	Res., SV	>440 000	100 000
Linnet	Field	Res., SV	936 000	1.75 million
Twite	G'n/wet	SV	130 000	
Bullfinch	Woodland	Res.	195 000	500 000–750 000
Yellowhammer	Field	Res.	2.2 million	2.4 million
Cirl Bunting	Field	Res.	906	?c. 1000
Reed Bunting	G'n/wet	Res.	440 000	450 000
Corn Bunting	G'n/wet	Res.	39 600 (m)	100 000

[a] G'n/wet, ground-nesting or wetland species. These categories are combined as most wetland species in farmland nest on the ground or in low dense ground vegetation.
Ground, wintering birds only: species that feed and roost on the ground in open country usually, but not always farmland.
Field, species that feed in fields but nest in hedges or trees.
Wd/field, woodland/field species. Those which mainly inhabitat woodland habitats but, in farmland, feed to an important extent in fields, defined as occurring in at least two-thirds of my field feeding counts at Oakhurst.
Woodland, species that occur in farmland mainly because of the presence of woodland, hedges or gardens. Many make some use of agricultural resources for food, particularly in winter, when those marked[d] feed regularly around farmyards and stockyards. Some summer migrants, e.g. Hobby and Whitethroat, might be more properly regarded as woodland/field species. Swallow, as an aerial species, has no category.
[b] Res., resident.
SV, summer visitor, taken as those birds that are summer visitors to farmland breeding sites but mainly winter elsewhere, e.g. waders in estuaries or upland birds in lowland farmland.
SM, summer migrants, taken as those species which migrate entirely out of Britain to winter.
(FI), numbers still affected by feral introductions.
WV, winter visitor, includes both winter migrants and residents with populations augmented by important winter influxes.
[c] m, males; f, females.
[d] See note [a] above.
Sources: The breeding estimates are based on the estimates given by Gibbons et al. (1993), adjusted to allow for the woodland populations of relevant species on the basis of the densities recorded in the CBC *Woodland Index* for the same period, or for population counts in other habitats outside farmland, where recorded. Where applicable, estimates have been updated by reference to Stone et al. (1997) and more recent surveys. Wintering populations are based on the estimates in Lack (1986), updated as necessary to take account of changes in breeding numbers recorded in the *New Atlas of Breeding Birds in Britain and Ireland* or of changes shown by Stone et al. (1997). The gull populations are for totals and an unknown but smaller proportion actually winters in farmland.

group of scarcer species, attracted by particular habitat features. This pattern is obvious in Figure 2.1; altogether 22 species with large populations account for over 80% of the total population of breeding birds. Many of these species are also common woodland species; 19 of them figure in the woodland index of the CBC and 16 of those are among the 25 most abundant woodland birds listed by Fuller (1982). They are highly adaptable and ubiquitous species and tend also to be common in suburban and urban habitats.

Only three ground-nesting species, Pheasant, Skylark and Meadow Pipit, figure among these 22 most numerous farmland birds. Although this is the most numerous category of breeding species (Figure 2.1), 57% have small populations, mainly because they have specialised habitat requirements. The entire group only comprises 15% of the total bird populations. Such patterns have presumably emerged because of the general difficulty of maintaining a ground-nesting habit in cultivated landscapes. Nevertheless an important proportion of the species nesting in trees, hedges and woods in farmland feed in fields or around farmsteads, thus exploiting farming processes, either habitually (field species) or to gather a significant part of their food there (woodland/field species).

Of the species inhabiting farmland that make little or no use of fields there seems to be little obvious difference in the proportion that are summer visitors and those that are residents. But, on the basis that Fuller (1982) suggested – that if a species occurred in half the sample censuses for a community, it was a typical member of that community – they are all typical woodland birds except Lesser Whitethroat. Some resident species in this group occasionally feed in fields and often around farmyards in winter but, as breeding birds, an important segment of the farmland avifauna clearly lives there because of the fragments of woodland habitats present rather than any attraction of agriculture. Such species in fact represent the most numerous category of breeding individuals (Figure 2.1).

Forty species are listed as summer visitors or migrants to farmland in Table 2.1 but 14, comprising 2% of total birds, are species such as waders which move out of farmland breeding sites into other habitats or areas within Britain to winter. The 26 summer migrants into Britain included here as farmland birds comprise only 9% of the total population. Murton (1971) noted that the comparative paucity of such birds reflected the general lack of a spring flush of surface invertebrates in farmland and the fact that residents were highly efficient at exploiting the available resources. O'Connor (1985) argued that such summer visitors to farmland were restricted to certain habitats unless severe winter weather reduced resident populations. This argument was supported by Fuller & Crick (1992), who found that migrant passerines were less widespread in all regions of Britain and Ireland and that resident ones used a greater diversity of breeding habitats.

O'Connor & Shrubb (1986) also noted that the extent to which hedgerow birds exploited the resources in fields may vary with the other resources available in any area. Comparing two farmland areas, Manydown in Hampshire with 62 ha of woodland included in the farmland area (11%), with my family's Oakhurst Farm in Sussex, with 2.5 ha (1.5%), showed that many more such species exploited fields in the breeding season in the latter. Observations I have made in Wales and those of Davis (1967) support this further. Nevertheless breeding birds in farmland characteristically use more than one of the major habitat elements available, many nesting in one habitat or crop type and feeding in another for example. Both elements are necessary and the loss of only one will have adverse effects.

Wintering populations

Table 2.1 also lists the species regularly wintering in farmland, with some indication of population size. It includes all the breeding species which winter regularly in farmland and all winter visitors which behave similarly. Altogether 82 species totalling $c.$ 177 556 000 individuals are involved. The basic structure of this population is shown in Figure 2.2.

Comparison of Figures 2.1 and 2.2 shows several important general differences between breeding and wintering populations in farmland. Wintering birds are more numerous. This only partly reflects the increment of the breeding season for substantial numbers of immigrants winter in British farmland. A greater proportion of wintering species has large populations; fewer have small populations. Although the proportion of species that are winter visitors is about the same as that for summer visitors, the number of individuals involved is much higher in winter. Many more wintering than breeding birds are ground-dwelling or field species ($c.$ 90 million vs. $c.$ 38 million). Much of the difference lies in the habits of seasonal visitors, with only one winter visitor, Goldcrest, making no use of fields for feeding, whilst nearly 30% of summer visitors do not. Finally wintering birds in farmland are strongly gregarious. Comparatively few truly colonial species breed in farmland. Flocking, however, is a characteristic habit of farmland birds in winter, because the food resources available tend to be distributed in scattered parcels which a flocking habit is best adapted to exploit.

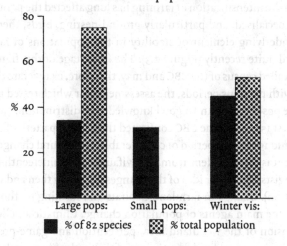

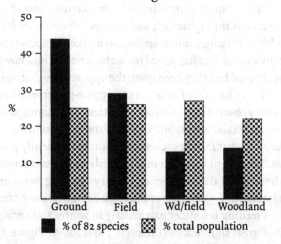

Figure 2.2. Summary of the basic structure of wintering bird populations of farmland. Definitions as for Table 2.1 and Figure 2.1. Both species and number of winter visitors includes residents whose local populations are augmented by significant numbers of winter immigrants, e.g. Lapwing, Skylark and Starling, as well as winter migrants such as Fieldfare. The actual number of winter visitors is unknown but presumably less than the gross figure.

Changes in bird populations

Although the intensification of farming has long affected the populations of more specialised, and particularly ground-nesting, birds, there was a strong underlying element of stability in the populations of farmland birds until quite recently (Figure 2.3). The final stage of this figure rests on the detailed census of the CBC and may, therefore, exaggerate the comparison with previous periods, the assessments for which rested more on subjective observations and a good knowledge of distribution. Nevertheless the first 15 years of the CBC confirmed the general pattern of stability into the mid 1970s, in a period of considerable agricultural change (Chapter 8) and it is quite evident from the avifaunas that nineteenth-century ornithologists had a clear idea of the changes occurring then and why (see Introduction). Change in farmland habitats and farming methods were often not the main agents of population change. Climatic amelioration, the expansion of the woodland area, persecution and game-preserving were all considered major determinants of this in the nineteenth century and the changes unequivocally ascribed to agriculture or habitat change arising from it were surprisingly limited. The extensive loss of more specialised habitats in the eighteenth and nineteenth centuries (Chapter 4) was always likely to change bird distribution markedly. But we should not automatically assume this happened for we have no real idea how far birds might have adapted had they been given the opportunity that was so often denied them by the barrage of collectors and game-preservers then operating. We can only observe that birds such as Marsh and Montagu's Harriers have shown remarkable adaptability to farmland habitats in the late twentieth century and that factors such as persecution frequently pre-empted habitat loss. The extent to which many grassland habitats remained unchanged throughout the nineteenth century has long been underestimated (Chapters 6 and 9), and the application of high farming created new habitats, and feeding resources particularly in winter and spring, which were often of great significance (Chapters 3 and 10). As Figure 2.3 shows the underlying pattern of stability has changed in the second half of the twentieth century and farmland bird populations are increasingly affected by changes in farming methodology and technology. Modern changes in grassland management, the timing of cultivations and the use of pesticides particularly have undermined the capacity of birds to adapt to farmland as a habitat.

Afforestation has also contributed to changes in farmland birds, having had some markedly contrary effects. Planting in the eighteenth and

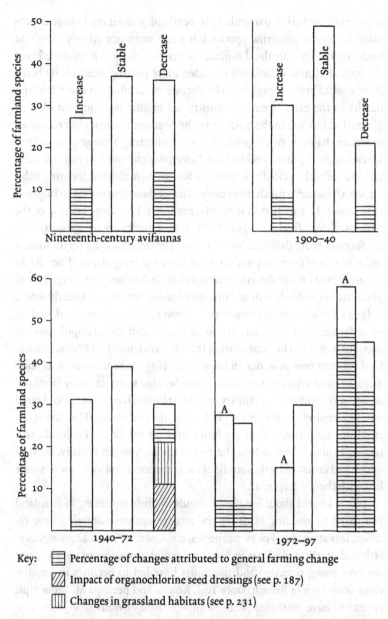

Figure 2.3. Changes in the breeding bird populations of farmland at different periods. A represents changes shown by a comparison of the two *Breeding Atlases* (see Chapter 8). (*Sources:* Nineteenth century, avifaunas; 1900–40, Alexander & Lack (1944); 1940–72, Parslow (1973); 1972–97, Marchant *et al.* 1990.)

nineteenth centuries, particularly in Scotland, encouraged range expansions in several passerine species but these were not strictly farmland birds. In lowland farmland nineteenth-century authors often noted that the spread of new woodlands provided more nesting sites, to the benefit of species such as pigeons. The decline of farming on some lowland heaths in the early twentieth century led to afforestation, and blanket plantation forestry in the uplands in the twentieth century has restricted moorland habitats to a significant extent, affecting a range of moorland species such as grouse and Golden Plover, which have lost nesting habitat. Species such as Golden Eagles and Ravens have also lost foraging habitat, which has been much more evident in Scotland and northern England than Wales (Marquiss *et al.* 1978, Mearns 1983). This arose because of the pattern of small freeholdings in Wales, which made extensive acquisitions for forestry more difficult. Welsh forests characteristically retain fingers and enclaves of farmland within them as well as irregular outlines. As an example Peers & Shrubb (1990) noted that, in Breconshire, of 151 km^2 of plantations established in 305 upland 1-km squares from the early 1960s, only 31 1-km squares had complete tree cover. They thus retained feeding areas for species such as Ravens, so that the Welsh Raven population has been little affected by afforestation (Newton *et al.* 1982). In Wales, too, upland forestry now provides an important refuge for species such as Song Thrushes and Whitethroats, which have become scarce in many farmland areas. In the nineteenth century small farm woodlands were also cleared quite extensively, often to supply fencing for enclosures (Hoskins 1955). Holdings long enclosed in my home area on the Selsey Peninsula still retained small woodlands at the end of the eighteenth century, most of which had vanished by the middle of the nineteenth and some species were lost with them (Chapter 4).

Less is known about long-term trends in birds wintering in farmland compared to breeding birds for the actual numbers wintering have received less attention. For some species, e.g. Golden Plover, Lapwing, Skylark and Starling, declines in breeding populations throughout Europe are now being recorded and this must have led to declines in populations wintering in Britain. Only Teal, Kestrel and perhaps Meadow Pipit appear to have increased (Teal) or shown little obvious change in winter, compared to declining local breeding populations. Wildfowl numbers have shown more increases than declines, mainly because of reduced shooting pressure but perhaps also because they have reacted favourably to changes in agricultural practice. Geese provide examples (Chapter 10).

Wintering populations of many common seed-eating passerines have undoubtedly declined in the past 25 years, with the decline of food resources (Chapter 10). Chaffinch and Brambling are probably exceptions. Winter thrushes also appear to be declining, at least in Wales (personal observation) but this may simply reflect the modern trend to milder winters.

Natural limitations

The distribution and abundance of farmland birds are not only governed by factors such as crop management and agricultural habitat changes. Underlying their distribution are the natural factors of geography, climate, soils and fertility and altitude and relief which act independently of the way land is managed. Each also influences farming.

Geographic position and climate

Bird distributions are affected by geographical patterns of climate. Whilst Britain lies on the northwest limit of many species' European ranges simply by virtue of her position as an island on the northwest edge of the continent, that geographic position strongly influences Britain's climate, which again influences bird distributions. The geographic ranges of many of our farmland birds cover a fairly wide band of latitudes throughout Europe. The maps in Voous (1960) indicate an average band embracing some 27–28 degrees of latitude for the species in Table 2.1, the bulk of which also breed across Europe from the Atlantic to the Urals. Clearly many tolerate a wide range of climatic conditions. Voous noted, for example, that the Lapwing bred in five different climatic zones. Nevertheless patterns in the distribution of some farmland birds illustrate more particular effects. Some northern species, characteristic of cooler climates, reach the southern limits of their European distributions in the British uplands, of which Merlin, Red Grouse, Golden Plover and Dunlin are included in the farmland list here. Declines in the southern populations of such species, which are particularly marked for grouse, Golden Plover and Dunlin in Wales and southwest England (Gibbons *et al.* 1993, Lovegrove *et al.* 1995) probably partly reflect the present warming phase.

Another suite of species reaches the northern limits of their European ranges in Britain. Farmland birds include Barn Owl, Stonechat, Chough, Cirl Bunting and Corn Bunting (and Red-legged Partridge but that is an introduced species). As these species are all residents, it is probably Britain's milder, more maritime winter climate that allows them to persist farther north there than further east on the continental mainland, although in Europe the Chough is a montane bird. The abundance maps in Gibbons *et al.* (1993) suggest similar climatic influence on the distribution of other farmland birds. Thus many resident species characteristically show a greater density in southern or southwest England, although breeding throughout Britain. Green Woodpecker and Long-tailed Tit are typical examples. As such species often decline markedly after severe winters, better winter survival in milder areas presumably contributes to this. But avian species diversity in Britain is strongly correlated with temperature, the geographical pattern of diversity changing seasonally in parallel with the change in temperature gradient, from south–north in summer to west–east in winter (Lennon *et al.* 2000). This fits well with the species-energy hypothesis, which postulates that higher temperatures support greater total abundance of life and that more species can persist where there is greater total abundance (see Gaston & Blackburn 2000).

A third suite of species has a limited distribution in Britain, usually in the south and east, although reaching much farther north in continental Europe. Farmland birds are Garganey, Marsh Harrier, Montagu's Harrier, Hobby, Turtle Dove, Green Woodpecker, Woodlark, Yellow Wagtail, Reed Warbler, Lesser Whitethroat, Garden Warbler, Chiffchaff, Nuthatch, Jay and once Wryneck and Red-backed Shrike. Of these, 75% are summer visitors and their British distribution is probably mainly restricted by the cooler and damper summer climate of north and west Britain. Some, e.g. Hobby, Reed Warbler and Lesser Whitethroat, are expanding their range in Britain today, in a warmer period. The increase in Hobbies may also be related to a decline in persecution (Chapter 12).

Other aspects of climate

During the period under review there have been four broad climatic phases in northwest Europe – a cold period which actually extended from *c.* 1250 right through to *c.* 1850 and was at its most severe from *c.* 1550–1700, a period known as the Little Ice Age, a period of marked climatic amelioration between about 1850 and 1950, a significant return to colder conditions until *c.* 1980 and subsequently a return to warmer conditions. Broadly again milder periods tend to be characterised by damp cool summers and mild wet winters, as Atlantic weather systems dominate our climate, and colder periods by drier, more arctic and continental conditions, as these become dominant (Williamson 1975, Burton 1995). Within these broad patterns, there have always been marked fluctuations (Manley 1952).

The mechanisms by which climate affects bird numbers and distribution fall broadly into two categories, the severity of the winters, which affects winter survival, and the warmth and dryness of the spring/summer, which influences food supplies. With farmland birds indirect effects may arise through the influence of climate on farming. I have already noted that climatic considerations may have encouraged the cultivation of heath in the eighteenth and nineteenth centuries. Williamson (1975) also remarked that the climatic amelioration after 1850 lengthened the growing season and enabled cultivation to expand northwards, which undoubtedly encouraged the northward expansion of birds such as Rooks and Lapwings. Similarly in the 1970s and 1980s a series of difficult springs but benign autumns may have contributed to the switch to planting most cereals in autumn.

Williamson (1975) and Burton (1995) discussed the effects of climatic change on birds in Europe in some detail. Table 2.2 summarises their findings for the list of farmland birds considered here and raises several points. An obvious one is the short list of species included under the early nineteenth century. Although this must reflect the general lack of detailed information at this period, distributions and habitat conditions were probably more settled when much colder conditions had been prevalent for long periods. Then the actual patterns of change are always less neat than this form of presentation suggests. Williamson (1975) remarked that birds make rather crude indicators of climatic change. The most marked increases during 1850–1950 tended to occur after 1900, particularly in the 1920s and 1930s, coinciding with a run of particularly dry warm summers, and some of these species were still declining up to the end of the nineteenth century. Similarly, many species which reacted favourably to climatic amelioration, were still expanding after that climatic phase had started to reverse in the 1950s.

For migrants climatic change outside the breeding range may be most important. The classic example is the Whitethroat, whose population crashed in 1969 as a result of drought in the Sahel region (Winstanley et al. 1974). Other farmland summer visitors have been similarly affected, for example Sedge Warbler, and continued fluctuations in the populations of such birds reflect conditions on wintering grounds far more than breeding.

For some species such as Bittern and Black-tailed Godwit decline is more usually ascribed to drainage of their breeding habitats for farming. Burton (1995) pointed out, however, that much apparently suitable

TABLE 2.2. Population changes in farmland[a] birds in Britain possibly or partly influenced by climatic changes

Contraction and decline, early nineteenth century	Expansion with amelioration, 1850–1950	Decline with amelioration, 1850–1950	Changes since 1950			
			+	−	+/−	−/+
(Bittern)	(Bittern)	Red Grouse	Red Kite	Garganey	(Bittern)	Mute Swan
Montagu's Harrier	Mute Swan	Quail	Hen Harrier	Red Grouse	Montagu's Harrier	Cirl Bunting
(Back-tailed Godwit)	Garganey	Stone Curlew	Hobby	Moorhen	Golden Plover	
Black-headed Gull	Shoveler	Golden Plover	Oystercatcher	Lapwing	(Black-tailed Godwit)	
Stockdove	Montagu's Harrier	Barn Owl	Black-headed Gull	Cuckoo	Turtle Dove	
Barn Owl?	Oystercatcher	Long-eared Owl	Collared Dove	Barn Owl	Woodlark	
Great Spotted Woodpecker	Lapwing	Wryneck	Reed Warbler	Wryneck	Twite	
Chiffchaff	Curlew	Wheatear?	Lesser Whitethroat	Song Thrush		
Starling	Black-headed Gull	Ring Ouzel	Chiffchaff	Grasshopper Warbler		
	Stockdove	Red-backed Shrike	Bullfinch	Sedge Warbler		
	Woodpigeon	Chough		Whitethroat		
	Turtle Dove	Tree Sparrow		Red-backed Shrike		
	Tawny Owl	Corn Bunting		Starling		
	Little Owl			Swallow		
	Green Woodpecker			House Sparrow		
	Great Spotted Woodpecker			Tree Sparrow		
	Woodlark			Corn Bunting		
	Swallow					
	Dunnock					
	Blackbird					
	Mistle Thrush					
	Grasshopper Warbler					

Sedge Warbler
Willow Warbler
Goldcrest
Blue Tit
Great Tit
Jay
Rook
Starling
Chaffinch
Greenfinch
Goldfinch
Linnet
Bullfinch
Cirl Bunting

a Species in parentheses are not considered farmland birds but are included as affected by major habitat changes caused by reclamation for farming.
b +, increase; −, decline; +/−, increase followed by decline; −/+, decline followed by increase.
Sources: Williamson (1975) and Burton (1995).

habitat persisted for both species, even in the Fens, until the mid nineteenth century, well after their main decline, and that both have recolonised England when habitat was far scarcer. He therefore suggested a climatic cause linked to the known period of severe conditions in the early nineteenth century. However these species were widespread in previous centuries, in conditions equally severe (Figure 6.1 for Bittern) and Burton failed to appreciate the malign influence of the collector on both (Chapter 12).

Similarly Burton linked the expansion of pigeons and doves to climatic amelioration rather than the expansion of rotation farming, arguing that it mainly occurred from about 1850. Figure 3.1, however, shows that the greatest expansion of such farming actually occurred from the late 1830s, coinciding quite closely with the expansion of Woodpigeons and Stockdoves. Contemporary authors often linked the expansion of these birds to the widespread establishment of new plantations, which provided new nest sites. White (1789, letter 39 to Pennant, 1773) comments on the reverse of this trend, noting a decline of winter flocks since 'our beechen woods were so much destroyed', a neat example of the importance of winter food supplies to these birds. It is most unlikely, too, that the Montagu's Harrier was primarily reduced in the nineteenth century by climatic change, rather than habitat loss and persecution. Burton also linked changes in the numbers of Tree Sparrows and Corn Buntings largely to amounts of rainfall, quoting Norris (1960) to the effect that their range is largely limited to areas receiving less than 100 cm of rain annually. This simply indicates that they are mainly restricted to areas of arable cultivation, also largely limited to areas of moderate rainfall.

A general point emerging from these examples is that changes in birds' ranges and populations are often not attributable to a single cause. Climatic trends provide a framework in which broad geographic changes in distribution will take place. On a more local scale they cannot overide the impact of factors such as habitat destruction and farming change. Nevertheless habitat degradation may well increase the vulnerability of species to unfavourable climatic change which would be tolerable in optimum habitat.

Williamson (1975) observed that a return to the pattern prevalent in the nineteenth century of three or four severe winters each decade would substantially alter the composition of our bird populations. This was supported by Greenwood & Baillie (1991), who showed that prolonged snow cover in winter reduced populations of most common passerines in farmland, although low temperatures alone were less important. The CBC has measured the impact of severe winters since 1961 and Table 2.3 looks at the impact of four winters between 1960 and 1990, in all of which mean monthly temperatures in Britain in at least one month during November to February were around 0 °C or below (see Marchant *et al.* 1990, Figure 3.2). The table looks only at those species included in the farmland index of the CBC. Several interesting points emerge. One is how few farmland species were significantly reduced in all the winters concerned – four out of 25 species reduced at least once. This underlines the point made by Marchant *et al.* (1990), that all such winters tend to differ in their impact. Secondly there was little obvious difference in the category of species affected at least once. They comprise eight ground-nesting birds, seven field species, four woodland/field species and eight woodland species; severe winters have a general impact. Thirdly, despite the extent of snow cover, only two finches or buntings were significantly reduced in 1963 when abundant stubbles and other winter food sources were still available, compared to seven in 1982, when snow cover was also severe and such food sources were much scarcer (Chapter 10).

This analysis adopts a fairly narrow definition of a severe winter. At least four other winters in the 1980s and 1990s included severe spells of frost and sometimes snow. Populations of Moorhen, Skylark, Wren, Dunnock, Robin, Blackbird, Song Thrush, Goldcrest, Long-tailed Tit, Linnet and Yellowhammer were all reduced significantly more than once after these winters and these are all species which figure in the lists of Table 2.3. This underlines Williamson's observation about the impact increasing frequency of such winters would have. The reverse also applies. The winter of 1962–3 was followed by a long series of mild winters, lasting until 1978. During this period the CBC shows many farmland species reaching the population peaks recorded by the census. We do not know how these peaks relate to the pre-1961 status of these species but this long series of mild winters must have contributed to the peak populations recorded. Similarly the present series of warmer summers is associated with population recoveries in several species, for example Hobby and Woodlark.

Altitude and relief

With increasing altitude higher rainfall, severer winters and a shorter growing season limit the farmer's options, ultimately to grassland and stock. For birds increasing altitude affects habitat availability. In considering this today, however, allowance must be made for the effects of

TABLE 2.3. *Significant percentage declines in farmland birds after four severe winters, 1962–3, 1978–9, 1981–2, 1985–6*

Percentage decline	1963	1979	1982	1986
78	Wren			
76				
74	Mistle Thrush			
72				
70				
68				
66				
64	Pied Wagtail			
62				
60	Moorhen			
58				
56	Song Thrush			
54	Lapwing			
52				
50				
48			Goldfinch	
46		Long-tailed Tit, Treecreeper		
44			Wren, Long-tailed Tit, Reed Bunting	
42		Wren		
40				
38				
36	Reed Bunting		Bullfinch	
34	Mallard		Skylark	
32	Linnet		Corn Bunting	Long-tailed Tit
30				Red-legged Partridge
28			Robin	Partridge
26		Partridge, Tree Sparrow	Dunnock, Mistle Thrush	
24		Reed Bunting, Corn Bunting	Pied Wagtail, Tree Sparrow	Wren
22			Partridge	
20			Moorhen	
18	Partridge, Skylark, Blackbird		Song Thrush, Linnet	
16		Red-legged Partridge, Bullfinch	Starling	
14		Skylark, Song Thrush	Blackbird, Blue Tit, Great Tit	

TABLE 2.3. (Cont.)

Percentage decline	1963	1979	1982	1986
12		Robin	Yellowhammer	Song Thrush
10		Blackbird		
8		Dunnock, Great Tit		Skylark, Blue Tit
6		Blue Tit	Chaffinch	
4				

Source: Common Birds Census farmland index.

historic habitat change and persecution. The greatest extent of semi-natural habitats that remains in British farmland is now concentrated in uplands and in one type of habitat – open rough grazings and moorland. This is significant to bird distributions. An important group of farmland birds, now considered upland species, such as Kite, Hen Harrier, Merlin, Black Grouse, Dunlin, Raven, Twite and possibly Ring Ouzel, was once widespread in lowland areas in moor, fen and heathland habitats (Holloway 1996). They are confined to the uplands today largely because that is where suitable habitat remains, their lowland habitats having been lost, or because, for raptors, the impact of persecution in more remote areas was never sufficient to exterminate them. Similarly, more widespread species once more numerous in lowland farmland, such as Meadow Pipit, Whinchat and Wheatear, are now increasingly upland birds because that is where most habitat persists for them. Altitude also affects wetland habitats. Rapid flows in hill streams render them unsuitable for species such as Moorhens or Reed Warblers, which need good emergent vegetation. Upland pools and lakes tend to be markedly less productive than lowland ones and thus support fewer species of water birds (Fuller 1982). Traditional farm ponds are rather scarce in upland districts, where stock can usually be easily watered at streams.

The approximate upper limits in the distribution of many farmland birds are shown in standard avifaunas and atlases. Several broad patterns are discernible. The number of species tends to decline with height, a well-known phenomenon with woodland birds (Hudson 1992). Such patterns have altered in recent years, as upland forestry has drawn many common woodland species uphill; altitude formerly limited the habitats rather than the birds. A surprising number of species, such as waders, appear to reach their upper limits in the range of 400–600 m. Several

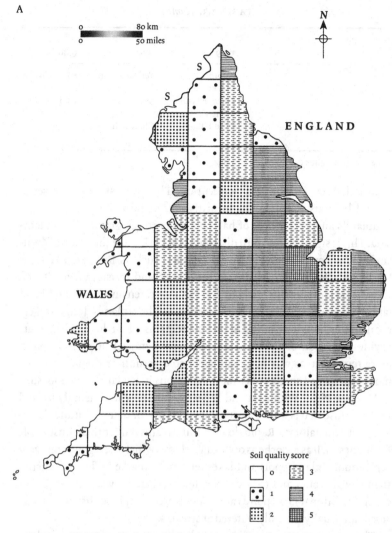

Figure 2.4. (A) Soil quality in England and Wales compared to the (B) abundance of common and widely distributed farmland birds during the breeding season. For sources and methods see text.

other upland species are most numerous at lower altitudes, such as Hen Harrier (mostly 200–300 m in Scotland), Red Grouse (mostly 300–600 m in Scotland) and Ring Ouzel (250–350 m in Wales and 250–500 m in Scotland). These points probably reflect the altitudinal distribution of particular moorland habitats.

B

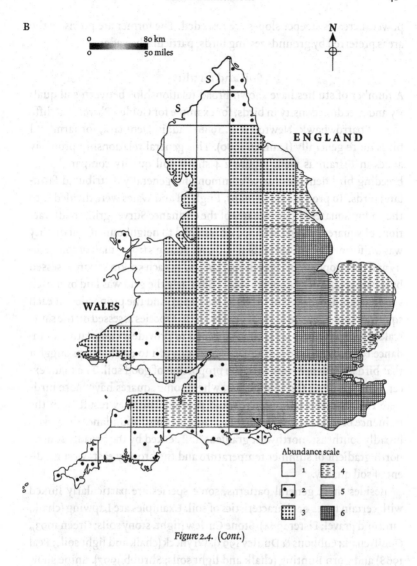

Figure 2.4. (Cont.)

Relief has a separate effect on farming because steep slopes inhibit ploughing. Thus on the southern Downs, unimproved grassland in many areas tends to be limited to steep slopes which have gone out of agriculture altogether. Unless grazed, scrub woodland develops rapidly on such sites, completely changing their character. Similarly, in the uplands, there is a strong tendency for the agricultural improvement of hill grasslands to be mainly a feature of plateaux and shallow slopes, although as tractor power increases steeper slopes are reseeded. The former are precisely the areas preferred by ground-nesting birds, particularly waders.

Soil and fertility

A number of studies have shown broad relationships between soil quality and breeding density in birds, for example for Golden Plover (Ratcliffe 1976), Sparrowhawk (Newton *et al.* 1986) and, in Denmark, for farmland birds more generally (Laursen 1980). This general relationship probably arises in Britain as well. Figure 2.4 shows soil quality compared with breeding bird densities for 56 common and generally distributed farmland birds. To produce these maps, England and Wales were divided into the 50-km squares or part squares of the Ordnance Survey grid, small fractions of squares being combined or assigned to neighbours if soil quality was uniform. This grid was then laid over Stamp's (1955) map of soil quality and the proportion of good-quality land in each 50-km square assessed by eye on a scale of 0 (none) to 5 (all). Similarly the grid was laid over each species abundance map in Gibbons *et al.* (1993) and the proportion of each square occupied by the highest density of that species assessed on the same scale. The bird scores were then summed to provide a cumulative abundance total for each 50-km square, on a scale of 1 to 6. The maps suggest that bird abundance rises with the proportion of good soil. The main exceptions are along the south coast, where some squares have more birds than their average soil quality would predict. This may result from the influence of climate discussed above. Summer bird abundance displays a broadly southeast–northwest gradient, influenced by the broadly south–north gradient of summer temperature and the broadly east–west gradient of soil quality.

Besides these general patterns, some species are particularly linked with certain types or characteristics of soil. Examples are Lapwing (chalk, sand and gravel; Lister 1964), Stone Curlew (light, stony soils; Green 1993), Quail (chalk; Gibbons & Dudley 1993), Wryneck (chalk and light soils; Peal 1968) and Corn Bunting (chalk and light soils; Shrubb 1997). Snipe show a preference for peat soils as peat retains moisture and remains easy to probe in summer; clay also impedes drainage but such species find clay substrates difficult to probe (Smith 1983).

3
Arable farming systems: high farming and before

There have been four basic systems of arable cultivation since 1750, the old two- and three-course systems typical of the open fields, followed by high farming, the three-year ley and modern chemical-based systems. High farming should be considered the first modern farming system. The period from 1750 to the present has seen a fairly continuous pattern of intensification in arable management for, even in recessions, agricultural development and experiment persisted. Farming systems also included a wider distribution of arable crops up to the 1870s than has existed since, because all farms needed some cereals for their horses and local communities depended more on local produce.

Any organism inhabiting arable farmland must adapt to the cycle of cultivation and harvest, which involves regular changes in habitat. Crop rotation adds to such problems by shifting habitats around the farm. Arable weeds cope with these difficulties by completing their annual growth and seeding cycle within the cropping cycle. Birds, being highly mobile, adapt to crop rotation by following particular stages around the farm. They can breed in one habitat and gather food in another at some distance, roost or winter elsewhere and are especially well-equipped to exploit temporary food sources, such as following the plough. Small mammals perhaps need hedges or ditches to facilitate any movements from field to field. Many invertebrates need settled conditions to complete annual cycles. These are lacking in continuous cropping systems but leys, hedges and ditch banks etc. provide habitats where annual cycles can be maintained.

Early arable systems

Little is recorded about farmland birds in the old traditional systems before high farming. Pollard *et al.* (1974) examining Laxton, Nottinghamshire, the only remaining open-field parish in Britain, noted that the bird population largely comprised field species such as partridges, Lapwing and larks and remarked that the area was rather a boring habitat for birds. But the habitat cannot be fairly assessed on the basis of a remaining fragment and without the web of other habitat elements such as commons, which was interwoven through farmland.

There was, I suggest, great diversity in the countryside in this early period. There was the mixing in many parts of lowland England of the landscape defined by Rackham (1986) as 'ancient countryside', typified by early enclosure, small fields, hedges, winding lanes and an intimate scatter of hamlets and farmsteads, and the open-field landscape, known as 'champion' or 'champaign' country to early authors. Hoskins's (1955) map of the Parliamentary enclosures indicates how much these were intertwined. Secondly there was the extent and distribution of habitats such as heath, downs, wolds, fens, marshes and mosses, largely covered by the old manorial term 'waste' and extensively managed by systems involving common rights, which were widely distributed through farmland, occupying at least 20% of the agricultural area of England and Wales before Parliamentary enclosure and probably significantly more (Hoskins 1955). They were also widespread in Scotland. They formed an integral and valued part of the agricultural area and rural economies exploited their natural resources (Chapter 4).

Nor were these marginal economies. Rackham (1986), for example, stresses the agricultural wealth of the pastoral systems of the Fens and noted that heathland was valued 'at not much less than arable'. The availability of such habitats influenced the way parishes were laid out. For example Tubbs (1993) noted that chalkland parishes in Hampshire were often long strips allowing access to the grazing of the Downs, to the coppice of the clay and the marshes of the river valleys. The same pattern can be seen in Sussex, around Lewes and along the Weald below the Downs in West Sussex. Spearing (1860) noted that farms in parts of Berkshire were similarly laid out and their steadings had become increasingly inconveniently placed. Tubbs (1993) also noted how the economies of chalkland parishes in Hampshire depended on the availability of downland and marsh grazing or water meadow, as well as the arable and enclosed pasture. The economies of the Suffolk Sandlings were similar and Armstrong (1973) noted that, with the decline of sheep in the early twentieth century, the heaths ceased to be grazed and were eventually afforested. Cobbett noted that the surrounding pastoral farms used the heath of Ashdown Forest, Sussex, for their young cattle (Chapter 9). The countryside was probably more open than today in many areas but openness should not be equated with lack of diversity.

In the Fens Rackham (1986) remarked that it was not local dissatisfaction that pressed for their drainage and enclosure but the desire of outside landowners and the Crown to make money from the application of new arable farming techniques. Thirsk (1953), writing of the draining of the Isle of Axholme, Lincolnshire, made a similar point, noting it as 'a scheme summarily embarked upon, without much prior investigation into the islanders' old way of life, or consideration of its merits'. Again she notes that 'Vermuyden's struggle with the islanders was not a struggle to create new prosperity in the place of poverty; its object was to substitute a new economy for the traditional one', which was perfectly satisfactory. Effectively a large part of the islanders' livelihood was confiscated. These are manifestations of a change in the philosophy of land management that became increasingly evident from the mid eighteenth century. Traditional uses and communities which had evolved them ceased to be regarded. Instead land conditions were everywhere adapted, by drainage, enclosure and the plough, to fit a chosen farming system. As noted in Chapter 1, there was genuine necessity. But their rights being inconvenient to such a philosophy, commoners were too often regarded as idle wastrels:

> Let those who doubt, go round the commons now open, and view the miserable huts, and poor, ill-cultivated, impoverished spots erected, or rather thrown together, and inclosed by themselves, for which they pay 6d or 1s per year, which by loss of time both to the man and his family, affords them a very trifle towards their maintenance, yet operates upon their minds as a sort of independence; this idea leads the man to lose many days work by which he gets a habit of indolence; a daughter kept at home to milk a poor half-starved cow, who being open to temptations, soon turns harlot, and becomes a distressed, ignorant mother, instead of making a good useful servant. (Bishton 1794)

It is the last part of that passage that spoils the effect. Nor does the Victorian literature provide evidence that the commoner's lot was improved as a farm labourer. Such moralising platitudes were frequently expressed. That they were exaggerated is shown by Evershed's (1869) description of the commoners of Cannock Chase, Staffordshire. The term 'waste' came increasingly to acquire the pejorative meaning we understand today. Later eighteenth-century writings on the subject are riddled with descriptions such as that of Claridge (1793) of the Dorset heaths: 'a most dreary waste and almost the only advantage derived from it at this time is the support of a few ordinary cattle and sheep, and the heath which is pared up by the surrounding villages for fuel'. Cobbett frequently makes similar comments but then inveighs as frequently against the social disruption which was the inevitable result of commons enclosure. The process of enclosure demanded capital and contributed in turn to the significant reduction of an independent yeomanry and peasantry. Farmers increasingly became tenants and their leases controlled far more what they could do than is so today. From this has stemmed the surprising uniformity of much of today's countryside.

A characteristic feature of the old two- and three-course systems was the fallow, left uncropped for a full year, grazed after harvest and ploughed at intervals, usually at Michaelmas, in late April and in late June, although there was much variation; for example five ploughings were common in Northumberland and Durham and six or more in Essex (Sturgess 1966). Such fallows still comprised 22% of tillage in the first decade of the nineteenth century (Grigg 1989). Fallows sometimes lasted longer than a year on some poorer soils. For example the Brecks of Norfolk and Suffolk, the Yorkshire Wolds and the sands of Sherwood Forest were often farmed on an infield/outfield system, where the outfield was managed in a cycle of cropping and grazing (Chapter 9). Such systems were an integral part of Scottish farming (Birnie 1955). On the clays tilled land was sometimes left to tumble to grass for two or more years as a crude ley to provide extra grazing, before being fallowed and brought back into wheat (Sturgess 1966). The main autumn-sown crop was wheat which, with a little rye, comprised about one-third of tillage in England and Wales in 1801 (Turner 1981), the remainder being spring cereals, roots and fallow. The crops were generally sown broadcast and crop density and yields were low. Weed populations were high everywhere and the state of the open fields was considered scandalous in the later eighteenth century. Genuine problems were undoubtedly widespread (Ernle 1922) but whether they should be considered quite as universal as implied is another matter. Thirsk (1964) makes the interesting observation that the regulations governing the management of the common fields were often 'at their most emphatic and lucid' in the seventeenth and eighteenth centuries, suggesting a system that was still developing. Thirsk (1997) also showed that good and innovative farming was practised in the open fields, whilst Fletcher's (1962) account of the agrarian revolution in Lancashire makes it clear that the prophets of high farming saw little good in any other system, no matter how obviously efficient and successful.

Two points about early farmland as a habitat are evident however. Firstly, food resources for seed-eating birds must have been abundant. Large finch populations were a major feature of the countryside in the eighteenth and nineteenth centuries. White (1789, letter 13 to Pennant, 1768) particularly commented that 'vast' finch flocks (mainly of Linnets and Chaffinches in his area) were found everywhere in winter farmland. Secondly the great amount of bare land probably particularly favoured ground-nesting birds. Habitats such as infield/outfield must have resembled the pseudo-steppes of Spain today and the early avifaunas confirm that they attracted many of the same species (Chapter 9). Although fallows were grazed the tillage pattern would usually have left time for ground-nesting birds to hatch and advance broods. The coarseness of the tilth and a scattering of weeds also improved the cryptic camouflage the eggs or incubating birds derived from broken ground, and provided shelter. Such micro-habitats have changed markedly over time with more efficient cultivation machinery. Species such as partridges, Quail, Corncrake and Skylark probably also benefited from thin crop stands. Tilling fallows provided summer feeding grounds for species such as Rooks, feeding on soil invertebrates, and seed-eaters, which find fresh seed brought to the surface at each cultivation.

Overall, however, it is far from clear in this period how far cultivated land was the main habitat for many birds now regarded as typical of farmland. As examples Table 3.1 examines the nesting habitats of a sample of 16 species of farmland birds, all widely distributed in the eighteenth and early nineteenth centuries, recorded in 46 avifaunas published before 1910. Most of these avifaunas are strong on historical detail and their authors had personal experience of the impact of commons enclosure and fen drainage, which feature prominently in their narratives. The data being rarely available, no assessment of numerical abundance within these habitat categories was made, each record of habitat use in each species account in each avifauna being simply scored as one. Habitats have been grouped as precisely as definitions and descriptive terms used allow. For some species more information is available on

TABLE 3.1. *Nesting habitat use in some farmland birds during the eighteenth and nineteenth centuries*

	Semi-natural habitats:[a] number of accounts recording the species in					Improved farmland:[b] number of accounts recording the species in					
Species	Heath/moor, commons, scrub, dunes, warrens	'Waste'	Downs, breck, wolds	Marshes, rough grass, fen and bog	Open hill, upland moor	Meadow, pasture (60%)	Fallow (8%)	Arable crops (40%)	Ley (4%)[c]	Hedges	Orchards, gardens, parkland
Montagu's Harrier	5		2	5	2				2		
Black Grouse[d]	22			14							
Quail	1		2	10		11		7	7		
Corncrake	1	1		7		28		11	11		
Great Bustard	3		10					3	1		
Stone Curlew	13	1	16				5	7			1
Lapwing	12	3	3	16		4	5	7			
Stockdove	12	1									6
Skylark	9		1	3		11		17			
Yellow Wagtail	5	1	3	10		23	6	15	2		
Whinchat	23	3	3	22	6			2		1	
Stonechat	34	1	2	2	6					1	
Wheatear	27	1	11	1	20	1		1			
Goldfinch	2	10		1							8
Linnet	36	7						1		4	4
Yellowhammer	10					2		1	11		

[a] See text for the extent of semi-natural habitats.
[b] Percentages of crops are of crops and grass (improved farmland) in 1808 from Grigg (1989).
[c] All clover or sainfoin except for Quail (1), Corncrake (4) and Yellow Wagtail (2).
[d] Black Grouse did not occur in any improved farmland habitats, except occasionally after breeding.

actual nest sites than habitat. This was particularly so for Stockdove and Yellowhammer.

Of 675 observations for these examples two-thirds are for semi-natural habitats. Obviously one cannot say that a greater proportion of birds was found in one habitat grouping than the other because numerical data are lacking. But, accepting that limitation, the analysis does provide an indication that these species were less likely to be found on cultivated land. As most are ground-nesting, such a bias is unsurprising. Accounts of the management of the open or common fields particularly, such as Thirsk (1964), leave the strong impression that they were too busy a habitat to be attractive to many such species. The overwhelming impression given by the records we have is that the large area of stable habitats such as heath or old pasture meant there was little need for many species, now associated with farmland, to adapt to cultivated land, which was perhaps an overspill rather than a key habitat. The Linnet provides a good example of this. Nineteenth-century avifaunas record it as moving right out of cultivation, where it wintered, to heath and scrub habitats for nesting (e.g. Haines 1907, Lilford 1895, MacGillivray 1837, Montagu 1833). Such patterns must have been more marked in the eighteenth century, when the area of semi-natural habitats was much greater. I return to this list of species in Chapter 4, when considering the impact of enclosure more fully.

High farming

High farming basically integrated crops and stock through fodder roots grown for winter feed, either to be folded and grazed off (mainly sheep) or carted to stalled animals (mainly cattle). The animals, in turn, fertilised the system through their manure from the sheepfold and the byre, making it largely self-sustaining. The development of high farming involved three fundamental changes from earlier farming systems. It introduced the regular, systematic and widespread use of short-term leys (grasses grown in arable rotations), usually of clover or clover/grass mixtures or, on chalk and sand especially, sainfoin. Although it is unclear how extensively, such ley farming was well established by the end of the seventeenth century, when the use of clovers was widely written about (Lane 1980); its origins were much earlier (Thirsk 1997). So this was a change in scale rather than the introduction of a new crop. Secondly the old annual fallow, land left uncropped for a full year, was replaced with root crops, mainly turnips or swedes but also mangolds later on the clays, grown to feed livestock in winter. Thus the land was cropped throughout the rotation cycle. Growing roots led to a third major change as roots altered patterns of livestock production, allowing far more animals to be kept over winter and fattened. This in turn raised the volume of manure available to fertilise the system. By 1870 roots occupied just over 18% of tillage (land that is cultivated annually) and rotation leys just under 25% of arable (land that is cultivated in rotation). Roots were the primary cleaning (i.e. weeding) crop of the rotation, the fields cultivated at intervals to kill weeds until planting in early summer and then hoed to suppress weeds in the growing crop, often into August. Clover and other short-term leys were established under spring-sown cereals (undersown) and were an important source of fertility in the system.

The success of the system in its early period led to its application throughout farmland, although it spread in two distinct phases. The key to the system was the successful cultivation of roots, for which clay soils in the eighteenth and early nineteenth centuries were unsuitable, being poorly drained and difficult to till sufficiently thoroughly. High farming bypassed the clays in this early period and farming there continued under the old three-course rotation until facilities for properly underdraining fields became more generally available (see Sturgess 1966). The expansion of the area of fodder roots provides some index of the spread of high farming during the eighteenth and nineteenth centuries and is shown in Figure 3.1. By 1800 c. 668 000 acres (270 000 ha) of fodder roots were being grown in England and Wales. Assuming a four-course rotation, this suggests that a quarter of the arable area was then devoted to high farming rotations. The proportion in Scotland was probably higher, for Smout (2000) stresses the scale and completeness of farming change there between 1750 and 1830; by 1856 460 000 acres (186 000 ha) of fodder roots were being grown there, 24% of tillage. The comparatively slow early spread of high farming suggested by Figure 3.1 reflected the limitations imposed by soils and also the demand for increased capital involved. The last particularly affected estates whose finances were encumbered by inherited debt or family settlements (Caird 1852). The very rapid spread of the area of fodder roots shown mid-century coincides with the great expansion of field drainage, particularly in the clays at this time (Chapter 6). But fodder roots were never so widely grown on the clays; the new farming there involving more extensive areas of green crops such as vetches and a greater dependence on purchased cake for animal feed (Thompson 1968).

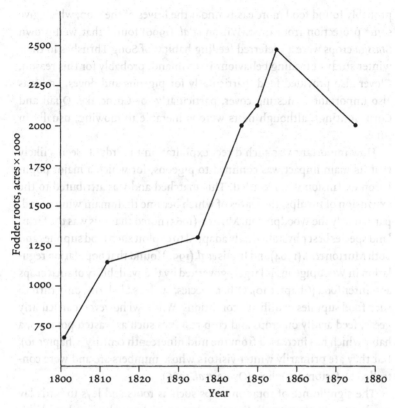

Figure 3.1. The increase in the area of fodder roots in England and Wales from 1800 to 1875. (Data from Trow-Smith (1951) and Grigg (1989).)

New crops and rotations and birds

For birds root crops were primarily a winter food source, both for grazers of the crop, such as Woodpigeon and Skylark, and for species seeking invertebrates in the crop, perhaps especially the caterpillars of the turnip moth (*Agrotis*) or of turnip sawflies (now extinct), known as jiggers. These particularly attracted plovers and larks (e.g. Borrer 1891), and presumably Stone Curlew and Great Bustards at one time, as both were recorded as feeding avidly in turnips in autumn. Borrer records that bustards feeding there on early mornings of heavy dew became too wet and heavy to fly. His father then used to course them, avoiding adult males as they could severely damage a hound! Finches, particularly Greenfinches and Linnets, were serious pests of turnip seed crops. Other species, especially gamebirds, used roots for cover and roosting and, in winter, ground-feeders probably found food more easily under the leaves of the crop, which give some protection from frost. Wilson *et al.* (1996) found that well-grown *Brassica* crops were a preferred feeding habitat of Song Thrushes in their winter study of feeding behaviour in farmland, probably for this reason. Clover also provided food, particularly for pigeons and doves, and was also important as nesting cover, particularly for Corncrake, Quail and Corn Buntings, although nests were vulnerable to mowing, usually in June.

How important was such direct exploitation to birds? It seems likely that its main impact was confined to pigeons, for which a major population expansion was recorded. This matched and was attributed to the expansion of turnips, the leaves of which became their main winter food, particularly the Woodpigeon. Murton (1965) noted that this was the farmland species best physiologically adapted to exploit such food supplies and both Murton *et al.* (1964) and Inglis *et al.* (1990) found that population regulation in Woodpigeons is largely governed by the availability of such crops as winter food (Chapter 10). Other species, such as Skylarks, only turn to such food supplies in difficult conditions. Wildfowl however, particularly geese, feed avidly on crops and crop residues such as wasted potatoes, a habit which has increased from the mid nineteenth century (Chapter 10). But they are primarily winter visitors whose numbers are and were controlled by factors elsewhere (Owen *et al.* 1986).

The significance of rotation crops such as roots and leys to birds lay much more in the overall diversity they brought to the farmland habitat. Most of the examples available to illustrate such relationships derive from modern studies and so say rather little about the use of fodder root crops, no longer widely grown. But the undersown ley was probably always the most important element of rotations, with at least two general advantages to birds. Undersown cereals could not easily be hoed, as the ley seeds were broadcast, not drilled. Although taller plants such as thistles could be pulled out, their stubbles were therefore always likely to provide weed seeds as well as waste grain; as the leys stood for at least a year without tillage, they also provided cereal stubbles as a source of food throughout the winter and spring. More recently they have also provided some check to the uninhibited use of herbicides. They are also essential sources of invertebrate food for many birds, largely because populations of important groups, such as sawflies, occur at much higher densities in traditional undersown leys than in cereals, where annual cultivations cause a significant proportion of total mortality (Potts 1986).

A feature of high farming was the great increase in manure applications, both from folding increased numbers of sheep and the greater volume available from stalled and yarded animals (e.g. see Thompson 1968). Farmyard manure was the main constituent but other organic manures, such as bone meal and guano, were increasingly used and, where readily available, town sewage. There can be little doubt that this increased soil invertebrate populations and both Shrubb (1988) and Tucker (1992) showed that annual applications of dung increased the attractiveness of fields to birds feeding on them. This factor was equally important in grassland, where there was a marked increase in fertility in the nineteenth century (Chapter 9).

Farmland birds characteristically use more than one crop habitat, perhaps an important adaptation to life in a habitat type subject to abrupt seasonal changes from cultivation and harvesting operations; the growth of crops also progressively inhibits many species feeding. Rotation offers alternatives. Such patterns may involve the use of more than one crop stage in one season or seasonal selection of different crops. The latter may particularly influence feeding behaviour, the former nesting. Or birds may show a strong preference for particular stages of a rotation and follow that around the farm whilst, for some, the rotation itself may be important.

The Corn Bunting provides an example of the last. Gillings & Watts (1997) found that Corn Bunting density was linked to crop diversity in the Lincolnshire Fens and Shrubb (1997) showed that this species' status in Britain has long been correlated with the availability of mixed arable rotation, rather than any particular crop. A study by Aebischer & Ward (1997) suggested why. The highest densities on their study area on the Sussex Downs occurred in areas combining ley and spring barley, which was linked to the much greater abundance of caterpillars in the undersown cereals compared to winter cereals and other crops. The Grey Partridge is another species for which such crop combinations are important, as the work of Potts and his colleagues has shown. The importance of arable rotations to partridges was also widely recorded in the nineteenth century, particularly in the west and north (e.g. D'Urban & Mathew 1895, Harvie-Brown 1906, McPherson 1892). The Rook is also influenced by rotational diversity. Brenchly (1984) noted that it reaches its greatest density in an optimum mixture of 44% tillage crops and 55% grass, with breeding density declining significantly as farmland moved away from that optimum in either direction. Finally, Wilson *et al.* (1997) found that the breeding productivity and population density of Skylarks on lowland farmland in southern

England declined with the loss of mixed rotations in favour of autumn tillage and intensive grass management. The birds needed to make two to three nesting attempts per season for self-maintenance which was possible with crop diversity but not in monocultures. Similarly Schlapfer (1988) suggested that crop diversity, by increasing the availability of secure nest sites, was a major influence on Skylark numbers in Switzerland. More recent studies, summarised in Wilson *et al.* (1995), have shown a strong preference by breeding Skylarks for rotational set-aside, a form of management that virtually replicates the old undersown leys used in high farming systems. Territory density ranged from 0.17–0.36 per hectare in set-aside, compared to 0.018–0.10 in winter cereals, and breeding success was six times greater. Interestingly wheat grown under an organic regime was as favoured and successful as set-aside, perhaps because crop density was more favourable for larks with reduced fertiliser and pesticide inputs. But that reduction probably improved food supplies and, again, such farming reintroduces many important features of high farming.

Such preferences also occur in feeding behaviour. I examined feeding patterns of birds over a complete year on two farms in Sussex, doing weekly or fortnightly counts of all birds observed on a fixed route. The counts were done in the early 1980s, when the farms still included one-year clover leys established under spring cereals. Overall the counts showed that permanent grass (all old unimproved pastures grazed by cattle), newly sown land and stubble were the most favoured feeding sites; clover leys were also important in spring and summer and pasture little used in autumn (Table 3.2). The table also examines the seasonal variations I recorded for major groups of farmland birds. It emphasises the lack of use of standing cereal crops and the importance of other features of the rotation, although plovers, larks and thrushes made much use of young cereals in autumn and winter. Wilson *et al.* (1996) obtained similar results in a study of winter feeding behaviour on five farmland areas in the Thames Valley in 1994–5 with a more rigorous statistical analysis. The main difference on their sites was that young cereals were avoided by all species, a difference that may have arisen from differences in rotations, particularly a lack of undersowing. Henderson *et al.* (2000) also found similar results in examining summer use of set-aside compared to crops and grassland; rotational set-aside was the preferred summer habitat for a broad range of farmland birds.

Davis (1967) made a similar survey in an area of mixed tillage crops with a small area of grass in Huntingdonshire in summer 1966. Again there

High farming

TABLE 3.2. *Habitats preferred for feeding by all farmland birds combined (overall) and different groups on two Sussex farms at different seasons*

Season	Pasture	Ley	Young corn	Standing corn	Spring cereals	New sown	Stubble
November–February							
Overall preference	+30	+17	−36	N/A	N/A	+55	−10
	Pigeons	Pigeons	Larks			Plovers	Pigeons
	Thrushes	Starlings	Thrushes	N/A	N/A	Pigeons	Larks
	Starlings	Finches				Larks	Finches
	Corvids	Buntings				Finches	Buntings
						Buntings	
March–May							
Overall preference	+30	+28	−91	−149	−135	+5	+12
	Plovers	Pigeons				Thrushes	Pigeons
	Pigeons	Larks				Corvids	Larks
	Larks	Finches				Finches	Thrushes
	Thrushes	Buntings				Buntings	
	Starlings						
	Corvids						
June–August							
Overall preference	+11	+35	N/A	−80	−98	N/A	+29
	Plovers	Plovers					Corvids
	Larks	Pigeons	N/A			N/A	Finches
	Starlings	Larks					Buntings
	Corvids	Thrushes					
		Starlings					
		Finches					
		Buntings					
September–October							
Overall preference	−34	−31	−23	N/A	N/A	+28	+148
	Thrushes	Thrushes	Plovers			Plovers	Plovers
	Corvids	Finches	Larks	N/A	N/A	Pigeons	Pigeons
		Buntings				Larks	Starlings
							Finches
							Buntings

Notes: The Preference Index was log(O/E) × 100, where O is the numbers observed in the crop habitat and E the expected number, if distribution was uniform in relation to area. Positive values indicate that the crop habitat was selected, negative that it was avoided. N/A, habitat not available. Groups are entered if the Preference Index was >+5.

Habitat definitions were: leys were all one-year leys established under a spring-sown cereal nurse, recorded as ley from 1 April to ploughing in the following autumn and as stubbles from harvest to 1 April; young corn was autumn-sown cereals from full crop emergence to 1 April; standing corn was autumn-sown cereals from 1 April to harvest; spring cereals were spring-sown cereals from full crop emergence to harvest; new sown was new-sown ground from planting to full crop emergence; stubble included undersown stubbles until 1 April (see ley), burnt and unburnt stubbles.

TABLE 3.3. *Preferred feeding habitats of birds on a Huntingdonshire farm during May to July 1966*

Pasture	Wheat	Oil-seed Rape	Peas	Beans
Skylark		Red-legged Partridge	Skylark	Blackbird
Blackbird		Woodpigeon	Starling	Whitethroat
Starling		Blackbird	Chaffinch	Dunnock
		Whitethroat	Linnet	Chaffinch
		Dunnock	Sparrows	Yellowhammer
		Yellowhammer		

Note: Preferences calculated as in Table 3.2.
Source: Data from Davis 1967.

was much variation in selected feeding sites but standing cereals were generally avoided, the small extent of pasture rather little used (perhaps a function of its limited area) and the rotation crops were most important, particularly for seed-eaters (Table 3.3). The study is particularly interesting for drawing early attention to the use of oil-seed rape by birds. Tucker (1992) looked at feeding behaviour by invertebrate feeding birds in winter in an area of mixed farming in Buckinghamshire. Permanent grass was the preferred habitat for the species he examined (which excluded seed-eating passerines). Permanent grass fields where farmyard manure was frequently applied were particularly favoured as were long-established pastures, especially by Golden Plover, Starlings, Rooks and Crows. Applications of farmyard manure increased the attractiveness of tilled fields as feeding sites. Patterns of stocking also influenced choice, with plovers avoiding sheep but favouring cattle, whilst Magpies and Jackdaws favoured sheep.

The classic example of a species using more than one crop habitat in one season is provided by the Lapwing which, in arable habitats, shows a strong preference for nesting on spring tillage (Shrubb 1990, Shrubb & Lack 1991). However Galbraith (1988) showed that breeding success depended on such fields being close to grass fields, where young could be moved quickly for rearing. Shrubb & Lack (1991) generalised this observation, showing a marked correlation between Lapwings selecting spring-tilled fields for nesting and the proximity of grass fields in England and Wales in 1987. Wilson *et al.* (2001) found this preference even more marked in 1998, when the population was continuing to decline sharply.

All these examples demonstrate that farmland birds benefit from a variety of crops and management and that they exhibit marked preferences, which may vary in different cropping regimes. Although deriving from modern studies there is no reason to deny their applicability to the high farming period. The patterns would have been similar but not necessarily precisely repeated. The significance to birds of farmyard and other organic manures, which emerges from some of these studies, would have been especially relevant in the high farming period. The fertility of soils was raised everywhere. It is worth repeating that high farming, by introducing favourable habitats and resources on a large scale, allowed species displaced by the enclosure of the waste to adapt to inhabiting farmland, an important contributory cause to the underlying stability and low level of extinction in farmland bird populations in the period. The impact of modern farming has been to remove many of these alternatives from farmland habitats, producing a type of farmland to which many species can no longer adapt (Chapter 8). The consequent decline in habitat diversity has been accompanied by declines in a high proportion of the species discussed (Figure 2.3).

Changes in the scale and timing of cultivations

A feature of the spread of high farming was an increase in the process of cultivation and marked changes in its timing. The basic cultivation is ploughing which reduces populations of soil invertebrates by inverting the topsoil, displacing both those animals living near the surface and those dwelling deeper. It thus has marked implications for the food supply of birds feeding on these animals, although availability and opportunity are important factors in their feeding patterns. Although soil invertebrates are most numerous in long-established pasture, birds often seek them behind cultivations where they are most exposed. Ploughing and cultivations also bring seeds to the surface and provide a valuable food source for many finches and buntings particularly. Ploughing also affects ground-nesting birds directly by destroying any nests present. In high farming the frequency of ploughing over the rotation's cycle increased by about 20% compared to older systems, the increase being mainly concentrated in the root course between late March and June. Thus the most significant change in timing was the loss of annual fallows, with their favourable pattern of cultivations to ground-nesting birds, to roots (Figure 3.2). The term fallowing came to mean preparing of land for roots, for which the soil was cultivated at intervals (of roughly a fortnight in my own experience) through the spring and early summer until planting. The roots were then hoed regularly to suppress weeds. The use of drills

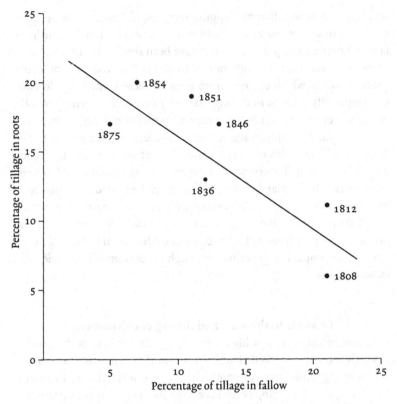

Figure 3.2. The percentage of tillage in fallow in relation to the percentage of tillage in roots in England and Wales during 1800 to 1875. Line fitted by eye. $r_s = -0.75$, $n = 7$, $p < 0.05$. (Data from Grigg (1989) and June Census Statistics.)

and hoeing to control weeds also spread such increased activity into cereal crops. Such factors contributed to the loss or decline of species such as Great Bustard, Montagu's Harrier and Stone Curlew as breeding birds in Britain (Chapter 4).

Although less vulnerable to such processes, Lapwings were also significantly affected by changes in cultivations. Many county avifaunas of the first half of the twentieth century stress the scale of nest losses in cultivated habitats as a reason for decline (e.g. Chislett 1953, Frost 1978, Smith & Cornwallis 1955), particularly where grassland was ploughed during the 1939–45 period. However, Shrubb (1990) found a rather more complex pattern in the BTO Nest Record Cards from 1962 to 1985. These showed that only 43% of nests on bare plough were successful and only 52% on all tillage other than cereals. By contrast, 70% of nests on spring cereals were successful and pairs hatched more young overall on tillage than grass because of that. The difference between this analysis and the county avifaunas reflected changes in the nature of spring tillage. The area of fodder roots and fallows declined by 47% between 1938 and the mid 1960s, whilst spring cereals increased by 193% with the adoption of the three-year ley system. The difference probably involved a substantial improvement in breeding success by Lapwings in tilled habitats in the mid twentieth century and Figure 3.3 illustrates the effect of this cropping change on the selection of breeding habitats by Lapwings at Oakhurst. Any benefits were short-lived. Shrubb (1990) showed that, in cereal-growing areas, the decline of Lapwings in the 1970s and 1980s was closely correlated with the decline of spring cereals in favour of autumn-sown crops.

Weed control

The introduction of high farming and its proper application in the period up to about 1870 led to a real reduction in weed populations in the countryside, perhaps most obviously in the reduction of thistles. Virtually every nineteenth-century avifauna stressed the point. The decline stemmed from two causes. The enclosure and cultivation of semi-natural habitats, such as heaths and commons, destroyed their herbaceous flora, a factor widely quoted to account for declines of the Linnet and Goldfinch. Secondly the application of drill and hoe to virtually all arable crops significantly reduced populations of arable field weeds (Chapter 7). Such methods control the abundance of these plants efficiently but depend on high inputs of labour and continuous application (Chapter 7). A combination of a series of wet and difficult years, declining prosperity, and social change and mobility for rural workers from about 1870 led to an overall decline in labour and a marked decline in the efficiency of weed control. This was also frequently commented upon, particularly the spread once again of thistles. This period of prosperity for farmland weeds persisted until the widespread application of herbicides which accompanied changed systems of rotation.

Recession

Major periods of recession in agriculture occurred between 1815 and about 1840 and between the late 1870s and about 1900. There was some recovery during the first decades of the twentieth century, particularly under the

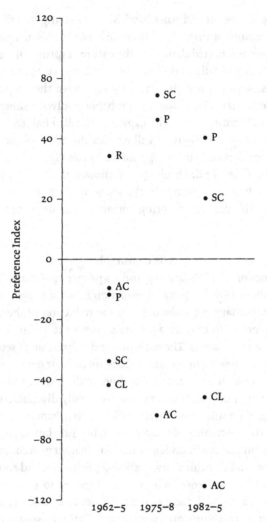

Figure 3.3. Habitat selection by nesting Lapwings at Oakhurst in three periods with different rotations (see text). Preference Index as Table 3.2; positive values mean habitat selected, negative avoided. Habitats available were: P, permanent grass; R, roots; SC, spring-sown cereals; AC, autumn-sown cereals; CL, clover leys. Field beans were also grown in the 1970s but never used by Lapwings.

stimulus of the 1914–18 War, but it was insufficient to reverse the trends set after about 1875, and there was a major slump from 1921 to 1939. It is important to realise that these periods of recession primarily affected the arable regions of the south and east. Farming in the north and west and in Scotland was always more resilient and profitable (Chapter 1). Over the past two centuries it seems extraordinary how much the prosperity of arable farming has depended on protected markets in one form or another or on direct Government subsidy. Such measures distort the financial and commercial disciplines and constraints in which farming operates. Artificially high prices, whether the result of war disrupting trade (the case particularly in the Napoleonic Wars), Corn Laws, deficiency payments or intervention, encourage arable farming to advance into increasingly marginal and unsuitable areas for it. Ernle (1922) observed of the period before 1815 that the inflated prices of the Napoleonic War had 'brought under the plough districts that, but for their stimulus, might never have been brought into cultivation, – areas that were forced into productiveness by the sheer weight of metal that was poured into them'. With recession arable withdraws but the environmental change wrought is often irreversible, a point made by Cobbett in the 1820s and Hudson (1900) when looking at the high downs in Sussex, Hampshire and Wiltshire, ploughed briefly for wheat in the early 1800s. Nevertheless, in the past recession in farming provided the environment with valuable relief from the pressures agriculture applied.

The major depressions of the nineteenth and early twentieth centuries had several common features. Wheat prices collapsed after both 1813 and 1875. Between 1816 and 1840 and in the 1870s a series of wet summers and poor harvests aggravated agricultural depression, through poor crops and by inhibiting cultivations. Maintenance of hedges and buildings was sharply reduced and, after 1875, drainage neglected. Land went out of arable, particularly on clay soils, stocking rates were reduced on grassland and land was abandoned altogether. Ernle (1922) showed that all these factors were important between 1815 and 1840 but comparing his account with that of Cobbett suggests that the major impacts of this period were particularly social – widespread bankruptcy, unemployment, personal hardship and pauperism – which both record, rather than extensive abandonment or dereliction of farmland which Cobbett rarely mentions, although he often notes good farming and crops. One interesting point was that this early recession did not slow the pace of drainage (Chapter 6). The size and volume of hedges certainly increased as maintenance was neglected (Chapter 5) which would have benefited hedgerow birds. Overall, however, we do not know whether this early recession had any marked general impact on farmland birds because information is scant.

The decline of arable farming

The main feature of the recession after 1875 was the expansion of grassland at the expense of arable land (Chapter 9). This theoretically represented both a loss of diversity and a contraction of resources, especially food supplies, for many bird species, perhaps particularly corvids, larks, finches, buntings, pigeons and some gamebirds. Although few census data are available, there is rather little evidence that this decline of food resources led to marked change.

There were perhaps two main reasons. First, although the area of arable declined by *c.* 41% in England and Wales and 16% in Scotland between the peak in 1871 and the nadir in 1938, arable farming methods did not change. High farming remained dominant, although new methods and techniques were being pioneered (Chapter 1). Coupled with this was a general decline in farming standards, largely stemming from the continual decline of labour, which resulted in a steady increase of weed populations, a fact that is well attested in contemporary county avifaunas. The mechanisation of tasks such as harvesting, threshing and crop handling also stagnated. Thus features such as binder, rick and threshing drum remained widespread throughout the farming system.

Secondly arable rotations remained widespread throughout farmland. Most farms needed to be self-sufficient in a period of financial stringency and tended to grow at least some of their own fodder crops, perhaps especially oats for the horses. Even in Wales, where pastoral farming has always been dominant, this applied, with patches of arable crops on nearly every farm, including hill farms. These were particularly exploited by Red and Black Grouse and their loss today may have contributed to the decline of the latter. Most of the farmland birds which exploit the food resources supplied by farming operations are flocking species, at least in winter, which enables them to exploit efficiently food resources distributed in such scattered parcels. The importance of a patchwork of arable fields in pastoral farmland in maintaining populations of some farmland birds has recently be demonstrated by Robinson *et al.* (2001). It is also possible that, at the peak of high farming, winter food supplies were superabundant, allowing considerable 'slack' in the system (Chapter 10).

Nevertheless, where more detailed data exist, they show that some farmland birds, at least, were influenced by this loss of farmland diversity. In Wales, for example, Grey Partridge bags fell steeply after 1900 (Figure 3.4), indicating a sharp population decline, associated in the avifaunas with the loss of arable farming, which was followed by a decline in

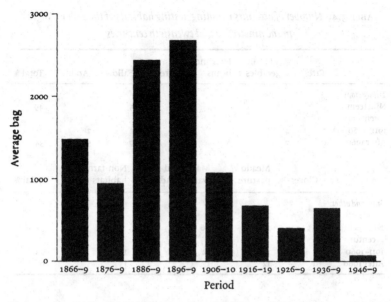

Figure 3.4. Average Grey Partridge bags on three Welsh estates from 1866 to 1949. Figures are the sum of five-year averages at Powis Castle, Voelas and Stackpole. (Data from Matheson 1953, 1957.)

gamekeepering. Similar patterns were recorded by county avifaunas elsewhere in western Britain for the same reasons. Rooks were also affected by the loss of diversity, as Brenchley's (1984) figure predicts. This was demonstrated by the sharp increase recorded between the 1930s and 1940s by Fisher (1947, in Sage & Vernon 1978), which was linked particularly to the recovery in cereal growing by Dobbs (1964). Corn Buntings also declined with the diversity of farmland (Shrubb 1997).

A particularly interesting species in this context is the Yellow Wagtail. Smith (1950) examined its distribution and population in detail. It virtually disappeared from Scotland in the first half of the twentieth century and declined markedly in southern England but it increased in the northwest. Nineteenth-century ornithologists expected to find nesting Yellow Wagtails in dry arable situations as often as grassland. Table 3.4 summarises the records for nesting habitats of this species recorded in 30 of the avifaunas examined for Table 3.1 (the other accounts gave no exact details) together with those in 17 giving similar detail for the period 1911–60 and 24, covering 27 counties, since 1960. As in Table 3.1 each record of each habitat use scores one and no information is available on relative abundance in these habitats.

TABLE 3.4. *Number of accounts recording nesting habitats of the Yellow Wagtail in the nineteenth and twentieth centuries*

	Tares	Roots and vegetables	Peas and beans	Cereals	Fallow	'Arable'	Total %
Tillage crops							
Nineteenth century	1	0	3	8	5	3	49
1911–1960	3	4	1	7	2	2	28
After 1960	0	7	6	5	0	5	34

	Clover	Meadow/ pasture	Marsh and water meadow	Non-farm habitats	Total %
Grass and other habitats					
Nineteenth century	0	12	6	3	51
1911–1960	3	19	12	14	72
After 1960	0	12	17	15	66

Mason & Lyczynski (1980) noted that 12% of nests in the BTO Nest Records were found in arable crops, 66% in grassland and 21% in other (non-farmed) habitats. Whilst these records may be biased towards finding nests in grassland and non-farm habitats, observers being less willing to trespass in growing crops, they do tend to confirm the continued declining use of arable sites. Most compensatory increase in habitat use in the twentieth century has not been in grass farmland but in non-farmland habitats – sewage farms, reservoirs, saltings, dunes and heathland. This decline in the Yellow Wagtail's use of tillage crops coincided with a marked population decline. Few of the nineteenth-century avifaunas consulted recorded any change in status but of the twentieth-century ones up to 1960 58% recorded a decline. That decrease has continued particularly in the pastoral west. Thus the change map in Gibbons *et al.* (1993) shows losses after 1972 in 109 10-km squares west of 2° W and gains in only 16, compared with 72 losses and 37 gains east of that line; the difference is statistically significant ($\chi^2 = 12.94$, df 1, $p<0.01$). The main concentration of the species in Britain now lies in the arable east of England, suggesting that the loss of tilled habitats in the west may well be a significant factor in the decline. Interestingly, Smith (1950) remarked that many pairs or groups of pairs of Yellow Wagtails bred only in particular crops, for example potatoes, and suggested that this was an inherited preference. In this case the major loss of tillage by the 1930s may have led to the total loss of breeding habitat for many pairs. The decline then would have resulted from a decline in the diversity of habitats available rather than changes in the scale of cultivation. Furthermore, in northwest England where Smith noted an increase, arable farming hardly declined in the recession, as there was a large increase in vegetables, a favoured nest site, grown for the industrial conurbations.

Otherwise the expansion of new crops, particularly vegetables and sugar beet, may have had impacts of which we now know nothing. They also retained arable farming over a larger area than otherwise would have been the case, helping to maintain the diversity of farmland. The spread of dairying was also of long-term significance in the management of grassland. Nicholson (1926) also suggested that the long-term expansion of orchards (which increased in area by 80% during 1875–1950) contributed to an increase in tits, Greenfinches, Bullfinches and perhaps Blackbirds and Starlings. Goldfinches probably also benefited and all the species which Nicholson mentioned were also recorded as expanding their ranges northwards during the period, largely as a result of the spread of trees in new plantings (Alexander & Lack 1944).

4
Enclosure

Enclosure was essential to the large-scale application of high farming because rights exercised in common in the open fields imposed limitations on such management. From the mid eighteenth century the economic advantages of high farming provided the impetus for the final enclosure by Parliamentary Act or Award of the areas still dominated by open field. By then this form of land tenure had become strongly concentrated in the English Midlands and eastern England from Norfolk north to the East Riding of Yorkshire and was progressively less important elsewhere (Hoskins 1955). Indeed, Thirsk (1964) noted that many pastoral districts probably never fully developed open or common field systems. As well as the remaining areas of open field, Parliamentary enclosure affected very large areas of common land and waste. For farmland birds this was a much more significant change. Parliamentary enclosure of the open fields peaked in the late eighteenth century, of common and waste in the early nineteenth (Figure 4.1).

Enclosure affected the pattern of landowning as well as habitats. Common grazings were particularly important to small farmers and smallholders, whose holdings were viable because they could graze their stock, which often included geese, on common land. Even today, many upland farms in Wales are similarly placed (Penford & Francis 1990). The right to graze stock and exploit other natural resources on the common was much more valuable than an additional patch of land or its monetary value, held to represent their interest, which is what such commoners received on enclosure. Enclosure also involved the expensive obligation to fence each holding within a year (Hoskins 1955). Costs for fencing were apportioned on the basis of the area of land allotted and were higher on smaller holdings as these have proportionately longer boundaries. Pollard *et al.* (1974)

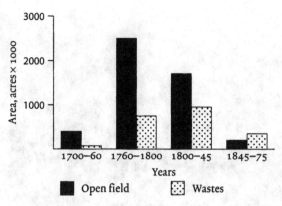

Figure 4.1. Areas of open field and of wastes enclosed in different periods in England and Wales; areas in thousands of acres. (Data from Hoskins (1955) and Hoskins & Stamp (1963).)

quoted figures for Warwickshire showing costs of £2 15s per acre for holdings of less than 50 acres (20 ha) but only £1 2s per acre for holdings of over 200 acres (81 ha). In today's money these costs ranged from c. £204 down to c. £82 per hectare.

The Award for the enclosure of Sidlesham Common, in my home area in Sussex, stipulated that 'any person to whom parts of land are allotted must accept their allotment within one calendar month and to inclose and fence same with a ditch three feet deep and five feet wide and bank and hedge to be made in good husbandry and workmanlike manner within nine calendar months' (Humpheryes 1981). Humpheryes noted that, as the cottagers with rights on the common were each allotted one-eighth of an acre (0.05 ha) as their portion, such a regulation shows why they sold their holdings and became labourers. Quite apart from the expense, such boundaries would have rendered their allotments pointless. Elsewhere commoners often exchanged them for a tenancy (Orwin & Whetham 1964). Consequently freeholders declined from owning 25–33% of farmland in England and Wales in 1690 to 15% in 1790 and 10% in 1873 (Grigg 1989). Landlords commonly imposed the forms of rotation to be followed, resulting in marked uniformity across farmland.

Perhaps Sidlesham was an extreme case but Awards typically specified that 'plots of land allotted by virtue of this Act shall be inclosed and fenced around with ditches and quickset hedges with proper posts, rails and other guard fences to such quickset hedges' (Pollard et al. 1974). Whether this was always comprehensively done is another matter. Acts

Figure 4.2. A sheepfold in turnips, Berkshire. Internal enclosures were not necessary for rotation farming unless for cattle or horses. (Courtesy of Rural History Centre, University of Reading.)

of Enclosure, in any case, included no regulations governing internal fencing within each ownership. A series of hunting prints, mainly from the early nineteenth century, illustrated by Nevill (1908), regularly shows fences but rather few associated hedges and those mainly of indifferent quality; nor are many ditches illustrated. This regulation perhaps varied with need; Read (1855a) noted that many such ditches were vestigial in Oxfordshire. Cambridge (1845), however, remarked on the frequency of pointless ditches in quite dry areas and advocated piping them in.

The legal process of enclosure, which involved extinguishing common rights and consolidating holdings into discrete parcels of land held individually, should be separated from the process of dividing blocks of land into fields. A field may be broadly defined as an enclosure or part enclosure devoted to a single crop, an enclosure as an area confined by a single boundary which may be divided into more than one field. Crop rotation does not need enclosures if simple and practical temporary fencing is available as needed for stock such as sheep (Figure 4.2); hurdles were widely used. But fences were necessary for cattle and horses and hedges were probably planted because they were the most efficient long-term solution available.

In lowland Scotland enclosure in the English sense did not occur, because land was almost entirely held by large estates. So consolidation of holdings in runrig (the Scottish equivalent of open-field systems) was within the landowner's fiat. Nevertheless the legal mechanism for consolidating strips held in runrig into single holdings was set up under the Runrig Act in 1695 and runrig died out by the end of the eighteenth century except in the northeast (Birnie 1955, Symon 1959). The abolition of runrig was followed by rapid improvements in agriculture which, as in England, involved extensive scrub clearance and reclamation of old damp rushy pastures and similar habitats. Smout (2000) stresses the sweeping and wholesale nature of this revolution in management from 1750. The new farms were laid out in larger and more conveniently shaped fields than in many parts of England and, as a result, hedges generally are younger and less varied in lowland Scotland (da Prato 1985). The Lothians and Berwickshire particularly became notable for their farming, of which Bradley (1927) has left an excellent description. In the Highland zone the Clearances had major long-term ecological effects, which are considered later.

A further source of changing field patterns in this period was the subdivision into smaller fields of parishes enclosed in previous centuries. Pollard *et al.* (1974) quoted the example of Buckworth in Huntingdonshire, fully enclosed by the 1680s into fields averaging *c.* 80 acres (32 ha), for sheep-ranching, and subdivided into fields averaging 16 acres (6.5 ha) by 1839, a process clearly in progress before the end of the eighteenth century. Such parishes do not figure in estimates of the area affected by Parliamentary enclosure, being already enclosed in the legal sense. But they added to the area of hedged countryside.

Enclosing the open or common fields

Figure 4.3 shows the basic distribution and extent of the area of open field enclosed by Parliamentary Act in the eighteenth and early nineteenth centuries. It is redrawn from Grigg (1989) but does not basically differ from the map of Hoskins (1955). Hoskins showed that only eight English counties had virtually no land in this tenure during the period – Durham, Lancashire, Cheshire, Monmouth, Devon, Cornwall, Kent and in Essex away from the chalk. Altogether between 1761 and 1844 Hoskins recorded over 2500 Enclosure Acts in England, covering over 4 million acres (1.62 million ha) of open field, with a further 400 000 acres

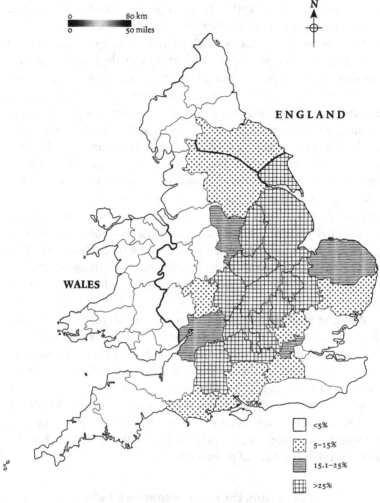

Figure 4.3. The percentage of agricultural area comprising open field arable enclosed by Parliamentary Act during the eighteenth and early nineteenth centuries. (Redrawn from Grigg (1989).)

(160 000 ha) dealt with before 1760 and 200 000 (81 000 ha) after 1845, for a total area comprising *c.* 21% of farmland. This was not the full area involved, for some enclosure was by private agreement or by the purchase of all rights in common in a parish or manor by a single person (Pollard *et al.* 1974), who suggested that the equivalent of only *c.* 3% of the area affected by Parliamentary Acts was likely to be involved because there was little incentive to use older methods during the period. But that increased the area

affected by some 150000 acres (61000 ha). These authors also noted that the pace of enclosure tended to vary in line with the prosperity of farming.

Whilst this process eventually produced a major landscape change in the areas of open field by considerable planting of hedges and trees, there is little evidence in the early avifaunas that it much affected farmland bird populations. Although a major change eventually was a long-term decline in the openness of the habitat, the common farmland species of open fields clearly adapted easily to enclosure. An exception was apparently the Quail, for which a number of authors noted a preference for the open fields rather than enclosures (see below). Early sources, such as parish or manorial records, noted the expansion of woodland birds into the new hedges of enclosure (Hoskins 1955, Murton 1971) but did not record the loss of open-field species. Similarly the nineteenth-century avifaunas record not the loss of common ground-nesting or field species to enclosing the open fields, but the loss of specialised species to the enclosure, drainage and ploughing of extensive heaths, commons and fens. Such losses were the result of habitat change (and persecution both of raptors and of scarcer species generally) rather than the change in the openness of the landscape. In fact enclosure did not automatically lead to rapid changes in this. Only the boundaries of allotments had to be fenced within a given time. Internal fences, being a private matter, were sometimes slow to appear which is obvious enough in old hunting prints, as Hoskins (1955) remarked. Overall, however, this apparent lack of ornithological consequences from what was a profound landscape change perhaps supports the idea discussed in Chapter 3, that the open fields never carried very important populations of breeding birds in an era when large areas of uncultivated land existed in the countryside.

Nevertheless the structure of bird populations within open field areas eventually changed considerably, as the new enclosure hedges came to support large numbers of common birds. Recent research has shown that these include a number of hedgerow specialists, so that hedges hold a distinct bird community, differing from but overlapping that of woodland (Fuller *et al.* 2001). O'Connor & Shrubb (1986) examined various habitat features of farmland and their relationship with the abundance of 57 common farmland birds. Openness of the habitat correlated significantly with the abundance of 25 of these species but it was a negative relationship for 16. On the other hand the abundance of 41 species was positively correlated with the presence of hedges with trees. However modern studies have shown that, unlike the overall abundance of birds, the number of species tends to decline rather little as landscapes become more open through hedgerow removal because of a gain in plains species (Bull et al. 1976, Murton & Westwood 1974, O'Connor & Shrubb 1986). We may reasonably assume that such factors operated in reverse during enclosure.

One important result of enclosing the common arable on the clay soils of the Midlands was a trend to convert it to pasture, a practice with a long tradition. Hoskins (1955) drew attention to the uniform field size of c. 10 acres (4 ha) which resulted. We can only speculate about the effects of this change on bird populations but the art of the grazier was to keep swards tightly grazed so that stock were continually grazing fresh springing grass. Paddock grazing was the preferred method, hence the uniform field size. Modern experience suggests that such grassland would have been of limited value as breeding habitats. Paddock grazing systems, with their continual shifts of stock, are rather traumatic habitats for ground-nesting birds for example, although Skylarks and Yellow Wagtails might have bred in young hedge lines. But many of these areas of pasture would have been good feeding sites for winter visitors such as plovers, thrushes and Starlings. Fuller (1986) remarked that many Lapwings, for example, feed in particular fields year after year, a comment that he based on his studies in the southern Midlands. Such traditional feeding sites may have been available for a long time there, for Broad (1980) shows that much permanent pasture in the southern and eastern Midlands of England dates well back into the eighteenth century and sometimes earlier.

Enclosure and woodlands

Enclosure demanded considerable amounts of fencing, a demand which was associated with a decline of woodland in what were already thinly wooded areas (Hoskins 1955). Fletcher (1962) gives scale to this demand. On the Hopwood estate in Lancashire nearly 20 miles (32 km) of fences were repaired and renewed between 1846 and 1848, requiring 66.5 tons of oak poles (I estimate a total of 8000–10000 poles) bought in, besides vast lengths found on the estate. The scale of use there was reckoned at 400 yards to the ton of poles (336 m/tonne). This was for repair and replacement rather than new fence lines, which would probably have demanded more; some of the fences illustrated in Nevill (1908) certainly suggest it. My home area in Sussex provides an example of the scale of woodland loss at this period. Yeakall & Gardner's map for the first Ordnance Survey, dated 1778, marks 36 small woods, varying in size from about 1 to 8 ha, widely scattered across the Selsey Peninsula, and more are marked to the north and east. They were a regular feature of farmland there and every holding had at least one. Most had vanished by 1846, some by 1824. Their disappearance coincided with the enclosure of part of Selsey parish and a string of commons stretching from Hunston to Wittering. Although I cannot prove this, it seems most likely that this surge in demand for fencing materials provided the finance to clear these woods for farming, instead of retaining them as coppice. Only four remain, now oak woods with remnants of coppice. Rackham (1990) remarked that the farmland created by such woodland clearance was of poor quality, a point readily discernible in my Sussex area. Woods were nearly always on poorly drained sites there; even today the outlines of those that existed on Oakhurst can be traced quite easily. I did counts in all the remaining woods in the 1960s. They then contained a total of 32 breeding bird species, of which Green Woodpecker, Nightingale, Blackcap, Willow Warbler, Chiffchaff, Spotted Flycatcher and Jay did not occur on my neighbouring farmland CBC area. Sparrowhawk bred in these woods in the 1950s and Long-tailed Tit and Treecreeper before the 1962–3 winter; all subsequently returned. Such woods would have added to the diversity of birds in farmland in the area and this must have been so everywhere. But if birds were lost from woodlands during enclosure, the nineteenth-century avifaunas record substantial new woodlands being planted elsewhere and hedges eventually provided an increasing area of new habitat as well. The impact of such changes, therefore, was limited largely to local fluctuations.

Enclosure of the wastes and commons

If enclosure of the open fields brought profound changes to the landscape, they occurred in even greater measure with the enclosure of the wastes and the loss of these habitats had marked effects on farmland birds. 'Waste' was a manorial term. The closest modern equivalent is perhaps the category 'rough grazings', although the waste provided much more than grazing to earlier rural economies. By and large such land occupied the areas unrewarding to cultivate by early methods, of infertile soil and/or poor drainage. It embraced habitats such as lowland heath, downland and limestone sheepwalks, fens and mosses and vast tracts of upland grazings and moor. Such habitats provided essential resources to early communities, mainly of grazing but also other important materials such as gorse (fodder and fuel), fuel (turf, peat, reeds and wood), bedding (sedge and bracken), thatch (heather, reeds and sedge) and fish and wildfowl in fens.

Much waste was subject to rights in common, a term which does not indicate common ownership, although it may once have done so. Such rights provided a measure of long-term stability and protection to the habitats. Commons were private property, however, and subject to a long history of restriction and encroachment (see Hoskins & Stamp (1963) for a detailed discussion). The Parliamentary enclosures were the culminating assault, fuelled by realising the profits possible with enclosure and the application of high farming for those with the necessary capital to invest. Many of these habitats comprised light and easily worked soils, sands, peats, chalks and limestone brashes which, with the limited power and technology available, lent themselves to the much more frequent cultivation that high farming demanded. Many were also free-draining, important in an era when poor drainage was a major bar to improving agriculture. Marling and high farming techniques solved the problems of infertility.

Most rough grazing today is in the uplands (32% of agricultural area compared to 6% in the lowlands) but in the eighteenth century every lowland English county included significant areas of heathland or similar habitats, and extensive marshes and fens were associated with most larger river valleys. Parishes were laid out to give access to such areas (Chapter 3) and commons often occurred in extensive blocks which were shared between parishes, rather than in discrete pieces attached to each. Thus in my Sussex area the old commons of Hunston, Sidlesham, Birdham and Wittering all lay in one continuous band of rather poorly drained soil to which each village had access.

In 1795 the Board of Agriculture estimated the total area of wastes in England and Wales as a little under 8 million acres (3.24 million ha). This seems quite accurate for the June Census for 1900 recorded 3.5 million acres of rough grazing in England and Wales, to which must be added 1.5 million acres of grazing common which still exists and *c.* 3 million acres of common and waste lost to enclosure (see below). Very large areas of open upland grazings also existed in Scotland, which embraced 75% of all agricultural land at the end of the nineteenth century and still occupy 70% (Figure 4.4). Hoskins & Stamp (1963) provided figures showing that 2.106 million acres (850 000 ha) of wastes were enclosed by a total of 1893 Parliamentary Acts and Awards in England in the eighteenth and nineteenth centuries, with 80% of this area enclosed between 1761 and 1844; in Wales the loss was proportionately greater, with *c.* 1 million acres (400 000 ha) affected, but these were mainly upland commons and are dealt with separately. Waste and common grazing in Scotland had a different legal

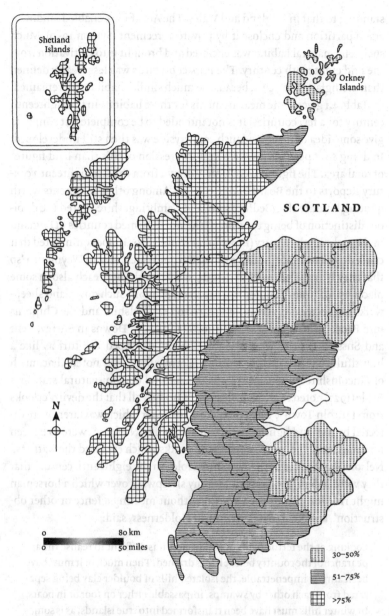

Figure 4.4. Rough grazings as a percentage of agricultural area in Scotland at the end of the nineteenth century by region. Note that Fife and East Lothian had less than 2% rough grazings and are left blank. (Data from June Census Statistics.)

standing to that in England and Wales. The Act of 1695 enabled landowners to partition and enclose it by private agreement (Symon 1959). Much such semi-natural habitat was enclosed and brought into cultivation from the mid eighteenth century. The impact on birds was less clearly defined than in England, however, because so much similar habitat still remained.

Table 4.1 gives some measurements for these habitats in the eighteenth century for a few counties. It is not intended to be complete but simply to give some idea of just how much countryside was then still undeveloped. In doing so it perhaps gives a better impression of this than bald figures of total area. The figures were largely drawn from the late eighteenth century Reports to the Board of Agriculture. Among other comments worth quoting here are that Bedfordshire and Cambridgeshire shared the dubious distinction of being considered the worst-farmed counties in England in the eighteenth century (Ernle 1922). In the 1720s Defoe commented that chalk downland sheepwalks extended from Winchester to Weymouth 'so that they lye near 50 miles in length, and breadth they reach also in some places 35-40 miles' (quoted in Furbank et al. 1991). Such downland sheepwalks also extended across Berkshire and Oxfordshire and the Chilterns into Bedfordshire and along the north and south Downs in Surrey, Kent and Sussex. On the Wessex chalk Cobbett described the turf as like a beautiful grey silk carpet. Smith & Cornwallis (1955) noted that much of Lincolnshire pre-enclosure was in a natural or semi-natural state and Ernle (1922) noted that 'from Sleaford to Brigg, "all that the devil o'erlooks from Lincoln Town", was a desolate waste, over which wayfarers were directed by the land lighthouse of Dunstan pillar. No fences were to be seen for miles, only the furze-capped sand-banks which enclosed the warrens.' Nelson (1907) noted of the Yorkshire Wolds in the eighteenth century that they were 'a desolate, grassy and stony sheepwalk over which a horseman might ride for 30 miles at a stretch without meeting a fence or other obstruction', and Reid (1885), writing of Holderness, said:

> Looking at the fertile fields of Holderness it is difficult to realise the appearance of the country before it was drained. Then much of it must have been almost impenetrable, the isolated hills of boulder clay being separated from each other by swamps, impassable either on foot or in boats. In winter hills must have been transformed into true islands, as is sometimes the case even in the present day.

Although Cobbett noted considerable enclosure in the Thames basin in the 1820s, he still recorded extensive heathland from Hounslow to Hungerford and from Kensington to Reigate.

TABLE 4.1. *Some areas of common and waste in eighteenth century England*

County or region	Date	Area	References
South and east England		470 000 acres of lowland heaths in eighteenth century	Tubbs 1985
London	1775	Not less than 200 000 acres of heath within 30 miles	
		96 000 acres in Surrey alone at end seventeenth century	James & Malcolm 1794
Hampshire	Eighteenth century	50 000 acres of pristine downland in eighteenth century	Tubbs 1993
		A total of c. 105 000 acres of wastes ex the Forests	Driver & Driver 1794
Middlesex	1798	17 000 acres common grazings	
Berkshire	1794	c. 40 000 acres of wastes. Windsor Forest extended to 24 000 acres in 1813	Pearce 1794 Spearing 1860
Buckinghamshire		c. 6000 acres of wastes and 92 000 acres of open fields	James & Malcolm 1794
Oxford	Early nineteenth century	Thousands of acres of wet common Kelmscote–Stanton Harcourt (map measurements suggest 15 500 acres). Otmoor 4000 acres	Aplin 1889
Cambridgeshire		17 500 acres of upland commons and 200 000 acres of wastes and unimproved fen	
Huntingdon	1793	44 000 acres of fen commons. Peterborough Fen, in the Soke, was a 6000–7000-acre common	
Essex	1794	c. 15 000 acres waste including Forests	Griggs 1794
Suffolk	1794	c. 100 000 acres of wastes, sheepwalks, common and warrens	Young 1794
Norfolk	1796	80 000 acres common but White's *Gazetteer* (3rd edn) stated that	Stevenson 1870

(cont.)

TABLE 4.1. (*Cont.*)

County or region	Date	Area	References
Lincolnshire		200 000 acres of common and heaths enclosed in 'the last 90 years'	
	1750s	'Wolds all warren for >30 miles from Spilsby to beyond Caistor'	Young in Smith & Cornwallis 1955
	1794	*c.* 200 000 acres of commons, wastes and saltmarsh, including 66 500 acres of fen commons, Axholme Commons *c.* 18 000 acres, Ancholme Levels *c.* 17 000 acres and extensive tracts of heath and open woodland on blown sand in the region of Scunthorpe, Brigg and Gainsborough	Stone 1794, Thirsk 1953, Straw 1955, Smith & Cornwallis 1955
Leicestershire		Charnwood Forest extended to 18 000 acres before enclosures	Moscrop 1866
Somerset	1760s	18 000 acres waste Quantocks, Mendip Hills 'uncultivated'	
	1797	50 000 acres of waste on Levels	
Dorset	1793	86 000 acres of wastes	Claridge 1793
Devon and Cornwall	1794	Many thousands of acres of wastes in Devon and waste estimated at one-third of total land in Cornwall	Fraser 1794
Worcester	1794	10–20 000 acres commons	Pomeroy 1794
Staffordshire		Cannock Chase exended to 25 000 acres before enclosure which started in 1820. Only part ever enclosed. Other forests were lost, e.g. Needwood (*c.* 10 000 acres, see below)	Evershed 1869
Yorkshire, East Riding	1750s	Bridlington–Spurn Point inland to Driffield, extensive areas of swamp and fen in the district of Holderness. Beverley to Barmston	Nelson 1907, Limbert 1980, Rhodes 1988

TABLE 4.1. (Cont.)

County or region	Date	Area	References
		Carrs occupied 11000 acres and Spalding Moor extended for 10 miles north of the Humber. The whole area of Thorne and Hatfield Moors and Potterick Carr by Doncaster embraced c. 70000 acres but that may include part of the Axholme Commons in Lincolnshire	
Northern England	1790s	Total of 1.5 million acres wastes in Lancashire, Westmorland, West Riding, Northumberland and Cumberland, of which 635000 acres considered improvable	Young 1770, Dickinson 1852, Ernle 1922
Wales		1450000 acres of grazing common before enclosure, 28% of land area	

Source: Ernle (1922) unless others listed.
Note: Areas are in acres as most so published; 2.47 acres = 1 ha.

The lowland mosses of Lancashire and Cheshire, extending into Cumbria, Staffordshire, Shropshire and Flint, embraced large areas of wet heather moorland and fen, mainly in glacial depressions. The whole extended to thousands of acres; in Lancashire alone Chat Moss was 6000 acres (2430 ha), Martin Mere 3600 (1457 ha), Rawcliffe Moss 1950 (789 ha) and Halsall Moss 1000 (405 ha), whilst in Cumbria Solway Moss covered 1700 acres (688 ha). Stevenson (1870) noted the eighteenth century in Norfolk as:

> a period when heath, warren and fen occupied in this county about the same proportion that cultivated land does now, when Norfolk did not produce enough wheat to feed its scanty population. When, even in the 'Enclosed' district, wide tracts of heath extended for miles through the inland portions, and an even wilder country as at Edgefield, Kelling, Weybourn and Salthouse, adjoining the coast, was divided only by the then undrained marshes from the sea-shore. When, on the east and

west the Broad and Fen districts, but little cultivated, were the fowler's paradise, and the Breck district with its heaths, warrens and sheep-walks – then, as now, the great stronghold of the Lapwing and Stone-curlew – presented a vast champain country with scarce a fence, fir-slip or plantation, over thousands of acres, divided only by mere balks to mark the rights of tenure.

It is often difficult to credit now, when driving along the north Norfolk coast, that this highly-farmed country was poor heathland in the eighteenth century: 'one blade of grass and 2 rabbits fighting for that' in the words of one contemporary.

Crown Forests generally included large areas of heath. They were, as Rackham (1990) commented, areas of multiple use, often under common rights and involved in the specialist activities of horse-breeding and pig-keeping. Sherwood Forest, in Nottinghamshire, for example, extended to c. 90 000 acres, largely of heath and breck (infield/outfield) in the first half of the eighteenth century (Chambers 1955). Rackham noted that most of the surviving Forests remaining towards the end of the eighteenth century were enclosed and destroyed and cited as examples Enfield Chase, enclosed in 1777, Windsor Forest in 1817, Hainault Forest in 1851 and Wychwood Forest in 1857. Others enclosed in the same period were Needwood Forest in 1801, Delamere Forest in 1812, Charnwood Forest in 1808 and Whaddon Chase in the 1840s (Coward & Oldham 1900, Hickling 1978, Read 1855b, Smith 1930–8). Other Forests, for example Parkhurst, Alice Holt and Bere, were converted to oak plantations in the early nineteenth century, almost as fundamental a habitat change as enclosure (Rackham 1990, Tubbs 1993).

Often the changes implemented by enclosure occurred over a comparatively short period, considering the extent to which the work was done by hand and horse power. Cobbett visited Lincolnshire and the East Riding in 1830 and the whole of the Wolds and Holderness were then described by him as the finest farmland; much of this change had been wrought in the previous 30 years (Smith & Cornwallis 1955). We have little real conception today of the scale of hard manual labour that enclosure involved. A common method of first clearing new land was to pare with a breast-plough, burn the turf and spread the ashes (Figure 4.5). The thought of breast-ploughing over 2 million acres (810 000 ha) even in 80 years is mind-boggling, although it is unclear quite how universal this practice was. Certainly from the mid nineteenth century steam tackle was being used for such work (e.g. Evershed 1869) and paring and burning came to be

Breast-plough (Mr William Ball, Ashwick, Gloucestershire)

Figure 4.5. Using the breast-plough. (Reproduced by permission of Cambridge University Press.)

banned in many areas from about 1820. In particular it exacerbated shrinkage in peat soils. Albery (1999) also opined that it was particularly destructive of the native flora.

Nevertheless much of the work had to be done by hand labour. Belcher (1863) noted that in clearing Wychwood Forest, Oxfordshire, for cultivation, after two years spent in felling trees and grubbing their roots:

> it was with the greatest difficulty that 4 strong horses drawing a large iron plough could break up half an acre a day, and many and long were the blacksmiths' bills for repairs where the plough was used. Some of the tenants tried digging, at a cost of £3 per acre; some used stocking-hoes and grubbed the ground 5 inches deep, carefully picking out the large stones, this plan cost 50/- per acre. On Potter's Hill farm breast-ploughing and burning was adopted and this course appeared to answer better than any of the others.

Potter's Hill Farm was 450 acres (182 ha) and was cultivated by the breast-plough for at least five years before it was considered worth thinking about using steam or horse ploughs. However Wychwood was perhaps an unusual case and converting downland or much heathland to the plough would not have involved the same problems. But another major input on light soils was claying or marling to improve soil structure and fertility. This was done extensively on the Norfolk sands and Williams (1970) noted that the peat moors in Somerset were clayed at a rate of 100–150 cartloads per acre. Fletcher (1962) quotes figures of 150–180 tons of marl per acre on

Lancashire mossland. In the Fens Clarke (1848) noted that claying was carried out by digging parallel trenches about 10–12 yards apart across the field, throwing out the clay to either side and backfilling with top soil. Clay up to 10 feet down was utilised by this process. Read (1858) noted that clay, marl, chalk and sand were all used in the Fens, at application rates of up to 300 cartloads per acre. Such fertilising agents had a fundamental impact on soils and vegetation. This together with the effect of paring and burning on vegetation on a very wide scale must provide an important reason why recession did not simply lead to reversion to older habitats (Chapter 9). Another method of making soil was warping, practised where rivers carried heavy silt loads, for example around the Humber marshes; land was embanked, flooded and the silt trapped to considerable depths (Creyke 1845).

Enclosure did not just involve clearing and cultivating land. A whole infrastructure of roads, steadings, housing and fences had to be constructed and comprehensive drainage was often necessary. Several sources noted the cost of such work, which averaged *c*. £10 12s per acre (£758 per hectare in today's money) before buildings were installed or preliminary cultivations done (e.g. Pell 1887, 1899), and up to £26 per acre (*c*. £1900 per hectare today) with buildings and cultivations included (e.g. Rowley 1853). Pell considered that some of the investment in land clearance and enclosure in the later years was ill-judged, an observation which Albery (1999) put firmly in perspective when he remarked that the periods of prosperity in lowland farming totalled barely 50 years between 1800 and 1939. Reading the nineteenth-century literature on the subject, however, one concludes that economic criteria were never considered when destroying natural landscapes and habitats for agriculture. It seemed always to be regarded as a desirable and necessary object of itself, irrespective of cost.

Commons enclosure and birds

Another way of estimating the significance and extent of the habitats of the waste is to look at the distribution of some specialised bird species for which they were important. Figure 4.6 illustrates the distribution recorded for Hen and Montagu's Harriers, Black Grouse, Great Bustard, Stone Curlew and Woodlark in the late eighteenth to early nineteenth centuries. These are all ground-nesting birds of wet or dry heathland, downland and fen and none was particularly rare within its range at this early date, even Great Bustard; Nelson (1907) recorded several hundreds in Yorkshire in the latter half of the eighteenth century, and similar

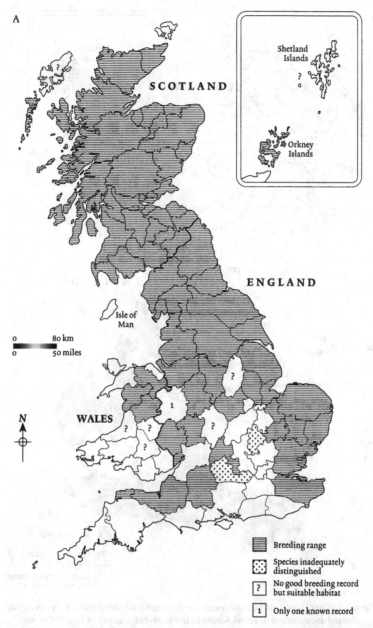

Figure 4.6. Breeding distribution of six typical species of the waste in the late eighteenth/early nineteenth centuries, by county. Main source early avifaunas, other sources shown as relevant. Shaded areas indicate breeding range. (A) Hen Harrier and (B) Montagu's Harriers: extensive lowland heaths and downland, fen (especially sedge fen), bogs, heather moorland. Note that the separation of these species is somewhat tentative as they were often confused at this period (see text). (C) Black Grouse: extensive heath, moorland and woodland edge, often damp. (Additional source Gladstone 1924.)

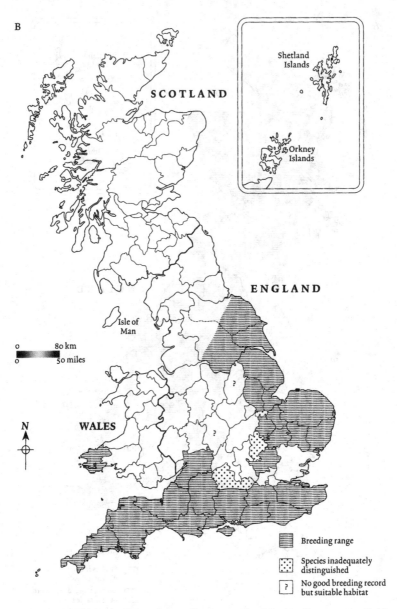

Figure 4.6. (Cont.) (D) Great Bustard: extensive dry grassland sheepwalks, very open. Said also to have occurred in Berwickshire in the sixteenth century. (E) Stone Curlew: extensive dry stony heaths and downland sheepwalks with scant vegetation, also dry fallows. Dates are for the last breeding in counties from which it disappeared during nineteenth and early twentieth centuries. In Yorkshire it persisted until 1937 but very rare after 1900. (F) Woodlark: wooded parks, hedged meadows with copses, sheepwalk and heath edge with trees (Yarrell 1837–43). By the end of the nineteenth century the main distribution was south of the line A–B (following Holloway 1996).

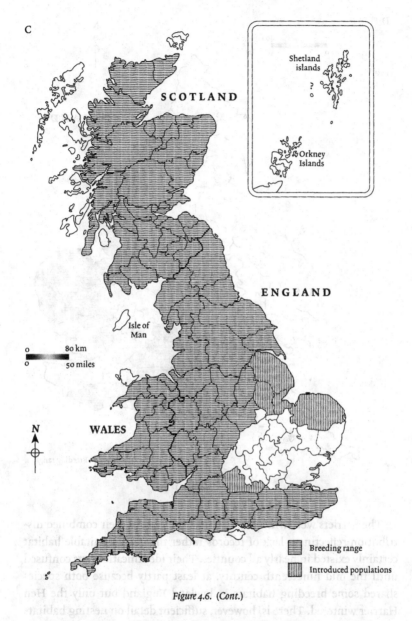

Figure 4.6. (Cont.)

numbers were still present in Sussex and in Wiltshire (Borrer 1891, Smith 1887, White 1789). Except for Woodlark these species also all tend to need large areas of habitat and they are, therefore, good indicators of how extensive their habitats were.

D

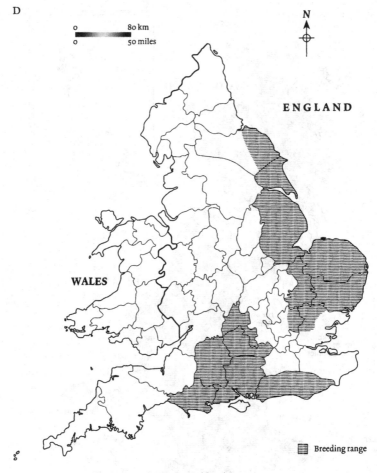

Figure 4.6. (Cont.)

The harriers were widely distributed, the gaps in their combined distribution reflecting a lack of records rather than birds; suitable habitat certainly existed in nearly all counties. Their identification was confused until the mid nineteenth century, at least partly because both species shared some breeding habitats in lowland England but only the Hen Harrier wintered. There is, however, sufficient detail on nesting habitats in the early avifaunas to suggest clearly that the Montagu's Harrier was the typical breeding harrier of lowland heathland and downland habitats in southern and eastern England, with the Hen Harrier much more confined to swamp, fen and marsh there (Aplin 1889, Christy 1890, Nelson

E

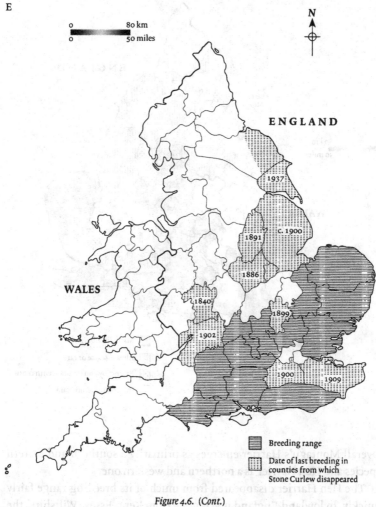

Figure 4.6. (Cont.)

1907, Stevenson 1866, Ticehurst 1909, Ticehurst 1932). Both species bred in the fenlands of East Anglia and south Yorkshire but Stevenson and Nelson both recorded Montagu's as the more numerous and widespread species. Figures 4.6A and B separate the breeding distribution of the two species on this basis and I have assumed that Montagu's was the breeding harrier on the Cotswold sheepwalks in Gloucestershire, that both species bred in Lincolnshire and that the breeding records of Montagu's for Durham and Northumberland are unreliable. I can find no substantive record of a nest.

F

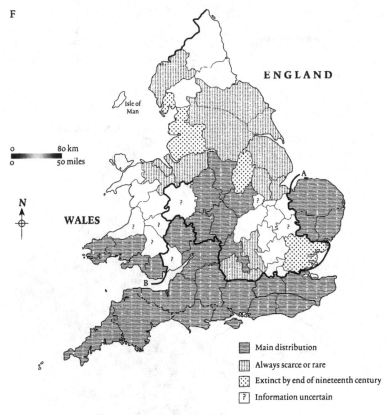

Figure 4.6. (Cont.)

Overall Montagu's Harrier emerges as primarily a southern and eastern species and Hen Harrier as a northern and western one.

The Hen Harrier disappeared from much of its breeding range fairly quickly. In lowland England it was gone from Kent, Essex, Wiltshire, the Forest of Dean and Leicestershire before 1825 and from Oxfordshire, East Anglia, south Yorkshire and probably Staffordshire by 1850. Watson (1977) noted that the great reduction in northern England and southern Scotland occurred between 1820 and 1850. There the loss of breeding places was insufficient to account for the scale of decline, the main cause of which was persecution by intensive gamekeeping. Very large areas of suitable habitat always remained in the uplands. However Watson records that draining traditional and very secure sites in lowland bogs, such as Billie Mire, Berwickshire, led to the Hen Harrier increasingly and fatally resorting to upland grouse moors. Unlike Montagu's Harrier, in lowland England it never adapted to nesting in crops, so that the enclosure and drainage of its breeding grounds was terminal; in Kent Ticehurst (1909) noted that it was pushed out of the Thames marshes by the spread of London. The timing of its extinction is a good indication of the speed and scale of habitat loss in much of England in the first half of the nineteenth century.

The avifaunas contain enough records to suggest that Montagu's Harrier was turning to crop nesting in the nineteenth century. Because its young are nidicolous, however, it was actually ill-adapted to do so and today this habit depends on protecting nests from harvesting operations (Clarke 1996). One can hardly envisage such protection being extended to a raptor in the nineteenth century. Sainfoin, widely grown on the chalk in the nineteenth century, was one crop it attempted to colonise. When left for hay, the crop was sufficiently undisturbed to allow laying and incubation; but it was cut in June. In other crops harriers' nests would have been quickly found and destroyed by weeding gangs. Nevertheless southern and eastern England always retained significant areas of its semi-natural habitats and Montagu's Harrier bred scarcely over much of its former range throughout the nineteenth century, although rarely successfully because of persecution. The latter must be held the main cause of the loss of this fine bird.

Black Grouse bred in every county in Britain, except in parts of the East Midlands and East Anglia (Figure 4.6C). It was introduced or reintroduced into Norfolk, Lincolnshire and Berkshire in the nineteenth century, and nineteenth-century populations in Hampshire and Surrey also stemmed at least partly from reintroductions. In Sussex Turner (1862) noted of Ashdown Forest that:

> it was hardly possible to ride or walk about it in any direction without disturbing some of them. At that time the forest was was thickly covered with heath, but since then this has been so generally cut and carried away that the black game, deprived of the food and shelter they so much delight in, have gradually disappeared, and in this locality are now very rarely to be met with.

The indigenous populations in north Wales, except Montgomeryshire, apparently died out in the early nineteenth century and the present population also stems from reintroductions (Forrest 1907). Its distribution gives a convincing picture of just how widely heathland and marginal habitats were distributed, particularly in England, and how extensive they must have been. Black Grouse are birds of the moorland and forest edge and need a mosaic of habitats, with open terrain for display, shelter for roosting and diverse food sources, with heaths such as *Calluna* needed in winter and shrubs and trees for feeding above the snow (Tucker *et al.* 1995). Yarrell (1871–85) noted that water was essential, particularly for chicks. Home ranges are large, exceeding several square kilometres (Angelstam & Martinsson 1990, in Tucker *et al.* 1995). The species has been declining in Britain since the mid eighteenth century and habitat loss was the major reason over much of England. The species had vanished from or become scarce or rare in most lowland areas by the end of the nineteenth century. However not every site from which it disappeared had been enclosed and ploughed. It seems likely that this species particularly was affected by the fragmentation of heathland habitats, with smaller scattered pieces lacking the necessary diversity of habitat elements (see below). The enormous amount of drainage carried out in the nineteenth century (Chapter 6) also had a general impact on water tables. Haslam (1991) pointed out that, with the general lowering of groundwater levels of the past 150 years, the abundance of springs has hugely declined. Such changes in the scale and dampness of remaining habitats must have contributed to this species' decline. The impact of enclosure in the uplands upon it is discussed below.

The distribution of the Great Bustard closely followed that of major areas of chalk and limestone sheepwalks (Figure 4.6D). The map shows the distribution by county and disguises its rather disjunct nature, on the Yorkshire and Lincolnshire Wolds, the East Anglian Brecks, the chalk heathlands on the Cambridgeshire/Hertfordshire/Essex borders, the Wessex chalk and along the South Downs. By the early nineteenth century, however, it was largely confined to the northern half of the Yorkshire Wolds, the Brecks, Wiltshire and East Sussex. It was gone by 1830. These last redoubts were remaining areas of very extensive sheepwalks and the main decline coincided with extensive conversion of open sheepwalk to arable. For example the Lincolnshire Wolds were almost entirely enclosed arable by 1830 (Smith & Cornwallis 1955) and thousands of acres of downland were broken up in Dorset in the late eighteenth and early nineteenth centuries (Fussell 1952). Similar changes were wrought in Hampshire, Berkshire and on the Cambridgeshire/Hertfordshire/Essex borders.

Nevertheless, as it does in the pseudo-steppes of Europe today, the species certainly occupied and used arable habitats in England, nesting in rye, wheat, clover and sainfoin and feeding regularly in turnips in autumn

(Borrer 1891, Nelson 1907, Smith 1887, Ticehurst 1932). But no one reading Nelson's account for Yorkshire can doubt that a major problem for this species was persecution in the early nineteenth century; in the final 20 years of its existence in Yorkshire it was simply exterminated. In the Brecks Ticehurst (1932) noted that once farmers changed from growing rye to wheat, drilled and hoed, the species lost too many nests to sustain itself. Smith (1887) made the same comment for Wiltshire, noting that hoeing especially led to nests being robbed. Bustards preferred very open landscapes and some authors suggested that enclosure, by breaking such landscapes up, hastened the decline. But many parts of the Downs and Wolds have always remained very open (see illustration on p. 282) and there seems no doubt that persecution arising from changes in farming methods was the main reason for its demise; the markedly increased activity in fields greatly increased its exposure to collecting. Rarity meant value. Whether the species could have adapted to high farming as a habitat to any extent is academic; the nineteenth-century avifaunas show that it was never given the chance.

The Stone Curlew (Figure 4.6E) was a widespread and numerous species primarily of dry semi-natural grasslands, extensively and traditionally grazed by sheep and often by rabbits, on stony heathland, warrens and downland and limestone sheepwalks. It also bred on cultivated ground in many areas, particularly on dry fallows on the Downs, Wolds and Brecks. It readily replaces lost clutches, an essential prerequisite for a ground-nesting bird on cultivated land (which Great Bustard may lack; Cramp & Simmons 1977). Bare open ground with very short vegetation is important and the grazing of very large numbers of sheep over its principal breeding range was an essential habitat element. Waldon (1982) gave an interesting insight into the mechanisms the species has developed to defend its nest against trampling. Of the species considered in this group, it should have been the least affected by enclosure.

Nevertheless it declined markedly in the nineteenth century. Figure 4.6E records final extinctions rather than the full pattern of decline but the dates for all these losses, except for Worcestershire, fall long after the main period of commons enclosure, after the peak of high farming and, indeed, after the great recession of the 1880s and 1890s. That possibly contributed in some counties as much land went out of arable (e.g. Albery 1999, Sheial 1976); such abandonment does not necessarily benefit Stone Curlews (Chapter 9). But the major cause must have been the much greater disturbance typical of arable fields in the nineteenth century. Amongst other things this led to increasing nest robbery; both Stevenson (1866) and Ticehurst (1932), for example, noted that declines in Stone Curlews partly stemmed from excessive egging, the eggs being taken and sold as plovers' eggs. Changes in the frequency and timing of cultivations in high-farming systems also tended to catch successive layings, particularly in fallows worked for roots, the bare broken ground of which exercised a strong attraction. Such habitats became traps (see Newton 1998). Several nineteenth-century authors, e.g. Read (1858), also comment on the high levels of nitrogen in soils resulting from high farming, and the effect this had on the growth and height of cereal crops. Thus autumn cereals, even in the nineteenth century, would have ceased to be suitable nesting habitats for Stone Curlews by early May, perhaps earlier, and spring cereals before the end of that month (see Green 1988), restricting pairs that lost first clutches to more dangerous fallows for repeat layings. Although little quantitative data is available, there can be little doubt also that the practices of rolling autumn cereals in spring, of rolling spring cereals after emergence, of working fallows at two- to three-week intervals and of hoeing and hand-weeding crops all contributed to a high level of nest losses and disturbance. Virtually all modern research confirms the significance of such factors (Easy 1962, Glue & Morgan 1974, Green 1988, Green et al. 2000). Ultimately all these factors should be regarded as effects of enclosure, pushing the bird into a habitat type where it was more vulnerable.

Like the Black Grouse, the Woodlark is primarily an edge species, needing a variety of habitat elements, short vegetation or bare areas for feeding, longer vegetation for nesting and trees and bushes for song posts (Tucker et al. 1995). Nineteenth-century avifaunas refer to it much more frequently as a bird of woodland edge than of heathland; grassy hillsides with bracken and scattered trees and bushes were also widely recorded, in the west especially. Yarrell (1837-43) noted that trees were essential to it. Ticehurst (1932) underlined this, recording that Woodlarks did not colonise the Brecks until the 1840s, probably attracted by the belts of firs which were by then well established on the sheepwalks. It never bred on tilled land. Figure 4.6F shows the distribution early in the nineteenth century. By the last quarter of the century it had become extinct or virtually so in several counties and was largely confined to some southern and eastern counties, which held the principal remaining areas of lowland heath and downland sheepwalk. Although persecution for the cage-bird trade was considered important in places, for example Pembrokeshire

(Mathew 1894), there seems little reason to doubt that the great decline of Woodlarks in the nineteenth century was largely due to habitat loss and changes in grassland management (Chapter 9). Many authors also noted the impact of the severe winter of 1880–1. However I am sceptical of the importance of climatic considerations in the long-term population decline of this bird, although they possibly had greater impact as habitats became more marginal.

These species were not, of course, the only ones affected by enclosure. The species listed in Table 3.1 provide further interesting examples. Lapwing, Stockdove, Goldfinch and Linnet, which bred mainly in semi-natural habitats, wintered abundantly in arable farmland, probably encouraging an increasing tendency to nest there as semi-natural habitats declined; Goldfinches, however, were increasingly affected by declining weed populations (Chapter 7). Besides Montagu's Harrier and Black Grouse, Stonechat failed to adapt to enclosure and cultivation. Wheatear and Whinchat should have been similarly affected but the nineteenth-century avifaunas actually record little change for either species, perhaps, as Holloway (1996) suggested, because of the much increased availability of other favoured habitats such as quarries and railway embankments. Corncrake, Skylark, Yellow Wagtail and Yellowhammer showed a more uniform spread across habitats, suggesting ready adaptation to farming habitats as the nineteenth century progressed. Corncrake, however, failed to adapt successfully to the mechanisation of haymaking, a factor which also affected Whinchat. The Quail was undoubtedly very much more numerous in Britain in the eighteenth and early nineteenth century than it is today (Moreau 1951). Both White in the eighteenth century and Lilford (1895) noted a preference for the open fields over enclosure and Aplin (1889) infers the same. Although Quail now make greater use of cereals, in the nineteenth century it was considered a grassland bird. In Norfolk (Stevenson 1866), Cambridgeshire (Lack 1934), Lancashire (Mitchell 1885), Oxfordshire (Aplin 1889), Suffolk (Ticehurst 1932), Kent (Ticehurst 1909) and Pembrokeshire (Mathew 1894) rough grassland and the fringes of fens and marshes were noted as particularly favoured. These accounts and that of Yarrell (1871–85) suggest that these were key habitats for the species. Moreau considered that the decline was attributable to human predation on migratory populations in North Africa rather than agricultural change. Nevertheless the sharpest decline in Britain actually coincided with the greatest loss of the preferred fen habitat in the first half of the nineteenth century (Chapter 6).

The Stockdove is something of an enigma. Holloway (1996) suggested, I think correctly, that there was much confusion over its identification in the nineteenth century. Its true distribution was thus underestimated and its expansion out of south and east England from about the 1850s exaggerated. Probably this partly arose through confusion over vernacular names; White, for example, called it the Blue Rockier and so did my father. Certainly Pennant recorded it as common in the Brecks in the late eighteenth century, whilst Eyton (in Forrest 1907) noted it as probably breeding in north Wales as early as 1838, MacPherson (1892) noted it as probably breeding in at least one site in the Lakeland counties by 1840 and Nelson (1907) recorded it as breeding on the Yorkshire coast by 1820. Many early accounts note it as a bird of heathland and warren and, particularly in the north and west coastal dune systems; it was known as the Sand Pigeon in Cheshire. It seems very likely that enclosure destroyed many of its preferred habitats and encouraged it to adapt to breeding more widely in the farmland habitats it already exploited in winter. Nelson in fact suggested this explanation for its spread in Yorkshire.

The group listed in Table 3.1 also involves species that have contracted northwestwards in the face of farming change and those which have contracted southeastwards. Characteristically the former were once generally distributed in Britain and, of the species in Table 3.1, comparisons between Holloway (1996) and Gibbons *et al.* (1993) show this shrinkage to the northwest clearly for Black Grouse, Corncrake and the three chats, whilst today it is also emerging with Lapwing despite losses in distribution in northwest Scotland. This pattern is, in fact more general in farmland birds and the same comparison also shows it for Hen Harrier, Ring Ouzel and Twite and suggests that it is emerging with Meadow Pipit and probably with Merlin, Golden Plover, Snipe and Pied Wagtail. How far the present warming trend may be contributing to this pattern is unclear. Species which have contracted south and east tend to be those which were already limited in distribution by other factors, such as soil or climate.

Such profound habitat change must have influenced many other species within and without farmland. What is remarkable, however, is how comparatively little long-term impact is evident from the nineteenth-century avifaunas. Despite its initial concentration in and spread at the expense of specialised habitats of heath and fen, the development of high farming led to few extinctions amongst birds. Ranges contracted but only Greylag Goose (in lowland England), Bittern, Marsh Harrier, Montagu's Harrier, Great Bustard, Black-tailed Godwit, Bearded Tit and ultimately

Ruff disappeared entirely and, of these, only the loss of the Great Bustard has proved permanent. Three of these are not really farmland birds and persecution was undoubtedly responsible for the final demise of Bittern, the harriers and Bearded Tit and made a major contribution to the loss of the others. Table 4.2 summarises the changes in farmland bird populations recorded by the nineteenth-century avifaunas. Bearing in mind that bird populations naturally fluctuate, very extensive changes in farmland habitats actually produced surprisingly limited changes in farmland bird populations. Of the 98 species in Table 4.2, 27 increased, 36 showed little change and 35 declined. Contingency analysis of the data suggested that fewer woodland plus woodland/field species declined than expected and field species showed most variation, but the differences were not quite significant ($\chi^2 = 9.19$, df 4, $p < 0.07$).

One difficulty in making such judgements is that nineteenth-century ornithologists lacked the sophisticated survey information available today. Estimates of population change rested on knowledge of distribution and subjective opinion but I doubt if significant changes went unnoticed (see Introduction). Furthermore assessing the impact of major habitat changes is obscured by what Nicholson (1926) anathematised as 'the Victorian leprosy of collecting' compounded by rabid game preserving. It is impossible to know now how well birds such as harriers or bustards might have adapted to high farming, when the nineteenth-century avifaunas make it only too clear that they were never given the chance. At least half the declines listed in Table 4.2 (all the raptors and corvids, Black-headed Gull, at least partly Great Bustard and perhaps Stone Curlew) resulted from persecution as much as or more than habitat change. The common wildfowl were heavily exploited for food, with such persecution particularly significant in spring, which undoubtedly reduced breeding populations. Even the Red-backed Shrike was commonly found on gamekeepers' gibbets because it preyed on pheasant chicks. Common sense may suggest that species such as Marsh Harriers must have declined with lowland enclosure. In many areas, however, they predeceased their habitats. Habitat loss was never total and the validity of such assumptions is further undermined by their behaviour today, when they have adapted to nesting in arable crops (Underhill-Day 1993). Species such as Stone Curlew were perfectly capable of breeding successfully in mixed high farming habitats. But the Victorian habit of collecting skins and eggs (Chapter 12) meant that such species came to have monetary value. Farm work was done in gangs and teams, ploughing, tilling, hoeing, weeding, haying etc., and

TABLE 4.2. *Changes in farmland bird populations in the nineteenth century*

Category	Increase	No change[a]	Decrease
Woodland	Whitethroat	Hobby[b]	Sparrowhawk
	Chiffchaff	Cuckoo	Great Spotted
	Goldcrest	Wren	Woodpecker
	Coal Tit	Dunnock	Jay
	Treecreeper	Robin	
		Lesser Whitethroat	
		Garden Warbler	
		Blackcap	
		Willow Warbler	
		Spotted Flycatcher	
		Long-tailed Tit	
		Blue Tit	
		Great Tit	
		Bullfinch	
Woodland/Field	Blackbird	Green Woodpecker	Tawny Owl*
	Mistle Thrush	Song Thrush	Wryneck
	Chaffinch		Red-backed Shrike*
	Greenfinch		
Field	Stockdove	Swallow	Red Kite
	Woodpigeon	Jackdaw	Buzzard
	Turtle Dove	Yellowhammer	Kestrel
	Long-eared Owl	Cirl Bunting?[c]	Barn Owl*
	Rook		Stonechat
	Starling		Pied Wagtail
	House Sparrow		Magpie
	Tree Sparrow		Chough
			Crow
			Raven
			Goldfinch
			Linnet
Ground-nesting/ wetland	Canada Goose	Mute Swan	Greylag (wild)
	Garganey*	Teal	Mallard
	Red Grouse	Pheasant	Shoveler
	Red-legged Partridge	Moorhen	Marsh Harrier
	Grey Partridge	Golden Plover	Hen Harrier
	Oystercatcher	Dunlin	Montagu's Harrier
	Curlew*	Skylark	Merlin
	Redshank*	Meadow Pipit	Black Grouse
	Grasshopper Warbler*	Yellow Wagtail	Quail
	Reed Warbler	Whinchat[d]	Corncrake*
		Wheatear[d]	Great Bustard
		Ring Ouzel[e]	Stone Curlew
		Sedge Warbler	Lapwing
		Twite	Snipe
		Reed Bunting	Black-headed Gull
		Corn Bunting**	Short-eared Owl
			Woodlark

[a] Species showing no change may have fluctuated markedly within a pattern of long-term stability.
[b] See Fuller *et al.* (1985).
[c] I believe that Cirl Buntings were overlooked rather than new colonists in the nineteenth century.
[d] For Whinchat and Wheatear see text p. 93.
[e] I have rejected nineteenth-century reports of Ring Ouzels breeding in lowland England.
*Change started in the second half of the nineteenth century.
**Sharp decline in the twentieth century started in the 1870s.
Source: Data from avifaunas published before 1910.

almost constant through the spring and summer. Rare or shy or proscribed species had little chance of escaping detection and destruction as a consequence.

I suggest three reasons why the impact of huge habitat changes in agricultural land in the nineteenth century on farmland bird populations was comparatively limited. Firstly, enclosure continued over a period of about 120 years, allowing time for adaptation. Secondly, for some species habitat loss was offset by the scale of tree-planting, in woodlands, plantations and ornamental plantings, which often accompanied estate improvement in the period. Holloway (1996) quotes this frequently as a reason for range expansion by birds in the nineteenth century. The new hedgerows eventually also helped replace older scrub habitats as nesting sites, for example for Linnets and Yellowhammers. More importantly many habitats that we would classify today as permanent grass, such as water meadows, hay meadows and old enclosed pastures, changed very little. The scale of drainage, ploughing and reseeding we have seen in them in the modern farm landscape is essentially a twentieth-century problem (Chapter 9).

Thirdly, and perhaps most important, was the scale of resources, particularly food supplies, which became available to birds within the new farming. Birds inhabiting farmland exploit crops, crop residues (stubbles), cultivations, operations such as haying and harvesting, stock feed and stored crops for food. High farming, combining stock and arable enterprises on one farm, provided a sequence of such feeding opportunities throughout the year, with grass and arable available, cultivations in every season, varied crops, stubbles left in winter and stock both stalled and folded over fields. Farm operations also took longer, making food available longer. The storage of cereals in ricks (Figure 4.7) was a major winter food source, which had a significant effect on the survival of seed-eaters in severe weather (Chapter 10). Rotations also influence feeding habits, which vary seasonally in line with farmwork. The adoption of such rotations virtually throughout farmland increased diversity within farms even if the overall diversity of the countryside had declined and must have had a favourable impact on birds' survival rates.

Fragmentation

Enclosure did not necessarily follow a logical sequence in any area. It was done by Private Act or agreement for each parish and that depended on the

Figure 4.7. A stackyard in Buckinghamshire in 1947, the work of Mr. William Chandler (at right), a master craftsman. (Courtesy of Rural History Centre, University of Reading.)

presence of landowners and/or tenants of enterprise, energy and capital. Consequently neighbouring parishes or commons might be enclosed at very different times so that extensive areas of semi-natural habitats tended to be increasingly fragmented, a process that has different effects on birds than simple habitat loss. Tubbs (1985) showed the relationship between declining area and increasing numbers of fragments for lowland heath from 1800 and that process affected downland and fen habitats similarly. The process has continued into the twentieth century, as Armstrong (1973) illustrated for the Suffolk Sandlings. Its impact remains evident in areas such as north Norfolk, where Kelling Heath, Walsey Hills and Salthouse Heath are the remaining fragments of what was once a virtually continuous belt of heathland east to Cromer, in the scattered fragments of heathland remaining in the Weald, and in Radnorshire, where the Maelienydd and Penybont Commons are the remaining fragments of the Great Common, which extended from Builth Wells 20 km north to Llanbister before enclosure in the 1850s and 1860s. Such remaining fragments were often the least fertile and left for the rural poor to gather fuel.

Fragmentation had an important general effect on heathland birds, for plot size is related to bird species diversity in heathland (Fuller 1982). Peers

(1997) recorded this relationship on a series of upland grazing commons in Breconshire. Reducing plot size increases the chance of any plot losing essential habitat elements. It increases the importance of edge effects and key species become more liable to extinction (Moore 1962). This probably particularly affected Black Grouse among the species illustrated in Figure 4.6. Harriers may have been similarly affected as breeding birds and probably Great Bustard. Thus as extensive areas of such semi-natural habitat were broken up in the nineteenth century the diversity of birds on the remaining parcels would have fallen.

Commons and livestock

The reports to the Board of Agriculture around the turn of the nineteenth century all condemned the management of livestock on commons, reporting that commons were overstocked and the stock of very poor quality and health. However accurate information about stock numbers in the eighteenth and early nineteenth century is limited. Sheep have always been the most numerous livestock and were perhaps the most important on many of the semi-natural grasslands of the waste. This subject is examined in detail in Chapter 9 but I estimate a total sheep flock in England and Wales of $c.$ 17.5 million head around 1800 and a total cattle herd of 3.5–4 million head. An increasing problem on common grazings in the late eighteenth century was that of 'surcharging'; that is, individuals outside a manor or parish, often from a distant township, buying a field within it to which common rights were attached and then abusing their entitlement (Ernle 1922). For the Somerset Levels Williams (1970) noted that the number of cattle brought into the 20 000 acres (8100 ha) of King's Sedgemoor alone for one year was reported to have reached 30 000 by 1775. That seems improbable (a stocking rate of 2.87 livestock units per hectare) but if stocking rates were so high on the Somerset Levels it perhaps accounted for the paucity of breeding birds there recorded by Smith (1869). Such figures may really refer to rights claimed, however, rather than animals present at any one time. Commons do accumulate extraordinary numbers of rights, which are not necessarily exercised. Penford & Francis (1990) record one of 7 ha on Mynydd Du Carmarthen, with 37 rights of pasture for 4071 sheep, 261 cattle, 118 ponies and 24 geese! Rights claimed and rights exercised are not the same thing and eighteenth-century claims of overstocking may rest largely on the former, which would be much easier to calculate than actual animals depastured at any particular time. Furthermore commons were often grazed by sheep and cattle, which replaced each other seasonally (e.g.

Fussell 1952). Although likely to be beneficial to them as habitats (Oates 1993), that could also lead to over-counting. One area where high stocking levels were well documented was Romney Marsh, Kent. Whitehead (1899) gave two reasons for this: the marsh was well drained and the salinity of the water in many of the dykes and fleets limited the presence of the snail *Limnaea*, the vector for liver fluke (see below).

Thus in terms of commons and wastes as habitats it is unclear how significant complaints of overstocking were. Nor do reporters say what they would have considered an appropriate level of stocking. Commons were also areas of multiple use. Complaints that commoners encouraged heather and gorse are not infrequent in the Victorian farm literature. But heather was valued as winter fodder for sheep in severe weather, and gorse as fuel, particularly for bread ovens (Evans 1956), and as cattle feed (Elly 1846). In Wales it was used as chaff for horses and attention paid to its management (the late Dilwyn Roberts personal communication). This hardly supports arguments for severe overgrazing. It indicates management and conservation of the resources available. Webb (1986) observed that when grazing was a major land use, the vegetation of heaths included a much greater proportion of grass, creating a habitat mosaic rather than the monospecific stands of *Calluna* we see today. He suggested that burning was not an important management tool before the nineteenth century, traditional practices and the scale of grazing maintaining the low nutrient status of the heath and its open plant community. This was probably, in fact, the best state for the birds. The decline of grazing pressure in the nineteenth century led to increased burning to keep the grazings open. That decline was the result of enclosure extinguishing common rights, so that farms became self-sufficient, maintaining their stock within themselves. Clarke (1848) provides another insight for a different habitat with his description of Littleport Fen, Cambridgeshire, where 'a large portion having been let at one shilling per acre, into which stock was turned amongst the reed and turf-bass, and not seen for days together'. Such a description suggests under- rather than over-use.

The most important point to which the reports to the Board of Agriculture drew attention was the state of the animals themselves on commons. The real problem in managing livestock on commons was the prohibition of fencing. All rights holders' stock ran together over the whole common, making the selection and breeding for improvement and the control of disease difficult or impossible. Harthan (1946) quoted a Mr. Duke of Bredon in 1794 as:

the mixture of property in our fields prevents our land being drained. Add to this, that although our lands are naturally well adapted to the breed of sheep, yet the draining is so little attended to that, out of at least one thousand sheep annually pastured in our open fields, not more than forty on an average are drawn out for slaughter or other uses; infectious disorders, rot, scab etc., sweep them off, which would not be the case if property were separated.

Sheep-farming in the eighteenth century was bedevilled by the 'rot', almost certainly liver fluke, which caused huge losses in frequent outbreaks. My impression is that this problem had increased by the eighteenth century, perhaps as a result of more wet years coupled with the deterioration and neglect of drainage systems. Ellis wrote of 1735, a very bad year, that 'the dead bodies of rotten sheep were so numerous in roads, lanes and fields that their carrion stench and smell proved extremely offensive to the neighbouring parts and the passant travellers' (Ellis 1735). Ernle (1922) provides numerous other quotations to the same effect and it is clear that the volume of carrion in the countryside in this period was very large. Williams (1970) noted that 10 000 sheep were said to have been rotted in one year on the moors of Meare parish in the Somerset Levels before enclosure and draining took place. Enclosure enabled farmers to drain or fence sheep away from undrained areas, the prime source of rot, control stocking, isolate flocks to improve disease control and breed for improvement. It led to a dramatic and well-documented reduction in the incidence of rot and the losses it caused, although major outbreaks occurred in 1830–1 and in 1879; the sheep flock in East Anglia never recovered from the losses sustained in the latter.

It seems highly probable that this result of enclosure, the loss of an enormous source of carrion, was an important factor in the rapid disappearance of Kites, Buzzards and Ravens from England. Table 4.3 sets out the pattern of decline in England for those counties for which there is reasonably exact information. It records the last decade in which there seemed to be a regular breeding population, albeit often small. The data are rarely more exact and most accounts make it clear that a marked decrease had occurred by the early nineteenth century for all these species. Of course, this loss of carrion was not, of itself, terminal but it must have led to significant decreases, which persecution then completed.

As the table shows the Kite disappeared most quickly, with the greatest loss before 1820. Hudson (1923b) noted that it last nested in London

TABLE 4-3. *The decline of the Kite, Buzzard and Raven in England in the nineteenth century (decades are the last in which at least some pairs still bred regularly)*

Before 1810	1810–19	1820–29	1830–39	1840–49	1850–59	1860–69
Kite						
Yorkshire	Cumberland	Wiltshire	Leicestershire	Northamptonshire		
Lancashire	Westmorland	Huntingdonshire	Hampshire	Lincolnshire		
Worcestershire	Devon	Nottinghamshire	Bedfordshire			
Cheshire	Somerset	Suffolk	Cambridgeshire			
Hertfordshire	Warwickshire	Derbyshire				
Norfolk	Dorset	Middlesex				
Gloucestershire	Oxfordshire	Buckinghamshire				
	Rutland					
	Essex					
	Surrey					
	Kent					
	Northumberland					
	Sussex					
	Berkshire					
	Durham					
	Staffordshire					
Buzzard						
Kent	Norfolk	Huntingdonshire	Durham	Gloucestershire	Herefordshire	Somerset
	Derbyshire	Sussex	Cambridgeshire	Staffordshire	Cumberland	
	Cheshire?	Northamptonshire	Lancashire	Bedfordshire		
		Oxfordshire	Northumberland	Yorkshire		
		Nottinghamshire	Leicestershire	Westmorland		
		Suffolk	Hampshire	Berkshire?		
		Middlesex	Essex			

		Warwickshire	
		Worcestershire	
	Surrey	Wiltshire	
	Yorkshire	Kent	Dorset
	Durham	Berkshire	Hampshire
	Huntingdonshire	Wiltshire	Somerset
	Cambridgeshire?	Norfolk	Derbyshire
	Oxfordshire?	Derbyshire	Devon
	Rutland	Suffolk	Essex
	Middlesex	Northamptonshire?	
	Worcestershire	Nottinghamshire?	Staffordshire
		Northumberland	Gloucestershire
			Sussex
Raven			
Lincolnshire	Cheshire		
Warwickshire	Surrey		
Leicestershire			
Buckinghamshire?			
Hertfordshire?			

Source: County avifaunas.

in 1777 and that it was remarkable for the speed with which it vanished in the early nineteenth century. The greatest loss of Buzzards and Ravens was in the 1820s and 1830s. Lilford (1895) noted (under Raven) for all these species that 'they disappeared before the advance of civilisation and the increase of the preservation of game'. He clearly meant before in the sense of time because he also said of Kite that it disappeared 'suddenly, not gradually diminished in numbers as though shot or trapped off by keepers'. Stevenson (1870) made the same point about the impact of persecution by keepers and Smith (1887) made similar observations, noting for Raven that, despite persecution back into the eighteenth century, it was not easily caught in the days of inefficient weapons. Smith and other authors noted that many ravenages latterly were in parks where they were often protected. Smith's point about inefficient weapons is an interesting one. Breech-loading sporting guns came into general use, replacing muzzle-loaders, in the 1850s and the three previous decades saw steady improvements in efficiency, starting with the use of percussion caps to replace flintlocks, always liable to misfire, in the 1820s (Hastings 1981). This development of sporting arms ties in well with the patterns of persecution of raptors (Chapter 12).

The enclosure of upland commons and its long-term effects

Substantial areas of upland commons were also enclosed from the second half of the eighteenth century. Unlike most of the lowland wastes, however, the land remained as pasture. Except for the great monastic estates, whose wealth derived from wool and which were already enclosed, the uplands in the pre-enclosure period were generally cattle-based or mixed pastoral economies. Transhumance was widely practiced. In Wales, for example, farmers and their stock moved up to a summer holding (hafod) from May to September, when cattle grazed to the highest ground, and down again to the main winter holding (hendre) from September (Davies 1984–5, Roberts 1959). The importance that cattle came to have in this economy is well attested by the scale of the droving industry, moving store cattle to the lowlands for fattening, from the fifteenth century in Wales and the seventeenth in Scotland (Sydes & Miller 1988). Stocking rates were comparatively low, set everywhere in the uplands by the capacity of the hendre or inbye to produce winter fodder for breeding stock and followers, although moorland hay was also regularly made to augment this (Roberts 1959). Most of the uplands of northern and western Britain were similarly managed (Rackham 1986, Sydes & Miller 1988).

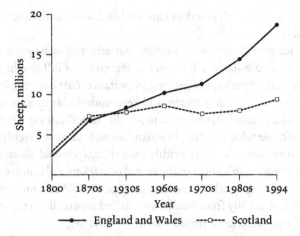

Figure 4.8. Sheep numbers in upland regions from 1800 to 1994. Figures for England and Wales are for Wales and northern England (Cumbria, Northumberland, Durham, Cleveland and Yorkshire) only, for Scotland for the whole country. Figures for 1800 in England and Wales are for 'hill pastures' (Fussell & Goodman 1930). Data for Scotland in the early nineteenth century are from Symon (1959). Figures for 1870s etc. are for the early years of the decade in each case.

The traditional cattle-based economy began to change from the middle of the eighteenth century as the upland common grazings were enclosed. Roberts noted increasing numbers of wether (castrated ram) sheep leaving Wales to be fattened on the Essex marshes and similar sites in England from about 1750. The change arose because enclosing upland commons produced individual self-contained hill units which lacked sufficient capacity for the production of winter fodder for cattle. They therefore increasingly concentrated on sheep which could be out-wintered and, in the twentieth century, particularly its second half, these changes have spread throughout the uplands. Numbers of sheep in the uplands of England and Wales have been rising since 1800. The change has been particularly marked in Wales, where hill cattle have declined steadily. In Scotland there was a large increase in sheep up to the 1870s, followed by a period of comparative stability for a century, before further increases (Figure 4.8). This pattern reflects the greater importance of hill cattle in Scotland and the importance of sporting interests there, both deer and grouse, which limit moorland sheep numbers. The composition and management of upland flocks has also changed during the twentieth century, with the decline of wethers (from 30% of Welsh flocks to <4%, for example), in favour of ewes, and the loss of traditional shepherding on the hills. This aimed to control grazing animals to utilise all the grazing fairly evenly and prevent overgrazing of areas such as heather, valued for winter keep (Bibby 1988,

Shrubb *et al.* 1997). Such practices have vanished with the decline of hired labour on farms.

These long-term changes in stocking patterns and management have had profound cumulative effects on the vegetation of hill grazings. The decline of cattle is perhaps of greatest importance. Cattle are the least selective grazers on upland pastures and keep undesirable species, such as *Nardus stricta*, in check (Grant *et al.* 1996); this is perhaps especially true of types of cattle which evolved in upland areas, such as the Welsh Black. Cattle also eat *Juncus* far more readily than sheep (personal observation). Treading by adult cattle checked the spread of bracken, as did harvesting it for winter bedding, another vanishing practice. The old custom of making moorland hay, usually from *Molinia* swards, led eventually to these being upgraded into reasonable *Agrostis* swards (Roberts 1959).

Wethers were at first the main constituent of out-wintered flocks on the uplands, kept for four or five years for wool and mutton and, like cattle, valuable in controlling unwelcome and invasive sward species, such as *Nardus* and *Juncus* (Roberts 1959). Roberts also noticed the importance of ponies, then widely grazed on the hills all winter, in managing swards. The modern change from wethers to ewes stems from consumers' increasing preference from the late nineteenth century for young lamb rather than mutton (e.g. Fletcher 1961). That and the loss of traditional shepherding practices have seriously reduced the proper utilisation of the upland swards. Ewes are very selective grazers and left to themselves concentrate on areas of sweeter grasses, creating lawns of heavily grazed areas in a sea of coarser herbage they largely ignore. If wintered on the hill they also need supplementary feeding, unlike wethers, which further degrades the vegetation around feeding sites where they tend to concentrate for long periods, rather than spreading out over the grazing.

The increased presence of ewes on the hill has resulted in a commensurate increase in the volume of carrion available, as dead sheep, dead lambs and afterbirths. This has led to a marked increase in predatory scavengers (Kites, Buzzards, Ravens and Crows) in some areas, particularly central Wales (Newton *et al.* 1982), a process which is the reverse of that described under commons and livestock. It has also supported a large increase in Foxes (*Vulpes vulpes*), originally triggered by the afforestation of open hill (Lloyd 1980).

Conversion of moorland commons to sheep farms also changed the seasonal pattern of grazing to one of grazing pressure throughout the year, although transhumance has reappeared today, with hill ewes being wintered on lowland pastures. This, in turn, has led to increased summer stocking rates (Lewis 1996). R. Davies (personal communication) has also pointed out to me that stocking densities are now more uniform, which again leads to changes in the impact of grazing on vegetation. In the past there were periodic fluctuations caused by serious outbreaks of liver fluke, usually in dry summers when the sheep finally found the only grazing available around springs and seeps, the prime source of fluke. Since about 1960 the availability of improved animal medicines to control fluke has removed this problem but, in the past, the effect was not unlike a vole cycle and the periods when sheep numbers were low saw recovery, for example, of flowering plants.

Fire has long been used as a management tool on upland grazings and moors. Carefully controlled rotational burning maintains a favourable age structure in heather for grouse and incidentally benefits other species such as Golden Plover, which often select recently burnt areas as nest sites, for example. Where sheep were or remain the main interest, burning has long been far more indiscriminate, largely to promote grassland at the expense of heath. Sydes & Miller (1988) noted that shepherds burnt large blocks of land annually, probably on some form of crude rotation. But, besides its impact on the vegetation, such frequent burning results in long-term degradation of soils, promotes gullying and erosion in blanket bog and lowers the capacity of the land to support vertebrate populations (Darling & Boyd 1964, Ratcliffe 1980, Ratcliffe & Thompson 1988). Birks (1988) also showed that industrial atmospheric pollution, particularly nitrate deposition, has played an important part in these ecological changes, especially in Wales, Galloway and the Pennines. Uplands in northern Britain and Scotland now receive 25–30 kg/ha of nitrogen annually from this source, sufficient, when combined with heavy sheep grazing, to eliminate heather moor in favour of grass (Smout 2000). This has led to the conversion of very large areas of dwarf shrub heaths to bracken and moorland grasses in the Derbyshire Pennines for example (Anderson & Yalden 1981).

Historically the impact of these changes may have been most marked in the western Highlands of Scotland. A combination of deteriorating climate and overuse from grazing and shelter there led to a marked loss of woodland from the sixteenth to the late eighteenth century (Smout 2000). But the woods were important for winter shelter for the cattle and to protect their grazings. The result was a marked decline in carrying capacity and sheep were increasingly favoured. In the early nineteenth century crofting communities there were widely evicted to make way for extensive sheep-farming. But Symon (1959) noted that towards the end of the century the fertility and carrying capacity of these moorland sheepwalks were declining sharply, some farmers maintaining that only two-thirds of the sheep possible in the 1840s could be kept in the late 1870s. By the 1890s the profitability of sheep-farming had collapsed. Symon suggested that the decline in carrying capacity chiefly resulted from the deterioration of the grazing once cattle, with their ability to control undesirable sward species, were removed. But management of the sheepwalks also involved repeated burning with the effects on fertility noted above. In response to declining returns estates increasingly cleared sheep and switched to letting moors for sporting rents, particularly for deer-stalking, tenants for which demanded exclusive use of the moors; the area of deer forest increased by 80% between 1883 and 1912 (Symon 1959).

This was hardly an ecological improvement. Habitat selection and grazing patterns of sheep and Red Deer overlap (B.C. Osborne 1984), so replacing sheep with deer does not necessarily ease the problems sheep create and, although sheep numbers in the Highlands have tended to stagnate since the nineteenth century (June Census Statistics), Red Deer numbers in Scotland have trebled since 1950 (Sydes & Miller 1988). Harvie-Brown & MacPherson (1904) refer under several species to the continued large-scale burning of heather moor in the interests of deer, leading to the final destruction of many grouse moors in northwest Scotland in the late nineteenth century. They noted that 'this was done without any consideration of the climate and peculiar conditions of the geology and surface soil, and the burning often was done down to the very roots of the heather-plants...and worst of all, the burned ground was not afterwards surface drained.' Gullying resulted and these authors go on to say that 'Miles of heather are being burned off to grant a few years deer grass, sweet and nutritious for a time; but in a few years it will become long and white and rank, and deer will not look at it unless burned again and again.' Infestation by bracken followed. In support of the idea that the decline in carrying capacity and general fertility of these hills as a result of this management history has affected bird numbers generally, Ratcliffe (1980) quoted O. MacKenzie (1924) that during the period 1860–1900 there were substantial declines in Red Grouse, Ptarmigan, Black Grouse, Golden Plover, Snipe, Lapwing, Dunlin and Partridge in Wester Ross. Such declines have continued and the abundance maps in Gibbons *et al.* (1993) show repeatedly how low densities for many bird species remain in this region.

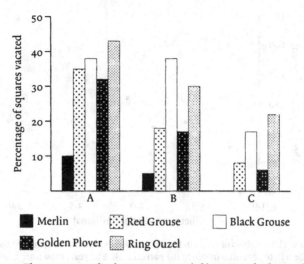

Figure 4.9. The percentage of 10-km squares occupied in 1968–72 by five species of upland birds but which were vacant in 1988–91 compared to the numbers of sheep per 1000 ha of total agricultural area in three regions of Britain: (A) Wales (7399 sheep/1000 ha), (B) northern England and southern Scotland combined (3350 sheep/1000 ha) and (C) northern Scotland (1152 sheep/ 1000 ha); Scotland is divided at the central lowlands. (Data from Sharrock (1976), Gibbons et al. (1993) and June Census Statistics.)

Hudson (1988) noted that habitat mosaics increase the bird species diversity of moorland and perhaps species density. Fuller (1996) gave a series of examples of the importance of vegetation diversity or mosaics to upland breeding birds. Different habitats may be used for feeding by adults and by chicks, for feeding and for nesting, for feeding at different seasons and so on. Such factors are essential features of the birds' ecology and the proximity of these habitat elements is often important. Such mosaics basically evolved from mixed grazing systems with an important proportion of cattle. But the overall trend of all the management changes outlined has been a long-term reduction in the diversity, quality and fertility of all open upland grazings and it has been exacerbated by a system of subsidies which favours sheep over cattle. Such increased intensity of use also contributes to the erosion of blanket mires, important to waders (Birks 1988). Many moorland areas have now become grossly degraded, dominated by unpalatable grasses or bracken, especially in Wales which holds by far the highest densities of upland sheep. Nor do these areas retain water as they once did. Virtually all upland breeding birds have been affected.

A broad association between declining numbers of upland breeding birds and increased grazing pressure by sheep is apparent. Figure 4.9

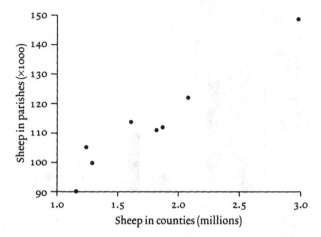

Figure 4.10. The numbers of sheep in five upland Welsh parishes compared to the total numbers in the counties in which the parishes fall. The years shown are those in Figure 4.11. The relationship is strongly significant: $r_s = 0.9$, $n = 10$, $p < 0.01$.

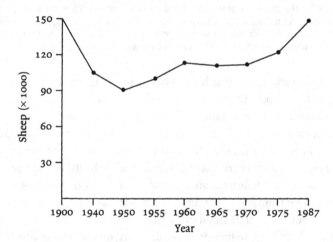

Figure 4.11. Numbers of sheep in thousands in five upland parishes in central Wales since 1900. Significant areas of hill in these parishes were afforested after 1945. (Data from June Census Statistics and Fuller & Gough (1999).)

shows this for three regions, which each have markedly different stocking densities, and for five typical upland species. The figure assumes, of course, that hill stocking densities follow the overall regional pattern. This seems a reasonable assumption and is supported by Figure 4.10, which shows a very strong correlation between numbers at intervals in the five Welsh upland parishes of Figure 4.11, which are predominantly open hill, and the counties in which these occur. Pain *et al.* (1997) reached a similar conclusion for a different suite of species. However their assumption that areas such as the Welsh uplands are occupied by low-intensity farming systems is unacceptable. The present stocking densities of sheep there are by far the highest in Britain and it is ludicrous to describe this as 'low-intensity farming'. It is because of this intensive land use that the relationship in Figure 4.9 arises.

Figure 4.9 only spans the 20-year period of the two *Atlases* (Gibbons *et al.* 1993, Sharrock 1976). However there is evidence that these changes have been continuing for a long period, although systematic long-term data are largely confined to the grouse and some of the waders. Thus for upland populations of Black Grouse there were general declines in the nineteenth century in Wales (Hope Jones 1989) and northern England except Northumberland (county avifaunas), and in Scotland in the early twentieth century (Baxter & Rintoul 1953). The only long-term environmental changes matching this long-term decline in upland Black Grouse populations have been the cumulative effects of the changes in stocking patterns and management outlined above, a relationship that has been confirmed by modern research. Thus Baines (1991) has shown that breeding success has been declining since at least 1950. Increased grazing pressure by sheep and deer has been of critical importance in this change. Baines (1996) found a 37% reduction in breeding success on heavily grazed moors compared to those only lightly grazed. The poor vegetational structure of heavily grazed moors led to sharp reductions in invertebrates important to chicks. Baines suggested that, with fewer large herbivores, improved habitat structure and increased invertebrate populations allowed Black Grouse chicks to survive irrespective of predator control, which did not influence breeding performance. Hope Jones (1989) noted that the inter-war period was one of population stability in Wales, which coincided with declines in sheep densities in the uplands (e.g. Caernarfonshire; Hughes *et al.* 1973) (Figure 4.11). Good *et al.* (1990) showed a relationship between the regeneration of hawthorn scrub and sheep densities in north Wales (Figure 4.12) and it seems very likely that the fall in sheep densities saw some recovery in the moorland vegetation important to Black Grouse. In all upland areas the decline was arrested temporarily after 1945 by afforestation, which saw much recovery of moorland shrub vegetation, fenced away from sheep, during the early stages of forest growth.

Red Grouse populations have shown similar long-term trends, with grouse bags declining throughout the twentieth century in England and

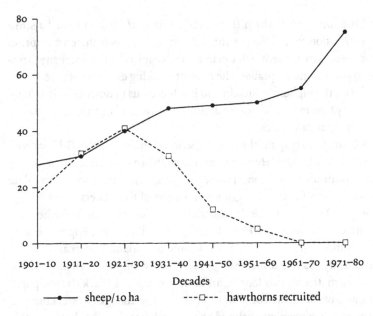

Figure 4.12. Relationship between regeneration of upland hawthorn scrub and June sheep densities in Snowdonia. (Data from Good *et al.* 1990.)

particularly Scotland (Tapper 1992). At the same time there has been a marked contraction of the species' range, stemming directly from the loss of heather moorland. In Wales this started in the mid nineteenth century. All the changes discussed so far in this section concern the impact of long-term changes in agricultural management on moorland vegetation and habitats. Underlying these are long-term strategic decisions about enterprises. One of the main reasons for the loss of heather moor from much of the Welsh uplands is simply that many estates divested themselves of upland holdings during the twentieth century, often to sitting tenants who were much more concerned with raising sheep than grouse. In the long term the loss of grouse management always leads to heather moorland giving way to grass sheepwalk (or forest), a point also noted by Gladstone (1910) for Dumfriesshire. Many examples of this change since the late nineteenth century can be found in Wales (Lovegrove *et al.* 1995). However, although the loss of heather moorland affects some other species, notably Hen Harrier, Merlin, Golden Plover and Short-eared Owl (Bibby 1988, Hudson 1992), its prominence in discussions about moorland habitats stems very largely from its importance to the upland economy through sporting rights. The Red Grouse is the only species confined to the habitat, many upland birds show no particular preference for heather and some, for example Dunlin and Skylark, avoid it (Fuller 1996). The diversity and structure of the habitat is more important and Fuller suggested that many birds may be more favoured by mixtures of heather and grass. Jenkins & Watson (2001) showed this to be so for Oystercatcher, Lapwing and Curlew on a moorland area in Angus.

Since the 1970s there have been marked declines in many upland breeding populations of waders (Gibbons *et al.* 1993). These have coincided with marked increases in stocking rates on upland pastures, especially of sheep, which stem directly from entry to the EU and its Common Agricultural Policy and the application of headage payments to upland stock farming. However in Scotland loss of upland wader populations is more usually ascribed to habitat loss from afforestation, which not only destroys moorland habitat but also limits the use by waders of areas adjacent to plantations, although agricultural improvement of moorland and moorland edge has certainly contributed. In Wales, and perhaps southwest England, forestry has been less important and the degradation of upland habitats in the face of large increases in sheep and declines in the standard and level of moorland management have been a major cause of losses (Lovegrove *et al.* 1995).

5
Some thoughts on hedges

Enclosure hedges

How much hedgerow emerged as a result of Parliamentary enclosure? Pollard *et al.* (1974) measured hedgerows on a series of aerial photographs of the late 1940s and estimated an average of 13 miles of hedge per square mile of improved farmland in England and Wales (for a total of *c.* 800 000 km of hedge in improved farmland or 81 m of hedge per hectare). This agreed well with a Forestry Commission survey of Britain in 1954–7, which estimated 1 526 400 km of field boundary, of which 992 160 km (65%) was hedge; most hedge was in the lowlands and more in England and Wales than Scotland. Areas of Parliamentary enclosure probably had less than 81 m/ha of hedge because enclosure aimed to produce more efficient field patterns (see Chapter 11). In the Lothians da Prato (1985) recorded a minimum of 80.2 km of field boundary before 1945 on 1579 ha of farmland (51 m/ha), which perhaps fairly reflects the larger fields of lowland Scotland in the nineteenth century and may give a realistic figure for arable areas of Parliamentary enclosure. Although earlier enclosures in arable areas tended to be divided into smaller fields, requiring more boundary, in some later enclosures no internal hedges may have been used at all (see below). In pastoral areas much more hedge emerged. In Leicestershire Pell (1887) gave figures which indicated 116 m/ha. I estimate that the arable : pastoral ratio within the areas of Parliamentary enclosure was roughly 75:25 and, if the examples given are typical, they allow an estimate of *c.* 170 000–200 000 km of new field boundary arising from Parliamentary enclosure in the lowlands to be calculated. To this must be added an unknown length of field boundary arising from the subdivision of older enclosures. In Scotland, da Prato's figures, if typical, indicate *c.* 82 600 km of field boundary arising from the creation of new field layouts in lowland farmland. Not all this field boundary was hedge. Walls were common in areas such as the Cotswolds and the Lincolnshire Wolds, and dykes in the Fens. In Scotland, Thom (1986) noted that hedges have never existed in many areas, field boundaries again being walls or fences; wire fences were being used in Perthshire by the 1840s (Dickson 1869). If the proportion of field boundary to hedge found by the Forestry Commission had not greatly changed, I suggest that the period of Parliamentary enclosure is unlikely to have resulted in more than around 200000 km of new hedge line. The uplands are excluded from these rough calculations.

Variations in field patterns

From around the mid eighteenth century, the size of the fields created tended to increase over time, a pattern obvious from field plans published by Taylor (1975) for example. In my Sussex area, for example, comparing a map of Selsey parish for 1672 (in Mee 1988) with the 1778 Ordnance Survey shows distinctly smaller fields in the north of the parish, enclosed by 1672, than in the south, enclosed a century later. By 1778, c. 4186 m of field boundary in the earlier enclosures had also been removed, increasing average field size there by 50%. Such rearrangments of field patterns over time were more frequent in nineteenth-century farmland than is often realised. On three farms on my Sussex study area, for example, this is evident from maps. In 1778 the area I examined on these farms comprised 132 fields or enclosures, with an average size of 2.41 ha. By 1846 15770 m of hedge/ditch had been removed and 1386 m put in, a net loss of 14384 m. By 1875 a further 3386 m had been removed and field sizes then ranged from 0.8 ha to 12.8 ha, with an average of 4.9 ha. Drainage has always been important for arable farming in this area and the original layout in small fields probably facilitated field drainage by open ditches, an idea confirmed elsewhere by Cambridge (1845). Much under-drainage (see Chapter 6) was done on these farms during this reorganisation but ditches remain a feature of all field boundaries there. The removal and making of hedge/ditches both increased field size and, more importantly, improved layout, producing a system dominated by oblongs. These farms were freeholdings and so not necessarily typical of nineteenth-century practice, when 90% of farms were tenanted. Nonetheless they provide a good example of the point made by Grigg (1989) that Victorian farmers were particularly exercised by efficient field layouts and similar examples have been described for Devon (Grant 1845), Hertfordshire and Kent

(Coppock 1958) and Lancashire (Fletcher 1962). Wilkinson (1861) gave the example of Joseph Blundell of Bursledon, Hampshire, whose first care on entering his farm was to remove every gate and hedge. Efficiency in this context meant what it does today, keeping the machine in work as long as possible without turning.

That such reorganisation was not more widespread was the result of the overwhelming influence of the estate in nineteenth-century land management. The constant grumbling of Victorian farmers about the frequency of hedges, the land they occupied and the inefficient working areas that so often existed clearly indicates that many landlords forbade the reorganisation of field systems. Farmers who complained about the land hedges occupied were unlikely to tolerate them otherwise. Baker (1845), Turner (1845), Grant (1845) and Read (1855b) all make clear the point that hedges and field patterns were in the control of the landlord. Read (1858) indicates that some Norfolk estates actually laid down when hedges should be relaid. Grant (1845) quoted Devon farmers thus:

> What is the use of our being told that we should study practical chemistry and copy the example of farmers in better cultivated districts, if we are not allowed to enlarge our fields by taking down those immense banks, or even to cut down the wood growing on them but once in seven years, and then only that which has not been previously marked for rearing? Look at what I lose from the shade of these hedges and trees, besides what they actually occupy. For nearly a ridge wide on each side the corn is hardly worth reaping.

As profitability declined in the later nineteenth century, landowners increasingly tended to regard amenity and sporting aspects of their estates more highly than agricultural investments and prevented removal (Orwin & Whetham 1964). McCall (1988) pointed out that field sports still had an important positive influence on estates' attitudes to features such as hedges. I have little doubt that the great surge of hedge removal since the 1939–45 War owes more to the decline of the estate and rise of owner-occupier farms to 75% of all holdings than to factors such as mechanisation, the availability of machinery specifically for the work or grants. Nineteenth-century farmers didn't need machines for such work (look how much they did by hand), had access to cheap loans for it if required and access to large machines (steam engines). Estates limited such work.

Maps for the first half of the nineteenth century in Coppock (1958), Pollard *et al.* (1974), Taylor (1975) and for my Sussex area show field sizes ranging from paddocks of 0.4 ha up to fields of 52 ha, with 16% of larger fields or enclosures (>8 ha) accounting for about one-third of the total area. This range of sizes and the proportion of larger fields or enclosures was probably fairly typical in arable areas then, as Cobbett and many authors of the county agricultural accounts confirm. We should discard the idea that there was any ideal enclosure size for practising rotation farming. Although pasture was probably always hedged, arable land did not need much internal fencing, for sheep were commonly hurdled (Figure 4.2). Read (1855a), for Oxfordshire, made this very clear: 'Very frequently...fences are not made, except by roads and the boundaries of properties. When the land is dry, and requires no ditches, outside fences only are necessary to each farm. Sheep are kept in hurdles...' Efficiency of working in convenient shapes, in many places coupled with field drainage, was probably always more important than size unless provision also needed to be made for fencing cattle and horses. Larger enclosures could be farmed as more than one field for the convenience of the rotation, a regular feature of our family farm in the past (see illustrations on p. 253 and p. 293). Interestingly the maps for my Sussex area show that virtually all the larger enclosures so farmed had once been two or more fields and the divisions still retained between crops almost invariably were made along the line of what had been a hedge; they were in fact ghosts.

Enclosure hedges as bird habitats

Whilst we know from incidental information in parish and manorial records that birds were colonising the new hedges of enclosure in the later eighteenth century, detailed information about numbers and variety of species is quite lacking. Initially, however, they were unlikely to have been large. Young hedges were kept closely trimmed or laid to encourage dense and stock-proof growth. Cambridge (1845) reckoned that a hedge should be able to withstand stock within four to six years of planting. He advocated trimming with shears in June and October, into a narrow oblong (Figure 5.1). Turner (1845) recommended similar management and such frequent trimming would have severely limited the value of hedges as both breeding sites and sources of food such as berries outside the breeding season. Hedges such as Cambridge and Turner described were clearly not unusual. Cobbett records them, they were noted as typical for the Lothians in the 1870s by Bradley (1927) and as fairly general in England in the mid nineteenth century by Newton (1896), who noted when writing of partridge-shooting:

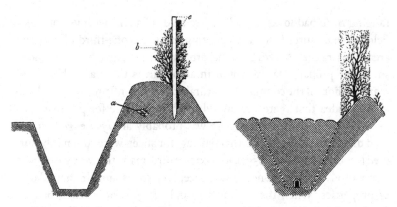

Figure 5.1. Typical enclosure hedge arrangement, with bank and ditch and guard fence, from Cambridge (1845). He advocated trimming with shears into an oblong about 1 m high by 0.3 m wide. Note that Cambridge also recommended that the ditch then be piped in. (Reproduced by permission of the Royal Agriculture Society of England.)

In the old days there were plenty of broad, tangled hedgerows which afforded permanent harbour for the birds, and at the beginning of the shooting season admirable shelter or 'lying' was found in the rough stubbles, often reaped knee-high, foul with weeds and left to stand some six to eight weeks before ploughing, as well as in the turnips that were sown broadcast. Throughout the greater part of England now the fences are reduced to the narrowest boundaries and are mostly trimly kept; the stubbles – mown to begin with as closely as possible to the ground – are ploughed within a short time of the corn being carried, and the turnips are drilled in regular lines...'

Authors such as Harting (1866) and Borrer (1891) stress the way that hedges were pared right back in the peak period for high farming and Coward & Oldham (1900) noted that 'old and tangled hedgerows, which afforded secure nesting places for Warblers and other birds, (have) been grubbed up and replaced by mathematically straight thorn hedges or wire fences'. Evershed (1856) and Read (1858) noted similar changes in Warwickshire and west Norfolk, as did Moscrop (1866) in the arable districts of Leicestershire. In grazing districts there the Leicestershire ox fence was typical, a double hedge allowed to grow to 10 m or more and laid every 15 years. These hedges were particularly managed to provide shelter for stock on farms which had few buildings.

Bradley (1927), writing of the 1870s in East Lothian, specifically noted that the hedgerows on farms there were poor habitats for breeding birds because they were kept so tightly trimmed, narrow and low. Cobbett made similar observations in the enclosures of the Fens and Lincolnshire Wolds in the 1830s, remarking on the lack of Song Thrushes, Blackbirds, Whitethroats, Linnets and Yellowhammers compared to the 'ancient' countryside of the Surrey heaths and Weald. Modern research confirms the accuracy of these observations, showing that low, tightly trimmed hedges hold rather few breeding birds, although Green et al. (1994) found that Whitethroat, Linnet and Yellowhammer preferred low hedges with few trees. Pollard et al.'s (1974) summary indicated average densities of two to nine pairs per 1000 m, involving up to ten species. Da Prato (1985) recorded similar results, particularly noting a lack of warblers. A second factor suggesting limited habitat value in early enclosure hedges was the lack of variety in the hedging shrubs used. The Enclosure Awards usually stipulated hawthorn (*Crataegus monogyna*) and this plant undoubtedly predominated. Elm (*Ulmus* sp.) was also frequently used and was favoured for standard trees. Pollard et al. (1974) noted that hawthorn hedges tended to hold more birds in total and more species than elm but O'Connor (1987) reviewed evidence showing that shrub-rich hedges hold more birds than monospecific ones and Green et al. (1994) also found that one-third of the birds they examined favoured shrub-rich hedges. However, enclosure hedges were normally planted on banks and very often were associated with ditches (Figure 5.1), many of which, according to Cambridge (1845), never held water and developed a considerable weed flora. Da Prato (1985) found that such factors increased both bird numbers and species, as did Arnold (1983) and P. Osborne (1984). Standard trees were also widely planted along new hedge lines, in time adding a significant dimension to hedge habitats; all studies of hedgerow birds have found that hedges with trees hold most birds. Thus, although initially enclosure hedges would have attracted limited populations of the most common farmland birds, in the long term they improved as habitats.

The significance of hedges as a farmland bird habitat has also varied markedly with the prosperity of farming; hedges expand as their management declines in recessions. That they became very overgrown in the recession between 1815 and about 1840 is clearly indicated in the early farming literature by a series of items in the *Journal of the Royal Agricultural Society of England* in the mid 1840s on 'the necessity for the reduction or abolition of hedges' and by accounts such as Clarke (1851), who recorded that the hedges and ditches of the Lincolnshire marshlands were then still badly kept.

The second great period of hedge prosperity was between about 1880 and 1916. Nearly all farming sources for this period describe a sharp reduction in hedge management, at least in arable districts. How far this was reversed during the food-production campaign of the latter half of the 1914–18 War is unclear. Sheail (1976), for example, says little about it. He described significant repair of field drainage systems, however, which could not have been done without attention to ditches and watercourses and associated hedges. The last great period of prosperity for hedges was between 1921 and the 1939–45 War and after, which built upon the growth of the late nineteenth century. Some of the hedges that emerged were enormous. I can remember one being cut back on a neighbour's farm in the early 1960s which had come to occupy a band around the field some 9 m wide. But Turner (1845) describes such hedges emerging as the result of the recession before the 1840s.

Such periods of prosperity for hedges were accompanied by other habitat changes, mainly the abandonment of land and the expansion of scrub. These are all parts of the same process of agricultural decline, which should be viewed as a whole in assessing changes in bird populations. Nevertheless extensive increases particularly in the height and volume of hedges, which occurred over a wide area, would have affected the size of farmland bird populations considerably. Accepting the measurement of just under 1 million km of hedge in Britain arising from the Forestry Commission survey of 1954–7, density figures given by Pollard *et al.* (1974) suggest a potential population of breeding birds ranging from a minimum of *c.* 2.5 million pairs to a maximum of *c.* 37 million. Of course hedgerows have never been all at either the best or worst end of the habitat spectrum and other factors would influence the presence of birds in them, but significant and large-scale changes along that spectrum must have affected large numbers of birds, to the tune of millions of pairs. Modern studies (e.g. Arnold 1983, O'Connor 1987, O'Connor & Shrubb 1986, P. Osborne 1984) indicate that it would have been populations of birds such as tits, robins, wrens, dunnocks and warblers that would have been particularly favoured; some field species, such as Skylarks and Corn Buntings, were probably reduced in number. The only species to which the nineteenth-century avifaunas consistently refer in this context, however, is the Red-backed Shrike. Eighteen of the accounts I examined discussed nesting habitats for this species and 17 considered tall and tangled hedges important; at least four suggested that cutting these back was responsible for reduced numbers. Nevertheless no general link between the growth of hedges and shrike numbers is apparent. Where the avifaunas discussed status changes before 1914, eight described declines and five increases, without any obvious geographic pattern.

The modern stock of hedges

Since 1945 hedges have not simply been cut back, they have been removed or have deteriorated on a scale that has had as much impact on the landscape as the enclosure movement. Haines-Young *et al.* (2000) examined field boundaries in Britain in 1998 on a random sample of 500 1-km squares. This showed that hedges, including remnant and relict hedges and lines of trees/shrubs, accounted for 43% (704 000 km) of all boundaries and, of these, 93% were in England and Wales and 7% in Scotland; 33% of hedges were classified as remnant, relict or lines of trees/shrubs, of which 89% were in England and Wales. Fences comprised 40% of the total boundaries (657 000 km). The survey did not classify hedges more exactly, not separating, for example, hedges with trees or with ditches, important habitat elements for birds; in fact ditches as field boundaries apparently were not considered at all. However, as an adjunct of the 1987 Lapwing survey, which required visits to one randomly selected tetrad in each 10-km square in England and Wales (Shrubb & Lack 1991), the BTO gathered some previously unpublished information on the nature of field boundaries in 49 tetrads in early 1990, which provides more detail and is summarised in Table 5.1. Boundary length was not measured, the analysis simply recording the nature of each boundary and counting each as one. The sample was rather small and biased towards midland and eastern England. Nevertheless it showed that, by type, 23% of hedges were in poor condition, of which nearly one-third were relicts, and two-thirds of the 2223 sound hedges were at the poorest end of the habitat spectrum for birds (low and closely trimmed). However 959 hedges (33%) were still hedges with trees and 305 (7%) were associated with ditches. Joyce *et al.* (1988) found that there were marked local differences in the occurrence of good or poor hedges in their central England sample and similar local variations were evident in the BTO data. All surveys agree on the deteriorating state of the remaining hedge stock and its scale.

The maps gathered by the BTO for the 1987 Lapwing survey also allowed some examination of the loss of field boundary over the previous 30 years. Altogether maps from 96 tetrads provided information on this subject and showed that a total of 270 393 m of field boundary had been

TABLE 5.1. *The nature of field boundaries[a] in 49 tetrads in England and Wales in early 1990*

	Percentage of 4358 field boundaries recorded								
No boundary feature	Fence only	Stone wall	Ditch or dyke only	Sound hedges[b]	Sound hedges with trees[b]	Relict/poor hedges[c]	Relict/poor hedges with trees[c]	Lines of trees[d]	Hedges with ditches[e]
4	11	10	9	32	19	9	3	3	7

[a] A field boundary is defined as a feature separating crops and is taken from intersection to intersection.
[b] Two-thirds of all sound hedges were low and well trimmed.
[c] Poor/relict hedges were defined as hedges with multiple gaps along their length (poor) or reduced to scattered shrubs along an old hedge line (relict).
[d] Most lines of trees were probably hedge/tree lines in the past and were regarded as such.
[e] Hedges with ditches are included in the main table.

Source: Unpublished BTO data reproduced with permission.

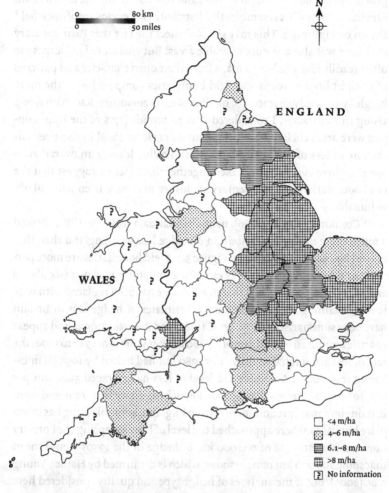

Figure 5.2. Removal of field boundary in England and Wales from 1950–1987 as metres per hectare of land area, by county. (Data from 96 random tetrads covered for the 1987 BTO Lapwing survey; unpublished data reproduced by permission of the BTO.)

removed, an average rate of 9.4 m per hectare of farmland area. The records showed that 68% of the boundaries lost were hedges, 11% hedge/ditch combinations, 17% ditches or dykes only, 3% walls and 1% lines of trees. Figure 5.2 plots the rate of loss by county and shows two interesting features. The extent of field boundary lost was strongly correlated with the proportion of improved farmland occupied by tillage in each county ($rs = 0.62$, $n = 27$, $p < 0.01$). Barr *et al.* (1993) also found that hedge loss was greatest in arable areas between 1984 and 1990. Secondly the distribution illustrated somewhat resembles the distribution of enclosure of open field shown by Figure 4.3. This may simply reflect the fact that Parliamentary enclosure was also a feature of arable areas. But enclosure landscapes are often readily identifiable on maps from their open character and patterns of straight lines in roads and field boundaries compared with the more higgledy-piggledy character of piecemeal early enclosure (Rackham 1986). Using this character, I considered that 26 tetrads (27% of the total sample) were areas enclosed by Parliamentary enclosure and in these tetrads the rate of loss of field boundary was 11.9 m/ha, leaving an overall average elsewhere of 5.23 m/ha. Taken together these factors suggest that the field boundaries of Parliamentary enclosure may have been particularly vulnerable.

O'Connor & Shrubb (1986), using habitat data from the CBC, showed a much more uniform distribution of hedge loss over England than that of field boundary suggested by Figure 5.2. Field boundaries are more permanent than hedges. They may exist for a long time as ghosts (see above) and Barr et al. (1993) showed that much hedge has been replaced with wire fences, retaining the field boundary. Estimates of hedge loss in Britain have been summarised in Joyce et al. (1988) and Barr et al. (1993), and appear to settle around an overall rate of c. 5000 km per year from 1947 to 1990; the rate of loss accelerated rapidly after 1980 but had halted by 1998 (Haines-Young et al. 2000). Assuming the level of loss has averaged 5000 km per year for 50 years means that 250000 km of hedge have been removed from Britain since 1947, predominantly from England. Some planting has taken place but has nowhere approached the level of loss. Accepting the Forestry Commission estimate of 962000 km of hedge in the 1950s would mean that some 700000 km remain today, which is confirmed by Haines-Young et al. (2000). But if the analyses of hedge type and quality considered here are valid, less than c. 200000 km is likely to be good bird habitat.

The deterioration of existing hedges is also of long standing. For example early maps show that the hedges in my Sussex area were mostly hedges with trees. Few trees remain today and not just because of the effects of elm disease in the 1970s. Over the last 200 years or so these hedges have slowly deteriorated to little more than remnants along open ditches. This represents a significant loss of habitat on these farms and gives force to the observation of Sturrock (1982) that simply making hedge removal subject to planning control to prevent change will not necessarily preserve hedges. Although on my three example farms a satisfactory field pattern for today's purposes had been reached by 1912, this has not preserved hedges, although the field boundaries remain as ditches. The loss of hedgerow trees has been more general. Once they were valued for their wood and timber, both within the farming economy and commercially (Rackham 1986). This changed from perhaps the mid nineteenth century and certainly during the twentieth as the farming economy declined and coal and iron replaced wood for many industrial uses. As a result hedgerow trees were no longer managed for wood and timber and their age structure altered markedly, many becoming senescent and ashes (*Fraxinus*), particularly, now senile and collapsing. Shrubb (1993a) linked this change to a greater use of tree cavities as nest sites by Kestrels from 1950.

Habitat loss on the scale described must have contributed significantly to the marked contemporary declines of farmland birds, although its precise nature in terms of numbers or species is obscure. The loss involved is also more complex than a simple loss of nesting habitat. O'Connor & Shrubb (1986) observed that the relationship between hedge density and bird density is not linear because of the impact of territorial behaviour by breeding birds. For some species at least, therefore, the highest bird density occurs at some intermediate level of hedge density, so that hedgerow bird populations may absorb some hedge loss without decline. Fuller *et al.* (2001) found that the needs of hedgerow specialists (see Chapter 4) vary and do not always coincide with the hedge types supporting most bird species; some prefer low tightly trimmed hedges. Thus management needs to maintain diversity in hedgerow structure (Figure 5.3). The loss of field boundary and hedges also means larger fields, which may also influence bird numbers if species avoid moving long distances across open fields. Territories then occupy long lengths of the hedge, which holds fewer birds than the hedge's quality as a habitat might predict (O'Connor & Shrubb 1986). Finally the loss of hedge species is no longer counterbalanced by increasing field species as habitats open, as field species are reduced by modern agricultural methods (Chapter 8).

The value of hedges as habitats outside the breeding season is still poorly understood. It can perhaps be summarised under two headings, shelter and food sources. Shelter may comprise roost sites or refuges from predators. The latter may influence choice of feeding site for field feeding species. Thus Arnold (1983) found that winter flocks of buntings and migrant thrushes strongly favoured grass fields with tall hedges, as did Tucker (1992), and I have made similar observations in Wales for thrushes, Starlings and finches. Field size would be a factor here if birds are more

Figure 5.3. A good diversity of hedge structures in a limited compass. (Courtesy of Rural History Centre, University of Reading.)

vulnerable to predators in large fields than small. Many birds tend to forage fairly close to the hedge, so increased field size is again unfavourable because it reduces the extent of safe feeding areas for them. Hedges provide large quantities of berries and seeds and Snow & Snow (1988) have given a detailed account of how birds exploit berries. Hedges and field margins, ditches and banks also provide important hunting sites for some raptors, particularly Barn Owls, Hen Harriers (in winter), Sparrowhawks and Kestrels; ditches are particularly valuable feeding sites for passerines (Arnold 1983). Finally hedges, as a major part of the permanent habitat skeleton of farmland, play a crucial role in providing havens for invertebrates in an environment subject to regular fundamental change during the year.

6
Drainage

Drainage is essential to farming. Land cannot be cultivated unless dry enough to create a tilth or carry machines, nor can pasture carry stock unless dry enough so that they do not churn it into mud. Improved drainage allows better plant growth, earlier cultivation or grazing, and better take-up of fertilisers, so increasing productivity. About 30–40% of improved farmland in Britain today is naturally free-draining enough to need little or no artificial drainage. In England, however, much of that area was waste in the eighteenth century (Chapter 4); elsewhere poor drainage inhibited agricultural improvement.

Land drainage can involve either arterial drainage of major wetlands by cutting new drainage channels or improving existing ones, or field drainage, usually today by under-draining with pipes or mole drains but in the past by ditches and surface drains; ditches remain important links between field drains and arterial channels. As with enclosure, of which they were often part, many major drainage schemes required authorisation by private Act of Parliament, not least because of conflicting and essential interests requiring different water levels, especially mills and navigation. Barge traffic penetrated quite small streams in the past (Haslam 1991).

Historically land drainage has affected two main habitat types of great significance to birds, major areas of swamp with reed-beds and permanent water, and wet or damp grassland flooding in winter/spring but allowing summer use for grazing or hay meadows. The draining of arable land has had less direct impact but, in the nineteenth century, allowed the expansion of high farming over large areas of poorly drained clay soils. This chapter considers the drainage of major wetlands and the development of field drainage. Factors affecting wet grassland are also dealt with in Chapter 9.

1750–1880

Drainage of major wetlands

Extensive wetlands remained a significant feature of the British landscape in the second half of the eighteenth and early nineteenth centuries. The largest areas were the Fens, the Somerset Levels, and Thorne and Hatfield Moors in south Yorkshire, totalling c. 400 000 ha (Marshall et al. 1978). There were also important wetlands in the Vale of Pickering and the Derwent Valley and in Holderness in Yorkshire, on the Lincolnshire side of the Humber, in the coastal marshes of Lincolnshire, East Anglia, Kent and Sussex, some of the river valleys of Sussex, Hampshire and Wiltshire, at the confluence of the Thames and its tributaries in Berkshire and Oxfordshire, in the fens and levels of the Severn and its tributaries and of the Carmarthenshire and Glamorgan coasts, the main estuaries of west and north Wales, the bogs, mosses and meres of the West Midlands and the low-lying coastal plain of northwest England and numerous similar habitats and wet river valleys and straths throughout Scotland and the Border counties. From the mid eighteenth century the bulk of these areas was drained; some major works started a century earlier. Marshall et al. (1978) surveyed the areas of England and Wales with extensive drainage channel systems and estimated the total area drained by these at c. 800 000 ha (c. 2 million acres).

The widespread nature of such habitats in the late eighteenth and early nineteenth centuries is indicated by the distribution of the Bittern and Marsh Harrier, typical inhabitants of extensive wetlands and areas of reedswamp (Figure 6.1). The Bittern was the more widespread, breeding in many areas of Scotland, where it was perhaps more widely distributed than shown; it was recorded in Sutherland in the 1630s, for example (Baxter & Rintoul 1953). It was a common species and, accounted a delicacy, was widely taken for the table, being recorded in the accounts of noble households as an item for banquets back to the late mediaeval period. Whilst there are no definite records of breeding in Kent, Hampshire, the coastal fens of Carmarthenshire or the mosses of north Wales early authors accepted the strong probability that they once did so because extensive areas of suitable habitat were present. However Bitterns apparently never bred in Somerset. The Marsh Harrier apparently never nested in Scotland, although it occurred, nor the West Midlands except for Cheshire. It may have been more widespread in southern England than the records show but I suspect that the nineteenth-century records of breeding

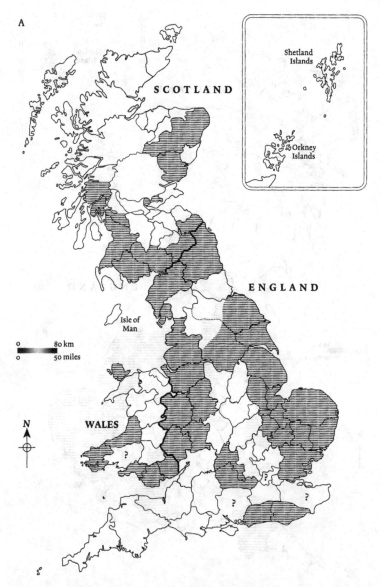

Figure 6.1. The breeding distribution of (A) (above) Bittern and (B) (see over) Marsh Harrier in the late eighteenth/early nineteenth centuries. ? indicates that breeding was considered probable, usually on the basis of the presence of abundant habitat. For Marsh Harrier I reject the Durham records (Hancock 1874) and Nelson recorded it as breeding in North Yorkshire/Cleveland, which is quite possible. Mather (1986), however, excludes these records. (Data from early avifaunas.)

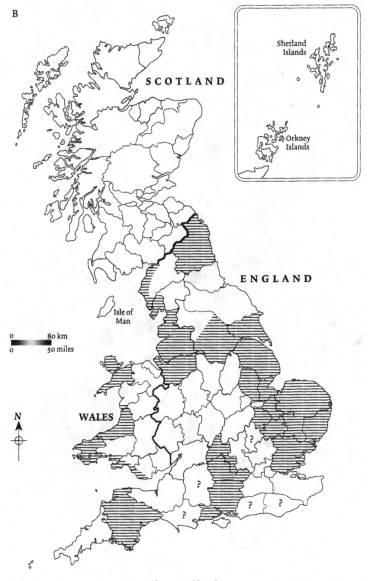

Figure 6.1. (Cont.)

on the Durham and North Yorkshire moors reflect confusion with Hen Harrier. Hancock (1874) recorded only one nest, on Wemmergill Moor, and the details he gave for that and his only Hen Harrier nest are the same. His claim that the Marsh Harrier was 'common on our swampy moorlands, where it bred', otherwise probably applied to Northumberland, which he also covered.

Other typical fen species were also comparatively abundant in this period. Spotted Crakes bred widely, if sparsely, in many wetlands in England and Wales (Aplin 1890, 1891) and probably bred regularly in southwest Scotland (Thom 1986). Bearded Tits were reported as breeding in the early nineteenth century in Kent, Sussex, Hampshire, Berkshire, Essex, Suffolk, Norfolk, Cambridgeshire, Huntingdonshire and Lincolnshire. Modern authors doubt some of these records, particularly for Hampshire and Essex, but suitable habitat certainly existed in the nineteenth century, a point confirmed by Montagu's (1833) account of watching Bearded Tits in Essex. Gurney (1899) added Devon as a county where the species bred regularly in the past. Savi's Warbler bred in the Fens in Norfolk and Cambridgeshire and possibly in the Thames marshes in Essex, and Black Terns were widespread in Lincolnshire, Cambridgeshire, Norfolk, Huntingdonshire, probably Suffolk, and in the Kent and East Sussex marshes.

All these species, and several others mainly confined to fen habitats in eastern England, for example Avocet, Ruff and Black-tailed Godwit, largely disappeared by the mid nineteenth century, and Table 6.1 summarises the loss of the principal species concerned. The Bittern was by far the most widespread and, as the Table shows, it vanished from Scotland and perhaps Wales, markedly earlier than from England. Some suitable habitat remained in Norfolk, around the Cheshire meres and probably in Suffolk, throughout the nineteenth century and breeding was re-established in the Norfolk Broads area in the early twentieth century. Thus drainage did not entirely account for its total loss in the nineteenth century and persecution came to be significant with increasing scarcity, as it did for the other species (Chapter 12).

Nevertheless the drainage of vast areas of wetlands was undoubtedly a major cause of the loss of all these birds. Table 6.2 provides a brief chronology for the drainage of a series of important sites. It is drawn largely from the early ornithological literature, which provides much incidental information on historic habitat changes. Whilst incomplete, it confirms that a massive amount of arterial drainage was carried out particularly from about 1770 to the 1840s. This comprised both new works and improvements of existing ones. The stimulus for this work was partly provided by a series of very wet years in the late eighteenth century and especially by the farming boom of the Napoleonic War period, when many improvement

TABLE 6.1. *The loss of selected fenland species as breeding birds in Britain in the nineteenth century*[a]

County	Bittern[b]	Marsh Harrier	Ruff	Black-tailed Godwit[d]	Black Tern[e]	Savi's Warbler[f]	Bearded Tit[g]
Kent					1840s		1845
Sussex	<1849				1843		<1860
Essex	<1820	c. 1830		<1820			<1800?
Berkshire	1780	1814					c. 1820
Oxfordshire	<1849	1814					
Suffolk	1836	1830s	1825	<1824	<1820		1885
Norfolk[h]	1850	1866	1844 F 1871 B	<1835	<1830 B 1850 F	<1860	not extinct
Cambridgeshire	1821	c. 1860	1840s	1829	?1824	<1850	1853
Huntingdonshire	1853	1853	<1810	c. 1829	1843		1853
Herefordshire	<1820						
Staffordshire	1863						
Shropshire	1836						
Lincolnshire[h]	<1835	1836	1825 F	<1820	1812		<1840
Yorkshire	<1820	1836	c. 1824	<1835			
Cheshire	c. 1854	c. 1820					
Lancashire	<1845	1860					
Durham			1825				
Northumberland	<1820	c. 1840	1853				
Cumbria	1804						
Glamorgan	?c. 1800						
Carmarthenshire	<1850						
Pembrokeshire	<1800	<1840					
Ceredigion	?						
Anglesey	?						
Berwickshire	c. 1830						
Roxburghshire	1793						
Galloway	c. 1830						
West Scotland	1782						
Clackmannan	<1820						
Kinross and Angus	1813						
Aberdeenshire	<1820						

[a] Dates are those of the last well documented breeding. <, before. Records simply given as early nineteenth century are shown as <1820.
[b] Macgillivray (1837–52) noted Bittern as of rare occurrence anywhere in Britain by the early 1850s.
[c] Ruff bred on the damp heaths of north Lincolnshire until 1882 and sporadically in Suffolk until the 1890s and in Norfolk, Durham and Lancashire similarly until 1922.
[d] Black-tailed Godwit bred sporadically in East Anglia until late nineteenth century.
[e] Black Tern bred sporadically in the Fens, with major floods, after 1850.
[f] Savi's Warbler was not found as a breeding bird until 1840, at least by ornithologists.
[g] Bearded Tit bred at Sudbury Essex in 1868.
[h] F, Fens; B, Broads; not extinct, still breeding at the end of the nineteenth century.

works were done by French prisoners of war (R. Leverton personal communication). It continued throughout the subsequent recession in farming between 1815 and about 1840. Increased understanding of drainage problems and better technology accelerated the process during the nineteenth

TABLE 6.2. *A chronology of drainage works in some major historic wetlands in Britain during 1750–1880*

Scotland and the Borders
Peak of drainage activity second half of eighteenth and early nineteenth century, with most Bittern sites lost before 1800 (e.g. Aberdeenshire, Roxburghshire) or between 1800 and 1820 but Lochar Moss in Dumfriesshire and Billie Mire in Berwickshire drained in the early 1830s. Loss of many flood and water meadows to drainage and conversion to arable in late eighteenth century. Selby (in Bolam 1912) noted that many bogs and mosses in Border counties, particularly Northumberland drained by 1830 but Prestwick Car, Northumberland, not drained until 1850s. Gray (1871), Muirhead (1895), Gladstone (1910), Bolam (1912), Baxter & Rintoul (1953), Symon (1959), Mitchell (1997).

Northwest England
Mosslands of northwest England mainly drained 1780–1830, particularly first 30 years of nineteenth century. The largest, Chat Moss and Martin Mere, finally drained 1849 and 1850, although former started in 1768 and latter in 1780s. The Abbey Holme in Cumberland drained 1840–59. Dickinson (1852), Mitchell (1885), Macpherson (1892), Fletcher (1962). Cheshire mosses drained later, not really under way until mid nineteenth century. Whitely Reed 'the wildest and deepest bog in Cheshire' drained 1852, Carrington Moss 1886. Probably spans period of activity. Coward & Oldham (1900), Phillips (1980).

Yorkshire
Hatfield Chase and Isle of Axholme originally drained 1620s but only partially successful (see text). Much of Hatfield and Thorne Moors relatively untouched and another drainage and enclosure Act in 1811. Holderness Carrs drained second half eighteenth century, Acts in 1764 and 1798. Doncaster Carrs drained same period, Act in 1764. Spalding Moor drained by 1829, the greater part of Thorne Moors by 1840 but 6000 acres (4050 ha) still undrained 1886, Pickering Carrs by 1846. Clarke (1854), Nelson (1907), Thirsk (1953), Sheppard (1958), Limbert (1978, 1980), Rhodes (1988). Hatfield Moors still extend to 1700 ha.

Wales and West Midlands
Main period of drainage similar to northwest England. Marshes and levels in the Severn, Avon and Teme Valleys drained approximately 1800–1840 but Longdon Marsh Worcester not drained until 1870. Forrest (1907), Smith (1930–8), Harthan (1946). Important mosses remain at Fenn Moss, Cors Caron and Cors Fochno.

Lincolnshire
For Fens see below. Ancholme Valley originally drained 1637, with subsequent Acts for improvement in 1767, 1802 and 1825. Straw (1955).

The Fens
At beginning of the nineteenth century greater part of Fens theoretically drained but many localities remained liable to inundation. Extensive meres remained, e.g. Ramsey and Ugg Meres (drained early nineteenth century), East Fen Lincolnshire (drainage Act 1801), Whittlesey Mere (drained 1851–3). A series of Drainage Acts were obtained in the first 40 years of nineteenth century to improve drainage but it was the rapid application of steam-powered pumps for drainage from c. 1820 which finally guaranteed success. By 1848 40–50 steam pumps operating of which 21, averaging 50 hp, were draining 123 300 acres (50 000 ha). By 1852 60 steam pumps working. Clarke (1848), Wells (1860), Darby (1968).

(cont.)

TABLE 6.2. (Cont.)

Norfolk and Suffolk
For Fens see above. Marshes of east Norfolk extensively drained during first decade nineteenth century, with drainage Acts for Cantley, Catfield, Fishley, Hassingham, Hickling, Ludham, Potter Heigham, South Walsham, Sutton and Upton between 1799 and 1807 but much inland from the coast still poorly drained in 1858 and described as almost valueless. Steam pumping had drained Buckenham Marshes by 1847. Salthouse Marshes drained 1851. Suffolk coastal marshes largely drained early nineteenth century, e.g. Minsmere in 1813. The Waveney marshes at Beccles drained in 1802. Read (1858), Lubbock (1879), Ticehurst (1932), Ellis (1965), Taylor *et al.* (1999).

Essex coast, Thames estuary and valley
In the Essex coastal marshes much reclamation of saltmarsh during 1780–1820 and some during 1850–80. Majority of the grazing in the Dengie and Crouch marshes converted to arable by mid 1850s. In the north Kent marshes much drainage in first half nineteenth century, a significant percentage of which was for urban development. In Thames Valley Kennet Marshes probably drained around 1815 and the wetlands in the Kelmscote–Harcourt area drained 1850–66 but much marsh still remained in Kennet Valley in 1889. Otmoor was inclosed in 1829 (date of Act) but still unimproved and subject to severe winter flooding to end nineteenth century. Read (1855a), Aplin (1889, 1890), Gillham & Homes (1950), Williams & Hall (1987), Williams *et al.* (1983).

Sussex and Hampshire
The swamps in the Arun Valley, known as Wildbrooks, drained in early nineteenth century, although still subject to major winter flooding. Other Sussex river valleys and marshes highly valued pasture at this period but extensive reed-beds in Pevensey Levels also cleared. Drainage of part of lower Test Valley in 1852. Knox (1849), Young (1793), Wilkinson (1861).

Somerset Levels
Main drainage activity 1770–1830 but many areas remained subject to severe flooding, although the new drains cleared flood water more quickly. Unlike many wetlands the primary landuse remained pastoral. Williams (1970).

century. The application of steam power to pump-draining the Fens was of particular significance.

As Table 6.2 shows, much drainage work in Scotland was completed rather earlier than in many important areas in England, which conforms well with the pattern of decline in the Bittern (Table 6.1). Baxter & Rintoul (1953) also noted that declines in the breeding population of Snipe caused by extensive drainage for agriculture were recorded as early as the 1790s. Symon (1959) gave a good summary of drainage activity in Scotland in this period, which confirms the brief chronology outlined in Table 6.2. He also stressed the much simpler legal basis of major arterial works in Scotland, where the need for drainage Acts rarely arose because of the extent of the landlords' interest.

Everywhere much of the work in this period involved a continuous process of improvement and setback rather than single schemes carried through to a conclusion. The draining of the Fens offers a prime example. Draining the peat fens in the seventeenth century was initially reasonably successful. Cornelius Vermuyden, in 1652, claimed that about 40 000 acres (16 000 ha) in the North and Middle Levels of the Bedford Level, Cambridgeshire, were sown with coleseed, wheat and other grain 'besides innumerable quantities of sheep, cattle and other stock, where never had been any before' (quoted by Clarke 1848). Large meres remained, as at Whittlesey and Trundle Meres, Ramsey, Ugg and Benwick Meres. At the turn of the nineteenth century the East Fen, Lincolnshire, was still 'quite in a state of nature, and exhibits a specimen of what the country was before draining. It is a vast tract of morass, intermixed with numbers of lakes, from half-a-mile to two or three miles, in circuit, communicating with each other by narrow reedy straits' (Southwell 1870/1). Such conditions led Sir William Dugdale (1662) to assert that wildfowl were little affected by the draining:

> 'As to the decay of fish and fowl, which hath been no small objection against this public work, there is not much liklihood thereof; for, notwithstanding this general drayning, there are so many great meers and lakes still continuing, which be indeed the principal harbours for them, that there will be no want of either; for in the vast spreading waters they seldom abide, the rivers, channels and meers being their principal receptacles; which being now increased, will rather augment than diminish their store' (quoted by Darby 1934).

Justifiable resentment at the way the inhabitants' rights and interests had been overriden by the Crown led to much destruction during the Civil War and Commonwealth (as it did in Hatfield Chase and Axholme; Stovin 1752), which was repaired under Charles II. But problems of soil shrinkage as the peat dried, exacerbated by the universal practice of paring and burning in preparation for cultivation, and 'blows' of the light fen soils, lowered the surface of the drained areas and progressively undermined the drainage systems, which were, in any case, very poorly maintained. Increasingly severe flooding became a feature of the eighteenth century, only partly met by the introduction of pump drainage by windmills, which in any case tended to start the cycle again. In addition the porosity of banks built of light moor soils allowed water pumped out by the windmills (which did not work in calm weather, a frequent complaint) to percolate quickly back. Because of these factors, many areas were close to reverting to fen, with floods increasing in area and persistence, by the late eighteenth century. Contemporaries also stressed the need for improved outfalls into the Wash. Young (1805) noted that 'The fens are now in a moment of balancing their fate; should a great flood happen within two or three years, for want of an improved outfall, the whole country, fertile as it naturally is, will be abandoned'. The accounts of farming in the Fens at this period hardly give a convincing picture of successful drainage by the end of the eighteenth century, with their reports of crops lost to floods, being harvested from boats or by men up to their waists in water, of cattle grazing up to their bellies in water and thousands of sheep lost to the rot; Young noted that 40 000 sheep rotted in East, West and Wildmoor Fens in Lincolnshire in 1793.

Nevertheless in dry seasons men tried again. Because it exacerbated shrinkage, paring and burning was discarded in the early nineteenth century in favour of claying, which gave body to the soil and increased fertility. Steam pumps were introduced from the early nineteenth century and, as drainage problems eased, farmers began to under-drain their fields with pipes to lower water tables further. This marked something of shift in land use, at least in the lower-lying districts, from pastoral to arable. Pastoral farming had always been the major feature of fenland agriculture and graziers preferred a high water table to maintain grass growth; full dykes also provided drinking water and obviated fencing, as in many marshland grazing areas. Arable farmers aimed to lower water tables by up to a metre (Clarke 1848). This probably influenced water tables over a wider area than that immediately drained (see Green & Robins 1993), which may have particularly affected species such as Black-tailed Godwits or those needing wet reed-beds, such as Bitterns.

Similar progressive patterns of drainage occurred in other major wetlands, which led Burton (1995) to suggest that climatic factors in the early nineteenth century, particularly a series of severe winters, were more significant in causing the loss of species such as Bittern and Black-tailed Godwit than inefficient drainage. However comparing Tables 6.1 and 6.2 shows a fairly robust agreement in the timing of species loss and drainage works. Furthermore seasonal inundation of what had become farmland represented significant habitat change or degradation, especially for fen species. As noted in Chapter 2, this may have increased fen species' vulnerability to unfavourable climatic change.

One other important development in arterial drainage at this period should be noted. The Land Drainage Act of 1861 set up a system of Drainage Districts and Commissions of Sewers, covering a total area of 100 000 acres

(40000 ha) with powers to improve and repair existing watercourses and outfalls, financing the work by a drainage rate on the land so improved. The principle of this system has remained the basis for maintaining arterial drainage works since.

Field drainage

Drainage in agricultural land normally leads to upgrading the agricultural operations practised there. In the nineteenth century the major change was an increase of 30% in the area devoted to arable farming, essentially high farming, largely at the expense of unimproved grazings and wetlands. The aim of most arterial drainage in this period was to convert the land to the plough. It was one part of a major drive in the nineteenth century to improve the drainage status of British farmland generally. The other component of this campaign was the improvement of field drainage. Early field drainage was by surface drains and ditches. Whether or not ridge-and-furrow was evolved as a drainage mechanism, it certainly came to be so used (Green 1975). Cobbett also refers to this on several occasions, e.g. in Essex, Gloucestershire and Hertfordshire. Green showed the close relationship between the distribution of ridge-and-furrow in Berkshire, Buckinghamshire and Oxfordshire and that of modern drainage. Both occurred in the same districts and soils, whilst free-draining land exhibited neither.

Under-drainage, putting drains beneath the surface of fields to carry water to ditches, was developed in the mid eighteenth century. Early drains were sod drains or filled with stones or bushes. During the nineteenth century claywork drains were progressively developed. Young (1770) noted farmers near Warrington using bricks laid in the form of goalposts to make the necessary conduit. Around 1800 the inverted U-shaped tile placed on a flat tile sole was developed and machines for extruding clay pipes for draining (known as draining tiles) were finally invented in the 1840s. Many estates built their own tileries. For example Fletcher (1962) noted at least 24 estates in Lancashire with such facilities by 1849 and there were at least 11 independent manufacturers there at the same time, whilst Bravender (1850) noted many in Gloucestershire. Nicholson (1943) pointed out that the period of agricultural depression after 1815 nevertheless saw major advances in the practice of field drainage, particularly the development of what was known as thorough drainage, the laying of drains at regular intervals and depths over the whole area of fields irrespective of the incidence of wetness. Clearly the investment was seen as still profitable.

Besides the prospect of worthwhile commercial returns from the investment, three factors encouraged a major surge in field drainage in the period up to about 1870. Draining tiles were exempted from tax in 1826, significantly reducing their cost (Trafford 1970). Secondly, with the repeal of the Corn Laws in 1846, a system of Government loans for field drainage and other improvements, available at $3\frac{1}{2}$% and repayable over 22 years, was introduced. It was particularly aimed at settled estates which otherwise found capital for improvements difficult to raise (Orwin & Whetham 1964). The success of the scheme and the demand that emerged led to the setting up of a number of public companies to raise money by the sale of stock to the public and lend to landowners for agricultural improvement. The loans were safeguarded by an obligation on landowners to have plans approved by the Inclosure Commissioners, who were already charged with administering the Government scheme. By 1872 some £10 million (£300 million today) had been advanced under these schemes, of which 72% was used for land drainage. Thirdly the invention of machines to mass-produce draining tiles in the 1840s lowered their price from £1 1s per 1000 to 6s (£30 to £8.70 in today's money), which, on figures supplied by Trafford (1970), would have reduced drainage costs per acre by 20–25%.

The area involved in tile drainage at this period has been estimated at c.12 million acres (4.8 million ha) in England and Wales by Robinson (1986). A substantial area was also drained in Scotland, where Orwin & Whetham (1964) showed that half of the £4 million loaned by the State by 1850 had been taken up; considerably more was spent by private capital. Symon (1959) assumed that, by the 1870s, most arable land in Scotland needing it had been under-drained and Smout (2000) gives a good picture of the scale of the work done there.

Robinson (1986) based his estimate on the evidence of old pipe drainage schemes recorded by Ministry of Agriculture, Fisheries and Food (MAFF) drainage officers during visits to approve post-1939 drainage work as eligible for grant. The bulk of these old works date from the nineteenth century, for little drainage work was done between 1890 and the 1930s (Nicholson 1943). Robinson's figure was supported by Trafford (1970), who produced a similar estimate based on figures for clay-pipe production. One factor undermines these estimates however. Many Victorian farming accounts make clear that much under-draining had to be redone as the nineteenth century progressed, particularly in Scotland, northwest England and the English Midlands. Murray (1868) noted that this happened because it was some time before landowners understood that drainage schemes had to be individually designed, a point also stressed by Caird (1852). The necessity for redraining especially undermines estimates of area drained based on tile production and the similarity of Trafford's and Robinson's estimates suggests that it affected the latter as well. Clearly, however, a larger area than the nineteenth-century estimates of $c.$ 3 million acres (1.2 million ha), based on the total amounts of Government and company loans advanced, was drained. Indeed Bailey Denton (1863) expected this when he assumed that 'a quarter of a million acres will, in future, be permanently drained per annum...' in Britain. Probably Government and public company loans widely supplemented private capital and the much greater availability of pipes and their steep decline in price must also have made a significant contribution to this surge in activity. Robinson produced a map showing that few areas in England and Wales had less than 25% of agricultural land under-drained by tiles prior to 1939, such low levels being found mainly on the chalklands of the south and southeast and the sands of north Norfolk. The highest levels were predictably found in the northwest, in Wales and in Devon, all areas with damp climates, and in the heavy soils of the Weald and the Midlands. Nicholson (1943) also noted that older methods of hollow draining with stone-filled drains continued in the nineteenth century in the eastern counties and that mole draining was also regularly used there and in the east Midlands in clay soils. The overall impact of this surge of drainage on farming systems is clear from Figure 3.1, with a virtual doubling of the area of fodder roots between 1830 and 1860, as the new arable methods spread onto the clays.

Although the loss of fen birds to arterial drainage seems straighforward, these were not really farmland birds. They were lost through the conversion to farmland of their specialised habitats. How far breeding populations of less specialised species, such as the commoner wildfowl, waders and passerines such as Reed Warbler and Yellow Wagtail were affected by the scale of nineteenth-century field drainage is far less clear cut. Alexander & Lack (1944) suggested little general change for the two most common wildfowl species of farmland, Mallard and Teal. But significant local changes did occur. Kear (1990) noted that an important factor in the demise of the duck decoy (Chapter 12) as the nineteenth century progressed was the decline of native Mallard brought about by drainage. Scarcer wildfowl certainly declined with the lack of a close season but started to recover at the end of the century, when one was imposed (Chapter 12).

Three of the common waders of agricultural land provided contrasting patterns. Snipe were most widely noted as declining, with half of the early regional and county accounts consulted recording historic changes in status, always declines as a result of drainage. The records usually referred to shooting bags in winter. Nevertheless the loss of wintering birds and sites is a valid indication of the impact of drainage and breeding populations must have been equally affected. That there was no geographical bias in counties recording declines suggests that the decline was general.

Changes in Lapwing populations followed a different pattern. Of the 42 avifaunas examined, 11 described a decline in the first half of the nineteenth century and two more a similar trend later, while five implied an increase at some point and 24 provided no evidence of historic change, although whether this always equated with population stability is debatable. Nevertheless there was a clear geographical divide in the records, with the bulk of declines occurring in eastern Britain, the area where drainage was accompanied by the most extensive conversion of old damp pasture to arable and egging was systematically practised. Stevenson (1870) and other authors considered that it was the combination of drainage and egging that caused the species' decline (Chapter 12). Lapwings also adapted to change. Gladstone (1910), for example, noted that the species had declined with the loss of 'wild pastures' in the early nineteenth century but by the turn of the century agricultural change had led to a change in habit, 'for they now nest as commonly on the cultivated land as uncultivated moor'.

The nineteenth-century avifaunas make it clear that the Redshank was unknown as a breeding bird in many inland and western counties before the second half of the century or later, being confined to counties on the eastern seaboard (Thomas 1942) (Figure 6.2). Stevenson (1870) noted that before 1825 it was considered more common than any other wader nesting in the Norfolk marshes but a sharp decline followed for the next 40 years. Other fenland counties recorded similar patterns, e.g. Cordeaux (1872) for Lincolnshire, Jenyns (1835) for Cambridgeshire (where Lack (1934) noted that breeding ceased for a spell in the 1850s) and Lilford (1895) for Peterborough. This timing coincided in the Fens with the introduction of steam pumping and of under-drainage for arable (Clarke 1848). The pattern of decline, however, reversed from the mid 1860s, well before the late nineteenth-century recession, when the species started to increase and spread again in East Anglia. In the 1860s and 1870s it had also spread into many inland sites in west and central Scotland (Gray 1871, Harvie-Brown

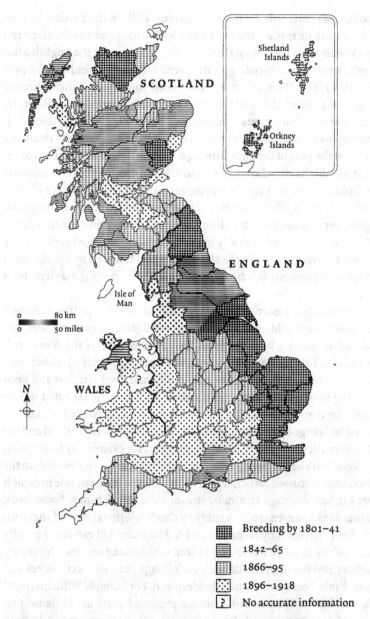

Figure 6.2. The breeding expansion of the Redshank in the nineteenth and early twentieth centuries by county. Note that the county basis inflates the distribution; for example there was only one breeding locality in N. Yorkshire in the 1840s. The species colonised Carmarthenshire after 1918, never bred in Cornwall and once only in Pembrokeshire, in 1955. (Data from county avifaunas and Thomas (1942).)

1906), it was breeding on the Lammermuir Hills in the Borders by 1876 (Evans 1911), increasing and expanding in Durham and Northumberland in the mid 1870s (Bolam 1912) and starting to penetrate the English Midlands, where it was breeding in the Trent Valley in Staffordshire by 1875 (Smith 1930-8). This expansion continued into the twentieth century (Figure 6.2). Much the most interesting point about this species is why its breeding distribution was apparently so limited in the early nineteenth century. As its population expansion was well under way by the 1860s and 1870s, at the peak of the high farming and drainage period, it is hard to believe suitable habitat did not exist and many nineteenth-century accounts of farming in various counties make it clear that it did (Chapter 9). This undermines too glib assumptions about the importance of drainage and agricultural factors to its breeding distribution in the nineteenth century. Climatic factors were probably important as Burton (1995) argued and a period of severe winters in the 1830s and 1840s, followed by a period with much severe summer flooding in river valleys (Clarke 1854), may have been important.

For farmland passerines there is little evidence of population changes stemming from field drainage. The Yellow Wagtail, for example, was considered as much a bird of arable land as grassland in the nineteenth century and little change in status was noted (Chapter 3). Other wetland passerines, such as Grasshopper, Sedge and Reed Warblers and Reed Bunting showed similar patterns. Alexander & Lack (1944) noted no evidence for marked change with any of these species, except for some increase and range expansion by Sedge Warbler in northern Scotland and the Northern Isles. Some local effects were noted, for example by Knox (1849) for Reed Warbler in Sussex, but these are all more adaptable birds than the fen species discussed earlier, with a wider range of habitats, able to exploit wet ditches, drainage channels, stream sides, canal banks, *Juncus* beds, plantations and even crops. A study by Catchpole (1974) showed that even the Reed Warbler is by no means tied to *Phragmites* and can breed equally successfully in other marsh vegetation, which many nineteenth-century authors also noted. Habitats such as drainage channels, wet ditches and canal banks became much more widespread. For example Williams (1970) showed how enclosure and drainage produced particularly dense patterns of drainage channels (rhynes) in parts of the Somerset Levels. Both Grasshopper and Reed Warblers in fact expanded their breeding range in Britain during the nineteenth century, the Reed Warbler from about 1820 (Holloway 1996). A limiting factor on Reed Warbler (and Sedge Warbler) numbers in major fenlands may have been extensive harvesting of reeds for thatching and for fuel. Graveland (1999) showed that both densities and breeding success of Reed and Sedge Warblers were significantly less in areas of cut reeds than uncut. Parslow (1973) noted that an increase in southwest England started in south Devon and benefited there from a cessation of commercial reed-cutting.

Four points should be made about the apparently limited impact of nineteenth-century drainage activity on farmland birds, outside the loss of specialist wetland and fen species to arterial works. It is important to bear in mind the limitations of the data in the early avifaunas, for many common birds (see Introduction). Secondly, a high proportion of the field drainage carried out was on land already devoted to arable farming; habitat change was not involved. Thirdly, a significant area of drainage was less successful and permanent than hoped; habitat loss was never total. Fourthly, an extensive core of permanent grassland, extending to some 13 million acres (5.3 million ha), remained in British farmland, much little affected by all this work and including habitats such as old pastures, hay meadows, flood plains and water meadows (Chapter 9).

In a lengthy paper dealing with drainage problems, Clarke (1854) showed that most of the major river systems of England and Wales then had abundant flood meadow, which was the major use of valley bottoms. Although the winter flooding was valued for its fertilising properties enhancing hay crops, prolonged flooding was regarded as damaging, as was summer flooding. Clarke wrote at a period of particularly severe flooding in the mid nineteenth century, hence the concern with improved drainage and flood protection. But Clarke, and other sources, shows that the competing interests of grain-milling and navigation (water transport was the principal method of moving bulk cargoes of grain for example) imposed considerable restrictions and difficulty on effective drainage on many rivers. Clarke used as an example the River Nene from Peterborough to Northampton, where there were 33 mills, 34 navigation locks and 11 staunches holding back the water along a winding course of about 96 km. The Drainage Act for this area (c. 1848) empowered the Commissioners to open any lock or mill gate 'upon any reasonable apprehension of a flood', and also arranged for a system of back drains to drain the meadows, leading drainage water to the back of the marsh and then round downstream of obstructions such as mills, instead of taking it straight to the river. An important point about this scheme was that it did not seek to prevent winter flooding but to reduce its duration and prevent summer floods. Clarke noted that the value and area of the land involved did not warrant paying to remove mills, as was done elsewhere. For example, under the Act of 1846 to drain the Rye and Derwent District in Yorkshire, the Commissioners were empowered to remove water wheels, replacing them with steam engines of comparable power and compensating mill-owners for the additional power costs. Wilkinson (1861) noted that all proprietors on the Hampshire Avon and Stour had agreed to remove all mills 'years since', so there were no artificial impediments to drainage 'as so often elsewhere'.

Victorian farming literature frequently complained about the difficulties that the competing interest of mills particularly, but also navigation and fishing weirs, posed to land drainage. Nor were they always grain mills. My wife's family once owned the Bulwell estate near Nottingham and a map of 1864 in her possession shows that this estate had at least six water wheels on about 2.5 km of the River Leen, powering two bleach works, a starch works, a cotton mill and two corn mills. Such a concentration of water-driven machinery owned by one estate points to a further problem. Conflicting agricultural and industrial interests owned by one proprietor must have often caused additional difficulties. Nevertheless corn mills were the most widespread problem. Smith (1949) noted that grain-milling in 1851 was still a largely rural industry based entirely on millstones not, of course, entirely water-driven, and in 1880 there were still 10 000 mills and 18 000 millstones. But grain-milling was revolutionised in the last quarter of the nineteenth century by its increasing concentration at ports to handle the growing level of imported bread wheats and by the switch to more efficient roller mills, mechanically powered. This coincided with the rise of railways to facilitate transport of bulk goods and a decline in industrial use of water power. By the time these changes became effective, interest in land drainage for agriculture was waning (Figure 6.3) but they led to the disappearance of major competitive commercial interests to drainage on many rivers, which has had an important influence on land drainage in the later twentieth century. It is important to remember that these commercial interests were essential to the economy of the time and usually required Parliamentary Acts to set them aside.

Nineteenth-century management of grassland had other features. Much pasture was certainly drained for improvement but Philip Pusey and others (e.g. Cadle 1867) described examples of good grassland spoiled by under-draining too enthusiastically, so reducing the quality of the hay. There was then far less reliance on reseeding, and water tables were manipulated. On his own meadows Pusey adopted 'the plan of damming

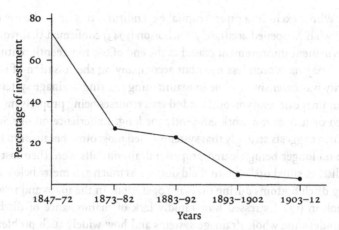

Figure 6.3. The percentage of capital investment in agriculture financed by Government or public company loans during the years 1847–1912 which was used for land drainage. (Data from Orwin & Whetham 1964.)

the streams in summer. The water in the land consequently does not escape from the land, whilst the stream finds its way up the drains and rises as in a sponge; so that this kind of sub-irrigation keeps the bottom cool and the surface green, while other meadows are scorched by the summer sun.' Such management appears to have been fairly widespread and Lincolnshire graziers considered that pastures for fattening bullocks degenerated into sheep land (an interesting derogatory term!) by too much drainage, whilst Read (1858) in Norfolk stressed the need to hold water tables up on poorer grasslands to make them worth anything at all. By and large valley grasslands and many other permanent pastures were used for fattening cattle and high water tables favoured to promote continuous growth. Sheep were kept off because of the danger of the rot, a clear indication of dampness. All these points, high summer water tables, dampness, cattle grazing, and generally rather low stocking rates (Chapter 9) indicate that plenty of good habitat for species such as waders persisted. Water meadows were a notable exception.

1880–1940

The prosperity of high farming collapsed in the 1870s and land drainage virtually ceased in England and Wales. Figure 6.3 shows the pattern of expenditure on land drainage as a percentage of all capital expenditure financed by Government or public company loans to agriculture until

1912. Whilst excluding private capital expenditure, this undoubtedly reflects what happened accurately. Nicholson (1943) confirmed that works of permanent improvement ceased at the end of the nineteenth century. There may have been less need but accompanying the cessation of new activity was extensive decline in maintaining existing drainage systems. Tile drainage depends on outfalls and watercourses being properly maintained or it does not work efficiently for long. Experience on our Sussex farm suggests strongly that systems failed most often because ditches were no longer being cleared properly; drain outfalls were then lost as the ditches silted up. We found old outfalls as much as a metre below existing ditch bottoms during drainage operations in the 1970s and 1980s. Nicholson (1943) stressed how rapidly lack of maintenance of ditches can undermine whole drainage systems and how widely such problems existed. The impact of poorly designed schemes was also probably exacerbated by declining maintenance. Many nineteenth-century drainage systems had placed the drains too deep to be effective for long (Symon 1959, Trafford 1970).

Thus, although Trafford estimated that, in the mid twentieth century, 5.3 million acres (2.15 million ha) of farmland in England and Wales were still well served by nineteenth-century drainage systems which had been adequately designed and maintained, a substantial part of the area underdrained in the nineteenth century needed redraining by 1940. As the agricultural recession affected arable counties most severely, it was probably there that most deterioration of nineteenth-century systems occurred (see Snipe, below). Sheail (1976) noted that the drive to increase food production during the 1914–18 War included draining 80 000 acres (32 400 ha) in about 18 months up to June 1918, although the work was hampered by serious labour and technical problems. Such improvements did not last and Nicholson noted that 'within a short period of the conclusion of hostilities progress ceased, to be followed by an interval during which this side of farm work fell more and more into desuetude; farming systems appeared in which it had no part and there grew up a section of the farming community which knew nothing of its practice.'

The major change resulting from the agricultural recession was a steady expansion of permanent grassland. As arable was abandoned and drains ceased to be maintained, substantial new areas of damp grassland occupied by a low level of farming emerged. However none of the later nineteenth-century avifaunas I have consulted, for example for breeding waders, makes any reference to the deterioration of drainage systems, although they all describe the effects of habitat loss and drainage in an earlier period. So the emergence of these habitats probably did not become significant until around the turn of the century. The history of the expansion of the Redshank supports this. Redshank were already penetrating many inland areas in the last 40 years of the nineteenth century, well before recession (see above), and the avifaunas then show a rapid increase in numbers within this range during the early twentieth century. This increase presumably resulted from the species exploiting the increasing availability of nesting habitat that the recession provided through the deterioration of drainage. A marked recovery in breeding Snipe populations also occurred but with an interesting geographical bias. The avifaunas for the period show that increase was marked in counties in southern and eastern England and Alexander & Lack (1944) noted also that Snipe had recolonised a large area of the south Midlands. But in those northern and western counties (except Somerset) for which contemporary avifaunas are available there was little evidence of change. Nor was there in Scotland (Baxter & Rintoul 1953). In northern England, Wales and Scotland this bird had, perhaps, always been most numerous on unimproved hill grazings and bogs. Thus the Snipe's changing distribution matched the decline of farming and drainage status, which was always most marked in arable areas in the English lowlands.

Although Alexander & Lack (1944) recorded an increase in Lapwings as a result of the 1926 Lapwing Act, two-thirds of county avifaunas covering the period 1920 to the early 1950s actually refer to continued declines and only one to a general increase (in Lakeland; Blezard 1946), although some range expansion was also noted in northern Scotland and the Isles (Baxter & Rintoul 1953). Although these population declines may partly have stemmed from the war-time ploughing campaign (Chapter 8), the avifaunas clearly show that numbers were still declining in many counties in the inter-war period, coinciding with the marked decline in farming, considerable expansion of grassland and deterioration in the drainage status of much agricultural land. As noted in Chapter 9, however, such changes do not always benefit Lapwing. Overall no particular general trend in farming emerges as clearly responsible for the long-term population decline in the species, which has been continuous, at least regionally, from the early nineteenth century. But one underlying factor may be that enclosure, drainage and ploughing in the nineteenth century led to the loss of many habitats with a long history of stability and settled usage, particularly in the grazing marshes of southern and eastern England and the sheepwalks and sheep/arable systems of the Downs and Wolds. Whilst they include no counts, the early avifaunas leave little doubt about how hugely abundant Lapwings once were in such sites, something of which we have no modern conception as, indeed, Stevenson (1870) also noted. The Lapwing may actually have proved to be less adaptable than we believe to the more disruptive agricultural systems introduced into its traditional habitats by high farming and its successors. Yellow Wagtails also declined (Chapter 3) and the Reed Warbler's range may have contracted somewhat by the 1930s (Holloway 1996). Otherwise populations of common wetland breeding birds of farmland were considered to be stable in this period. Local increases were recorded for Teal and Mallard and it seems unlikely that similar changes did not occur with species such as Moorhens, Reed and Sedge Warblers and Reed Buntings but they were not recorded.

The avifaunas also give rather little indication of change for common wintering wildfowl and waders. Shoveler and Pintail certainly increased, probably largely as a result of properly enforced close seasons, and it is likely that other ducks also benefited. But flock sizes recorded in inland areas remained moderate. Plovers were clearly abundant, although Golden Plover had declined locally, with the loss of traditional sites, often to development. Snipe had almost certainly increased markedly and flocks of 100–200 were frequent.

What did not happen with this recession was the re-emergence of any substantial areas of fen. Green (1976, Figure 1), for example, actually shows an increase in arable area in the Fens during the years 1918–1937, the deepest period of the recession. Arterial drains were undoubtedly neglected, which exacerbated the decline of field drainage. A good example was provided by our family farm, where all the large area of low-lying meadows flooded extensively in winter in the 1930s and 1940s because the channels in Pagham Harbour, through which this water had to go, were badly silted up. As a result these meadows were of limited agricultural value although they always provided summer grazing, but they had been arable in the late nineteenth century, indicating a very different status then. Sheail (1976) noted other examples when discussing drainage status of farmland during the 1914–18 War. However, neglect of arterial drainage was limited by the existence, from the 1860s, of public bodies charged with the responsibility of maintenance and repair of watercourses and outfalls and financed by rates levied on the land. From that date also, there has been a series of Land Drainage Acts largely aimed at maintaining arterial drainage. Sheail (1976), for example, noted that the Land Drainage Act of

1918 was passed to ensure that powers invoked under wartime regulations to improve drainage remained in force after the War. The Act made it easier to set up drainage boards and extend the areas covered by existing ones. Williams & Bowers (1987) also noted that, between 1921 and 1926, over £1.2 million (say £20 million + today) was spent on sea defence and arterial drainage of agricultural land under the Unemployment Grants Committee of the Ministry of Agriculture. What could have happened is vividly brought to life by Ennion's (1949) description of the Adventurers' Fen in the 1920s and 1930s. Here land that had been fully farmed during the period 1910–20 reverted to true fen in the space of less than ten years, once the drainage systems collapsed. Such sites were the exception.

Nevertheless some fenland birds re-established themselves, mainly in Norfolk and Suffolk, where suitable habitat had always persisted. Bitterns bred in the Norfolk Broads in 1911 and increased and spread slowly in Norfolk until the 1950s (Taylor *et al.* 1999). Breeding occurred in Suffolk in 1929 and 1930 and perhaps in other years (Ticehurst 1932) and the species bred in Cambridgeshire in 1938 (Ennion 1949). Marsh Harriers also re-colonised the Broads area from 1911 and bred regularly from 1927. Bearded Tits were breeding again in Suffolk and Norfolk, Black-tailed Godwits made four or five breeding attempts during the years 1930 to 1942, mainly in the south Lincolnshire fens, whilst a marked increase in passage birds also occurred from the end of the nineteenth century (Holloway 1996). Parslow (1973) suggested that there was a surge of records for Spotted Crake between 1926 and 1937. The re-emergence of these species probably mainly reflected increased protection rather than any change in land management.

1940 to the present

The recession in farming ended abruptly with war in 1939. The immediate result in agriculture was a campaign to return grassland to arable (Chapter 9) and to repair drainage systems. Murray (1955) provided figures showing that a total of 163 936 farm drainage schemes were approved for grant aid during 1940–4, altogether affecting 4.04 million acres (1.64 million ha) of farmland in England and Wales. Of these schemes 70% were for the improvement and repair of ditches and watercourses, 20% for tile draining and the remainder for mole draining. By area the proportions were 85%, 5.5% and 9.5% respectively (Nicholson 1943). The greatest emphasis in this work was cleaning out ditches and watercourses and Nicholson makes it clear that this recovered in working order nineteenth-century pipe drainage systems that had ceased to work as a result of ditches silting, something confirmed by experience on our family farm. Moling was particularly done in clay soils and the extent of tile drainage was limited by cost and lack of labour. Besides financial assistance, much of the actual work was done by Agricultural Departments, which built up a formidable fleet of drag-line dredgers and trenchers for the purpose. In Scotland, where Murray (1955) thought that natural drainage made artificial drainage less necessary, the main effort was concentrated on arterial works, improving particularly some of the smaller rivers and watercourses in Perthshire, the Lothians and Lanarkshire. Symon (1959) made the same point and Green (1974) observed that drainage work continued in Scotland throughout the period 1880–1939. So it is fair to assume that maintenance would have done so also. The difference in scale in remedial work necessary in Scotland compared to England and Wales again underlines the much more resilient nature of Scottish agriculture in the recession.

Progressively through the 1950s the State's direct involvement in drainage was reduced to zero and the work switched to private contractors. But land drainage continued to be grant-aided, at a level of 50% of cost until 1980 and reducing thereafter to nil by 1985; grant aid at 15% from the EEC continued to be available after that in some circumstances. As grant aid depended on prior approval by the Agriculture Departments, the State retained a highly significant level of influence in land drainage throughout the period.

Figure 6.4 shows the extent of tile draining in England and Wales from 1940 to 1980 by five-year periods. Altogether the figure shows that a total of 1.91 million ha was tile drained over the period. The drainage status of farmland was probably largely restored to the Victorian position by the early 1970s, when 886 000 ha had been drained and the area of tillage as a percentage of crops and grass was being maintained at the same level as the early 1870s. Adding that figure to Trafford's (1970) estimate that 2.15 million ha of land were still well served by nineteenth-century drains suggests that a total of 3.04 million ha (7–8 million acres) were actually drained in the Victorian era, rather than the 12 million acres estimated by Robinson 1986 (see p. 138). The repair of drainage systems and the reversion of substantial areas of grassland to arable undoubtedly led to some contraction in the breeding bird populations that had expanded in grassland during the recession but the general status of these species was not fundamentally damaged (Chapter 9).

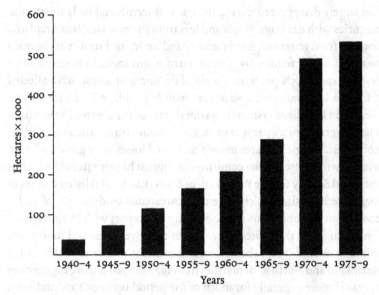

Figure 6.4. The extent of tile draining in England and Wales in each five-year period from 1940 to 1980. (Data from Trafford (1970) and RSPB (1983).)

That is no longer true. From the late 1960s a shift in emphasis in drainage and land improvement began to emerge more strongly. A feature of drainage works from this point has been the extent to which they have affected permanent grasslands, particularly those where nineteenth-century operations were limited, the habitats we categorise today as lowland wet grassland. This process appears to have accelerated rapidly from 1970. For example Green (1976) drew attention to the increasing amount of tile drainage in hill pastures and rough grazing apparent in the early 1970s. Carter (1982) reported that just under 10% of drainage (72 500 ha) done in England and Wales during 1971–80 was of new drainage of wetlands. Williams & Hall (1987) showed a steep increase in the conversion of grassland to arable in the North Kent and Broadland marshes from 1970. Lovegrove *et al.* (1995) estimated that, of *c.* 100 000 ha of damp pasture drained in Wales by 1992, the highest rates, *c.* 3000 ha per year, occurred from 1970.

The remaining extent of lowland wet grassland has recently been quantified for England at *c.* 219 410 ha by Dargie (1993). Newbold (1998) estimated a further 40 000 ha of flood plain, mainly occupied by wet grassland, in Wales. In Scotland, however, Mitchell (1997) noted that the habitat had largely disappeared during the late eighteenth and early nineteenth centuries with drainage. Roots and leys from the new farmlands replaced bog-hay from water and flood meadow and undrained straths as the main source of winter fodder. Dargie estimated recent losses in England as averaging about 1.5% per year, a scale of decline that would have affected c. 100 000 ha from 1970. Losses were mainly to tillage in the east, to reseeding in ley/tillage systems in pastoral areas or to industrial/urban uses. They ranged up to 4% per year in the east but down to as little as 0.2% in the northwest. Dargie's figures mainly measured losses after grant had been withdrawn in 1985. Earlier comparisons suggest higher rates of loss. Thus the Royal Society for the Protection of Birds (RSPB 1983) listed a series of major wetland sites which were then threatened by drainage, often for conversion to arable farming. Comparing this report with Dargie's (1993) figures suggests these sites lost c. 30% of their wet grassland in 10 years. Williams & Hall (1987) showed a similar extent of loss to arable in the Broadland and North Kent marshes from 1970 to 1981. If these higher rates operated more generally for much of the period between 1970 and 1985, they suggest an overall loss of c. 300 000 ha of wet grassland in the period. As losses in some areas have also been continuing over a longer period, the total loss since the War must have been greater. For example Williams & Hall found that the greatest loss in the east Essex marshes had been between 1940 and 1950 and over 40% of the total area of Romney Marsh was drained by 1970 (Sheail & Mountford 1984).

It is less clear from Dargie's survey how much wet grassland has been altered by drainage whilst remaining agriculturally improved grassland rather than being converted to arable. But this has been very widespread, perhaps affecting at least as great an area as the losses to arable already considered. In Wales, for example, agricultural improvement of wet grassland has been the major habitat change in flood plains such as the Tywi Valley (personal observation) and, in fact, the single most important change throughout Welsh farmland (Lovegrove et al. 1995). Changes may be quite subtle. Green & Robins (1993) showed how simple adjustments to the management of existing pumps and sluices on the Somerset Levels, operating them at lower threshhold water levels than in the past, had significant effects on the wet grassland habitats and their breeding bird numbers. They noted that, although conditions may then still permit breeding to begin in April, such increased drainage may not allow suitable conditions to persist long enough for it to be successfully completed. As a result they found breeding Lapwing numbers declined by 54% between 1977 and 1987, Snipe by 69%, Curlew by 38% and Redshank by 20%. Wetlands everywhere in pastoral districts have been subject to similar management changes and other case studies showing similar impacts on breeding waders were reported in northern England by Baines (1988) and Buckinghamshire by Fuller (1994), for example. Dargie also showed a high level of fragmentation in lowland wet grassland, recording a total of 2197 blocks with a mean area of 100 ha but a median of only 28 ha; 1024 blocks were less than 25 ha in extent. Fragmentation was rather more evident in arable areas than pastoral. As in other habitats, fragmentation leads to declining species diversity within remaining areas and, in this habitat, lowering water tables to accommodate changing use in one part makes it difficult to retain sufficiently high water tables in the remainder, again affecting wetland species, a problem particularly noted by Williams *et al.* (1983) in the North Kent Marshes and Green & Robins (1993) in the Somerset Levels.

Although drainage and conversion to arable leads to the loss of marshland breeding species on that site, it is in pastoral districts that the greatest general declines in breeding grassland bird populations have been recorded. In pastoral districts drainage is just one of a series of major changes in the process of intensifying grassland management and breeding birds may be affected as much or more by changes such as reseeding and increased nitrogen use. The whole subject of changes in grassland management and its impact on breeding bird populations is discussed in more detail in Chapter 9.

The drainage of wet grasslands also directly affects wintering birds, perhaps mainly surface-feeding ducks, grassland plovers, Snipe and possibly winter thrushes, although the amount of information on the latter is scrappy. However the evidence that extensive drainage has led to important general declines in the wintering populations of these species in Britain is ambiguous. One can say at once that the draining of individual sites leads to the loss of wintering birds. An example within my own experience is provided by the Arun Valley in Sussex, where extensive winter flooding on the Amberley and Pulborough Brooks until the mid 1960s supported regular flocks of up to 4000 Wigeon and 3000 Teal and smaller but significant flocks of Pintail, Shoveler and sometimes Pochard and Tufted Duck; large flocks of grassland waders also wintered there. Major draining and flood prevention works in 1967 led to sharp declines of all wintering wildfowl. Counts became much more often of a few hundreds than in thousands and significant flocks were present for much shorter periods (**Sussex Bird Reports; Shrubb 1979**).

Such patterns are common following significant changes in water tables and flood incidence. But the Wildfowl and Wetland Bird Survey counts record that, whilst drainage of wet grasslands for agriculture has accelerated since the mid 1960s, wintering numbers of the common surface-feeding ducks have been steadily increasing (Pollitt et al. 2000). Outside estuaries, the principal sites Pollitt et al. list are mainly industrial/man-made sites, such as reservoirs or gravel pits, or nature reserves, and Owen et al. (1986) show that protection has been important in these increases. Wildfowl are adaptable and gregarious and quick to exploit new wetlands such as gravel pits and reservoirs which have become increasingly available. How far such sites are used only as daytime roosts, with the birds fanning out in the surrounding countryside to feed at night, is difficult to assess. But nearly all these species, like geese, exploit the crop residues available in farmland (Chapter 10) and such behaviour would facilitate use of the new sites.

Changes in the distribution and numbers of grassland plovers in winter is much more difficult to assess. I suspect that there has been some quite significant desertion of inland sites. This has been very marked in Wales for the Lapwing, for example. Lack (1986) showed a wide dispersal of winter flocks of up to 430 birds away from the coast and and some larger flocks. Few of these flocks can be found today. The Welsh Bird Report has reported flocks of 50–300 in only eight such inland sites during the 1990s and few of those regularly. Significant flocks are now virtually confined to estuaries and neighbouring coastal grasslands. Golden Plover were never so numerous inland in Wales, although the Tywi Valley flood meadows at Dryslwyn attracted up to 3000 in the 1970s (Lovegrove et al. 1994), but inland flocks today are even scarcer than those of Lapwing. These changes do not just result from drainage but also from changes in the whole pattern of grassland management. In particular they probably reflect changes in stocking patterns. Tucker (1992) observed that both plovers avoided pastures with sheep and selected those with cattle. Sheep have increased markedly at the expense of cattle in many pastoral areas, especially Wales, where there is also increased wintering of hill sheep flocks on lowland dairy farms, for example in the Tywi Valley.

In considering the impact of drainage on wader species, one general point is important. Callaway (1998) noted that a problem with flooding newly created sites on arable land was that it killed or expelled terrestrial soil invertebrates that had colonised old flood plains and that it may take a period of years for such sites to regain a wetland soil invertebrate fauna.

Ausden (2001) noted that flooding dry grassland reduced earthworms by over 90% and leatherjackets by over 80%. Nevertheless waders may have actually benefited by remaining food becoming more available and Ausden noted that wet grasslands with high breeding densities of waders contained only very low densities of soil invertebrates. Shallow pools of flood water may, however, contain high populations of aquatic invertebrates by early summer, which are important food sources for chicks. It must be supposed that these processes also work in reverse, causing major upsets in food resources with drainage.

Finally it is worth stressing that little that has been done in wet grasslands is irreversible. Callaway (1998) was describing the restoration of wet grassland habitats in the Pulborough Brooks in the Arun Valley, which has led to the recovery of major wintering and breeding populations of waders and wildfowl there. With agriculture in decline there seems little reason why this cannot be repeated elsewhere.

7
Weeds, weeding and pesticides

The control of weeds has always been a prime concern of the farmer, for weeds compete with crops and limit yields. The weed floras of arable fields and pastures, however, are major sources of diversity in farmland habitats. They are important to birds both as a direct source of food (seeds and seedling leaves) and as habitats for invertebrates upon which they feed. Salisbury (1961) made the point that the weed flora of cereal fields had shown little qualitative change until comparatively recently but this has accelerated with the widespread use of herbicides since the 1960s. By contrast quantitative changes have always been more noticeably linked to cropping patterns, husbandry techniques and the relative prosperity of farming. Potts (1991) noted that 131 species of flowering plant had been found in cereal crops on the Partridge Survival Project's study area in West Sussex. He also suggested that 200–300 species could be associated with the national cereal acreage; many are now very rare. Some crops were associated with a particular weed flora. Flax (*Linum usitatissimum*), for example, widely grown in the past for fibre, was associated with a range of species with closely similar seeds, which disappeared when traditional flax cultivation was abandoned in Europe; it has not reappeared with the modern interest in growing flax as linseed (Wilson 1992). Rye also had a rather distinctive weed flora in Britain, perhaps largely because of the poor sandy soils typical of its cultivation.

Despite the variety of the arable flora, however, comparatively few plants in any area may be exploited for their seeds by birds. Newton (1967) listed 76 farmland weed species as food plants for nine species of finch on his Oxford study area, of which 37 were more typical grassland species. But his observations for Greenfinch, Goldfinch and Linnet show that a much smaller range of species formed the bulk of the diet over the year.

TABLE 7.1. *The main food plants (listed in order of frequency) of three species of finch in England at two periods*

	Greenfinch		Goldfinch		Linnet	
	1920s	1960s	1920s	1960s	1920s	1960s
	Stellaria	Arctium	Plantago lanceolata	thistles	Sinapsis	Sinapsis
	cereals	brassicas (wild)	Stellaria	Senecio vulgaris	Senecio vulgaris	Polygonum persicaria
	Rumex acetosa	cereals	Centaurea	Arctium	Rumex acetosa	Chenopodium
	Cerastium	Senecio vulgaris	Senecio jacobaea	Dipsacus	Stellaria	Stellaria
	turnip seed	Polygonum persicaria	Taraxacum	Taraxacum	Senecio jacobaea	Taraxacum thistles
Percentage of total recorded diet[a]	96	59	97	79	87	74

[a] The total numbers of plant species recorded by Newton were Greenfinch 31, Goldfinch 24 and Linnet 24.
Sources: Collinge (1924–27) and Newton (1967).

Collinge (1924–7) recorded a similar pattern (Table 7.1). Not too much can be read into the differences between the two periods for the data are not strictly comparable and only two of the plants recorded by Collinge altogether were not recorded as taken by Newton, *Chyrsanthemum segetum* and *Galium aparine*. Figure 7.1 shows that the same pattern occurs seasonally. It is important to remember that all these species take a much wider range of plants overall, which provides flexibility and adaptability in their diets. Historically finches have modified their diets to take advantage of or in response to changes in the availability of weed species. The studies concerned were also for relatively limited areas and other studies may show emphasis on different plants. One genus not included, *Galeopsis*, was found important for Skylarks in eastern England and Grey Partridge in Finland for example (Green 1978, Pulliainen 1984). Nevertheless the figure shows clearly that a rather small group of plants was of primary importance for farmland seed-eaters – cereals, brassicas both cultivated and wild, *Stellaria*, *Chenopodium* and *Polygonum*.

Some characteristics of farmland weeds

Newton (1967) observed that finch diets were ecologically well separated in woodland but not in farmland. Presumably finch species coexist or coexisted successfully in farmland because of the abundance of seeds available.

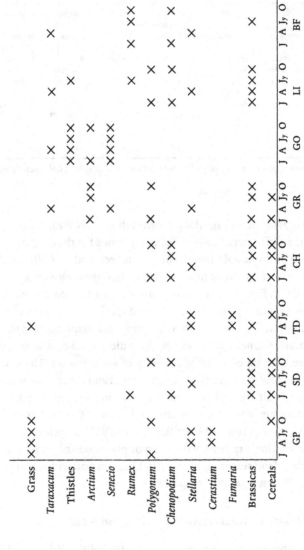

Figure 7.1. The main seeds taken by some farmland seed-eaters at different seasons. Species are: GP, Grey Partridge; SD, Stockdove; TD, Turtle Dove; CH, Chaffinch; GR, Greenfinch; GO, Goldfinch; LI, Linnet; BF, Bullfinch. Seasons are: J, January–March; A, April–June; Jy, July–September; O, October–December. (Data from Murton 1971, Table 15.)

TABLE 7.2. *Average seed production per plant and dormancy of some important finch food plants in Britain*

Species	Average seed production per plant	Period of dormancy of viable seeds
Sinapsis arvensis	1000–4000	many years
Cerastium	6500	40 years
Stellaria	2500	10 yrs, 22%
Chenopodium	3000	?
Polygonum	200–1500	20–60 years
Rumex	5000	39 years, 5–6% (docks)
Senecio vulgaris	1100	?
Taraxacum	2000	?

Source: Salisbury (1961).

Salisbury (1961) provided some figures for the average seed production of some arable weeds that are important finch foods (Table 7.2). Plants are also present in abundance as seeds in the soil's seed bank, which is an integral part of weed populations in farmland. Seed banks arise because many farmland weeds have a marked capacity to lie dormant in the soil until conditions are suitable for germination, usually when cultivations bring them near the surface, even after long periods under grassland. Some examples are shown in Table 7.2. In arable soils Roberts (1981) noted that total seed numbers held in seed banks averaged more than 4000 per square metre of topsoil (0 to 25 cm depth) in European studies; in very weedy fields numbers may reach 70–80000 per square metre, whilst up to 496000 has been recorded in Denmark and 86000 in Britain. Cereal fields in the English Midlands typically had 1800–67000, with a median of 5500. The main contributors are annual weeds and a fairly small number of species typically account for 80% or so of seeds present in any area – in the English Midlands sample above *Poa annua, Polygonum aviculare, Stellaria, Chenopodium* and *Aethusa cynapium*. Both haying and cereal harvesting and threshing help to ripen, scatter and spread seeds. So do grazing and the spreading of dung, as the seeds of many farmland plants are effectively scarified by being passed through the gut of herbivores.

Arable weeds

The seeds of arable weeds vary markedly in their capacity to lie dormant in the soil. Species that lack this capacity are poorly represented in the soil's seed bank and may be particularly vulnerable to improved seed-cleaning techniques. Corncockle (*Agrostemma githago*), now very rare, is an example. Species also typically germinate at different seasons and within seasons some are early and some late (Wilson 1992). Thus the increased concentration today on autumn cultivation and then the shift towards ploughing and sowing cereals in September/October rather than October/November has contributed to changes in the range of weeds present and population declines. Minimal cultivations (Chapter 11) particularly favour grass weeds, which are readily controlled by ploughing (Edwards 1984). The capacity for dormancy and fairly precise germination seasons interact to allow weeds to adapt to rotation farming, with its varied seasons of cultivation and leys. Many cereal-field weeds benefited from the thin crops and poorer soils typical of early farming systems. Modern cereal varieties are much more competitive and benefit from high applications of nitrogen fertiliser. Both factors suppress associated plants (Figure 7.2). But some arable weeds, for example *Galium apinare* and the grasses *Bromus sterilis* and *Alopecurus myosuroides*, thrive in high nitrogen regimes and are now the most important cereal-field weeds. Some weeds, such as couch grass (*Elytrigia repens*), reproduce vegetatively as well as by seed, surviving ploughing by regrowing from fragments of root.

Grassland weeds

Unlike arable weeds, which are mainly annuals, grassland weeds are largely perennials, able to withstand grazing, seeding before or during haying and intolerant of regular ploughing. The modern practice of frequent reseeding has reduced the flora of old grasslands even more dramatically than modern methods have reduced arable weeds. The grassland flora is also susceptible to the high nitrogen applications now common in grassland management (Chapter 9). On the basis of samples provided by Salisbury (1961), grassland weeds tend to have less copious seed production than arable weeds but the difference is not statistically significant. Flowering seasons and seeding patterns in grassland are also strongly influenced by grazing regimes. Salisbury made the point that traditional management of pasture and hay meadow favoured different groups of plants. Such differences must have declined in importance since the early nineteenth century, with the increasing tendency to rotate meadow around the grass (Chapter 9).

The seeds of grassland plants are probably important to birds at two points. Newton's (1967) figures for Greenfinch, Goldfinch and Linnet suggest that they were most important in spring and summer, when arable weed seeds were least available. Secondly hay has always been a prime

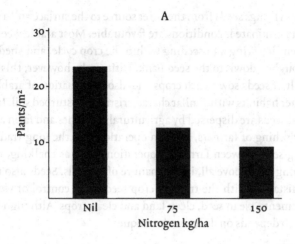
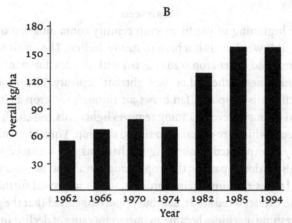

Figure 7.2. The impact of nitrogen fertilisers on weeds. (A) The total numbers of plants per square metre of 14 species recorded at harvest in cereal plots subject to different rates of nitrogen. (Data from Wilson 1992.) (B) The increase in nitrogen usage in cereals at intervals since 1962. (Data from Surveys of Fertiliser Practice.)

source of winter food for seed-eaters in pastoral land (Chapter 10). It may have been particularly important to buntings because of the high level of grass seeds present. Today intensive silage production allows few plants to seed, so that these food sources have virtually disappeared.

Weed control in high farming

The main sources of weeds in crops in the past were their seeds sown inadvertently in crop seeds by the farmer, and the soil's seed bank. Every cultivation brings seeds from the latter source to the surface and a proportion will germinate if conditions are favourable. Most arable weeds complete their flowering and seeding within the crop cycle, and shed seed is then ploughed down to the seed bank. Ultimately, however, this store is the result of seeds sown with crops and also colonisation of arable fields from other habitats with similar characteristics of disturbed soil. In arable farmland seeds are dispersed by agricultural machines and farm animals, on the clothing of farmers, by farm operations, by the long tradition of changing seed between farms, by operations such as chalking, marling and claying and, above all, in the manure of animals. Seeds also migrate longer distances with the trade in crop seeds. The control of weeds has three elements, clean seed, clean land and clean crops. Although interrelated, each depends on different techniques.

Clean seed

Since the beginning of the nineteenth century contamination of cereal seed by arable weed seeds has been in steady decline. The efficient cleaning of cereal and other crop seeds started with the development of winnowing machines at the end of the eighteenth century. The principle of such machines is simple. A fan blows air through the crop as it moves across and through sieves. Fanning removes light seeds and chaff, sieving heavier seeds which are different in size to the crop. Threshing machines eventually incorporated winnowing mechanisms and included the further sophistication of passing the crop through an adjustable rotary screen of vertical wires running against brushes, which improved further their capacity to extract weed seeds. Salisbury (1961) suggested that the general use of threshing machines became an important cause of decline in several cereal-field weeds, including *Agrostemma githago*, *Burpleurum rotundifolium*, *Centaurea cyanus*, *Chrysanthemum segetum* and *Melampyrum arvense*. Modern seed-cleaning machinery is extremely sophisticated, capable of removing well over 90% of all impurities from seed grain. The increasing efficiency of seed-cleaning machinery has been progressively reinforced by legislation, to protect the users of seed and to lay down guaranteed standards of quality, starting with the Adulteration of Seeds Act 1869.

Clean land

Until the development of herbicides in the early twentieth century cultivations were the main method of weed control. The only reference I have found to chemical means of control in the nineteenth century was to the use of salt to help control couch. The most important stage for weed control in high farming rotations was their starting-point, the root course. Ploughing was the basic cultivation used but the nineteenth century saw the development of a variety of cultivators and scarifiers for this work, particularly dealing with couch. An important development in the high farming period was that greatly improved drainage allowed a marked increase in autumn ploughing and weeding to be fitted into the farming year. This was particularly aimed at perennials and especially couch. If time allowed the land was ploughed, cultivated, harrowed and rolled and then forked over by hand, then cultivated again and picked over by hand. Seeds of annual weeds were encouraged to germinate by these cultivations and seedlings killed by subsequent ones. Roots were planted in May and June, which gave a further prolonged period to repeat these cultivations in spring, attacking those weeds which were spring germinating.

Such work was not, of course, confined to preparation for roots, although this was the most important. Stubbles were also forked over by hand to eliminate couch and winter ploughing of stubbles continued as long as weather allowed. Details varied between farms, districts and soils but the main point about all such work was that it was remarkably successful in controlling weeds if conditions were favourable, which meant good weather and abundant labour. Wet weather severely limited cultivations, allowing weeds a chance to recover if prolonged, which meant that it was never possible to eliminate them.

Clean crops

The efficient weeding of crops was impossible until the use of the seed-drill became widespread (Chapter 11). The main weeding tools were hoes, hand and horse-drawn, the use of which requires crops to be planted in rows (Figures 7.3 and 7.4). Once drilling cereals became usual, hand-hoeing them became a regular part of farm routine. Baker (1845), for example, noted that in Essex the winter wheat was hoed twice, in March and April, and beans were hoed three times. Hennell (1934) noted that different widths of hoe blade were used for wheat and barley, indicating that spring cereals were also hoed. Both Dutch hoes and draw hoes were used. Horse-hoeing cereals was also practiced but required the crop to be planted in rows at least 1 foot (0.31 m) apart. Roots were horse-hoed at intervals from crop emergence and hand-hoed at least twice. Standing root crops were also forked between the rows in the autumn to remove couch. The use of the hoe was supported by hand-weeding. Annuals such as charlock and

Figure 7.3. Hand-hoeing turnips, Perthshire 1937. In the nineteenth century most arable crops were hand-hoed for weed control by such gangs, highly labour intensive work. (Courtesy of Rural History Centre, University of Reading.)

Figure 7.4. Tractor hoeing sugar beet, 1951. Horse-hoes did exactly the same job. Planting in rows was essential. (Courtesy of Rural History Centre, University of Reading.)

poppies were hand-weeded from cereals, and thistles and ragwort from grassland. Jonas (1846) considered that strict attention to hand-weeding could and did eliminate quite severe infestations of charlock from cereals. In cereals the increased density and height of crops contributed to weed control by smothering weeds more efficiently and the high levels of soil nitrogen that accumulated may also have contributed.

The striking point about all these operations was the sheer scale of hand labour involved, often casual labour by women and children. Little information on the numbers employed is available but Verden (2001) quoted figures of 6000–7000 women and children employed in 'public gangs' (i.e. their labour was subcontracted) in the eastern counties and an unknown number, thought to be much greater, were employed in private gangs direct by farmers. Bradley (1927), for example, noted the gangs of Highland girls employed on such work at Fentonbarns, East Lothian. Verden made the interesting observation that such labour did not displace regular employees. Contemporary accounts recorded instead that the work would not have been done without them.

There can be no doubt of the capacity of such methods to control weeds, to suppress them in crops and to depress populations. I can attest to this from my own experience and observation. Newton (1896), discussing Goldfinch, noted particularly the 'extirpation' of the Compositae under high farming and the nineteenth-century avifaunas frequently confirm the decline of conspicuous weeds in the same conditions. Roberts (1958, 1962, 1968) and Roberts & Stokes (1965) described the effect of introducing intensive vegetable cultivation, using methods similar to the high farming techniques described here, on weed populations of arable land used previously for cereals and short-term leys. Initial weed populations averaged 163 million seeds per acre ($40000/m^2$) in the top 6 inches (15 cm) of soil. Populations declined by about 45% per year during the first four years when conditions for cultivation were good, recovered in the wet year of 1958, declined again in 1959 but recovered once more in 1960, another wet year. The experiment showed clearly that such intensive cultivation and weeding severely suppressed weed populations but confirmed that they were unlikely to eliminate weeds because good conditions for cultivation cannot be guaranteed.

Impact on birds

The nineteenth-century literature provides little evidence that the increased weeding efficiency of high farming much affected farmland birds.

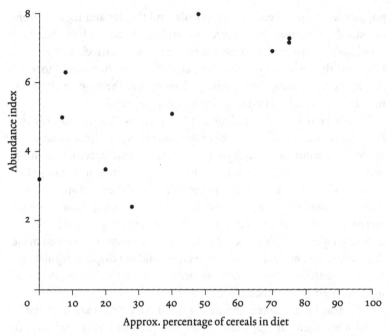

Figure 7.5. The abundance of 10 farmland sparrows, finches and buntings (Twite and Bullfinch are excluded) before 1910 in relation to the aproximate percentage of cereals in their diet. The abundance index is the sum of the abundance scores on a scale of 0 (absent) to 8 (abundant) in each pre-1910 avifauna divided by the number of species accounts providing information. The relationship is significant: $r_s = 0.73$, $n = 10$, $p < 0.02$. (Data on diet from Collinge (1924–7) and BWP.)

Seed-eaters should have been the species most influenced. That they were not, with one notable exception, probably derived from five factors. Many farmland seed-eaters make important use of wasted cereal grains, the availability of which increased with the area and yield of cereals. Whilst mechanisation tended to reduce loss, some grain was shed at every handling (cutting, stooking, carting, ricking, threshing). Read (1858) also noted that high farming led to high nitrogen levels in many soils, which caused increasingly frequent lodging in cereals (crops beaten down by bad weather). This would both have increased the amount of waste grain available and allowed birds easy access to unharvested crops. Figure 7.5 compares the abundance of ten species of farmland finch and bunting with the maximum percentage of cereals in their diets, suggesting a positive correlation between the two. Secondly, not all high farming was of a uniformly high standard. Caird (1852) noted that up-to-date methods were still then found alongside poor farming; wet seasons also reduced the effectiveness of cultivations. Thirdly, although the plants declined, their capacity for dormancy allowed the seeds to remain available in the soil. Most seed-eaters readily exploit the seeds brought to the surface by cultivations (e.g. Newton 1967; personal observation), a particularly important food source in spring. The experimental work of Roberts and his colleagues, summarised above, showed that seed banks in the soil would have persisted despite significant reductions, because of intermittent replenishment. Fourthly, undersown stubbles provided guaranteed feeding sites throughout the winter and spring. Finally stackyards and stockyards provided important winter food resources, particularly in severe weather; the value of cereal stacks is obvious but hay fed to stock was always another important source for seed-eaters (above and Chapter 10). Morris (1851–7) gave a timetable for the feeding ecology of Chaffinches in farmland, with stubbles in autumn/early winter, stack- and stockyards from January to March, spring sowings in March/April, invertebrates during summer and ripening cereals in August and September; they were also pests of early-sown horticultural crops.

The Goldfinch was the exception. Rather little on its status in the eighteenth and early nineteenth centuries is available beyond general statements that it was common. However some glimpses suggest that it was extremely numerous. Thus Cobbett in Gloucestershire in September 1826 saw a flock he estimated at 10000 individuals occupying about half a mile of roadside thistle banks, and he remarked on several other very large flocks in the same autumn in the same district. I doubt if anyone has recorded such a flock since. The species declined for much of the nineteenth century in virtually every county, remaining fairly common only in a few southern English counties and over much of Wales (Figure 7.6). The decline appeared to start in the 1830s and 1840s, thus coinciding with the major increase in high farming techniques (Figure 3.1) and enclosure of the wastes (Figure 4.1). It reached a nadir in the 1870s and 1880s and some recovery was noted from the turn of the century. By the mid 1950s the species was described as at least fairly common in 60% of the avifaunas consulted and increasing in all but two. Once again these population changes coincide well with changes in farming efficiency.

A species particularly of the waste, the Goldfinch lost vast areas of habitat with enclosure and was unable to adapt to the changing ecology of farmland. It takes a much narrower range of farmland weed seeds than other cardueline finches (Newton 1967) and one arable weed important to

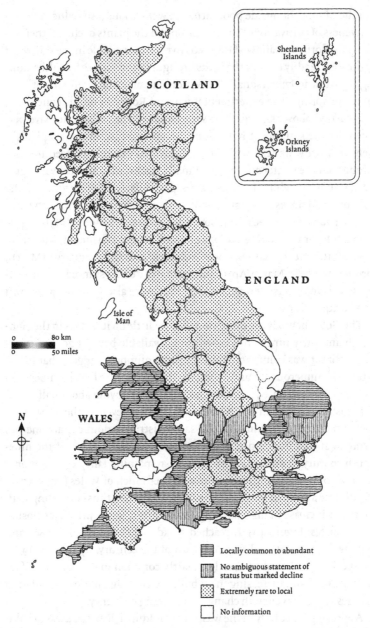

Figure 7.6. Breeding distribution of the Goldfinch in the late nineteenth/early twentieth centuries. (Data from county avifaunas.)

it, *Senecio vulgaris*, was widely noted as becoming rare in high farming, for example in Devon, Norfolk, Shropshire, Lancashire and Somerset. It also feeds much less on the ground than for example the Linnet (Newton 1967) and does not feed on seeds brought to the surface by cultivation (personal observation). Nor was it noted as an important component of flocks resorting to stackyards in winter in the nineteenth century; it takes few cereals. These limitations made it particularly vulnerable to the changes wrought by high farming. The distribution shown in Figure 7.6 shows the weight of the population in the mainly pastoral southwest of Britain, with good populations also in important orchard counties. Orchards were frequently recorded as favoured nest sites and thistles were probably more readily available in old grassland. However another factor in this species' decline was persecution for the cage-bird trade (Chapter 12). One factor, however, argues strongly that bird-catchers were not the primary cause of the decline. The scale of destruction of House Sparrows was even greater (Jones 1972) but there is little evidence that such operations had any impact on a species universally described as a too-abundant pest because of the availability of abundant habitat and food.

1870–1940

The period of the arable recession after the mid 1870s saw a decline in the standards of arable farming and weed control. As stressed above efficient weed control in high farming depended on high inputs of cheap labour. But the labour employed, both full-time regular employees and casual labour, declined sharply from the middle of the nineteenth century because of its rising cost in a period of falling returns and because the pool of labour available was shrinking (Chapter 11). For weeding the most significant loss was that of the casual labour of women and children. Verden's (2001) account indicates clearly that their loss led to much weeding no longer being done. The demand for labour also declined with the fall in arable area but this decline in demand still left a gap averaging *c.* 30% between labour employed and what was probably needed to maintain farming standards throughout the period from 1880 to 1930.

This period saw marked recovery in weed populations. Bradley (1927), for example, remarked on this in southern England as early as the 1880s and many of the avifaunas from around the turn of the century comment upon it. Such an effect was to be expected as labour and the frequency of cultivations declined. For birds, however, the significance of such changes remained limited. If weed populations increased in arable land, the arable area itself declined markedly between the 1870s and 1940. In general there is rather little evidence of major change in the populations of common farmland seed-eaters at this period. Changes in bird populations that did occur tended to reflect the overall loss of diversity in farmland and the recovery of specialised habitats (Chapter 3).

Herbicides

Methods of weed control changed dramatically from the mid 1940s with the development of selective herbicides based on the phenoxyacetic acids, particularly MCPA (2-methyl-4-chlorophenoxyacetic acid) developed for the selective control of dicotyledonous weeds in cereals. Chemical control of weeds had been the subject of experiment and limited practical application from 1911 (Murton 1971) but the phenoxyacetic acids were the first successful selective chemicals available. They were also safe to use, unlike some earlier herbicides. Research and development of chemical herbicides has been a major preoccupation of agricultural science since and Figure 7.7 summarises the increase in chemicals and capacity in cereal and root crops since 1955. First-generation herbicides such as MCPA were probably no more efficient at reducing weeds than high farming techniques, properly managed and applied. Potts (1986) observed that the numbers of weed species per square metre on cereal fields of the Game Conservancy's partridge study area more than halved between the 1930s and 1970 and that similar declines were noted in Europe over the same period; in Canada susceptible weeds were reduced by 70% by long-term applications of MCPA. These are similar levels of reduction to those recorded by Roberts and his colleagues in intensive vegetable cultivations (see above). Nevertheless the new herbicides had the advantage of being less weather-dependent than cultivations, so that they maintained pressure on weed populations and they were outstandingly successful at controlling or eliminating some troublesome annual weeds such as *Sinapsis*. This disappeared as a significant weed in cereals on our family farm, for example, about five years after regular herbicide use began.

However the main advantage and value of these chemicals at this period was that they dispensed with the need for expensive and frequent cultivations and hoeing. They also facilitated continuous cereal rotations, which otherwise promote weed populations (Rademacher *et al.* 1970). They had significant limitations. Several weed species important to birds, such as

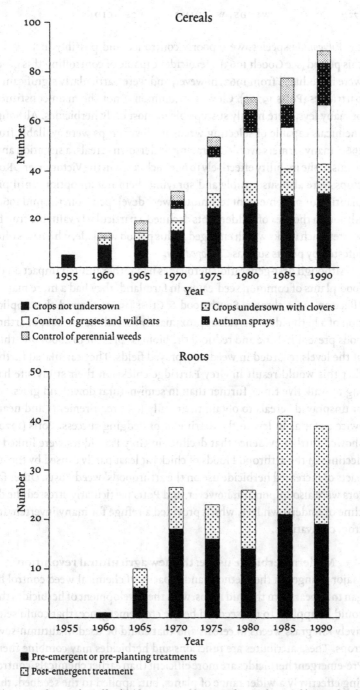

Figure 7.7. The number of herbicides approved by ACAS for weed control in cereal and root crops 1955–90. Root crops include potatoes. (Data from ACAS lists.)

the *Polygonum* species, were poorly controlled and possibly increased in this period (see Gooch 1963). Herbicides capable of controlling this group were introduced from 1962, however, and were particularly significant to partridges (Potts 1970a). Clovers, a common if not the main constituent of many leys, were highly susceptible to most early herbicides. Although chemicals capable of selecting weeds in clover crops were available from 1960, many farmers avoided spraying undersown cereals, a situation analogous to the inability effectively to hoe such crops in the Victorian era. Root crops were also susceptible and spraying them not an option until pre-planting or pre-emergent techniques were developed from the mid 1960s. Although the area of fodder roots declined a particularly valuable crop for winter finch flocks which emerged at this period was kale, which was often infested by plants such as *Chenopodium*.

Although first-generation herbicides had rather limited impact on the food plants of common seed-eaters in farmland, they had a more marked effect on invertebrates. Southwood & Cross (1969) showed that application of a herbicide to cereals approximately halved the number of arthropods present in June and reduced the biomass present to about one-third of the levels recorded in weedy unsprayed fields. They calculated further that this would result in Grey Partridge chicks on their study site having to walk five times further than in semi-natural downland grass, hay or unsprayed cereals to obtain their daily food requirements and nearly twice as far as in ley, to the detriment of fledging success. Potts (1970b) showed further evidence that declines in Grey Partridges were linked to declines in the arthropod foods of chicks at least partly caused by the impacts of increased herbicide use on the arthropods' weed hosts. Other factors were also important, however, and Potts particularly stressed the decline of undersown leys, which provided a refuge for many invertebrates from cultivations.

Modern herbicide usage: the new agricultural revolution
Major changes in the methods and capacity of chemical weed control began to appear from the mid 1960s, with the development of herbicides that could be applied to the seedbed before crop emergence, that could selectively kill grass weeds in cereals and that could be used in autumn-sown crops. These attributes are functions and herbicides may combine them. Pre-emergent herbicides are more efficient than earlier chemicals, controlling effectively a wider range of plants, but, applied to the seedbed, they also prevent weeds competing with the crop during establishment. This enables well-established cereals to tiller strongly in response to spring

nitrogen dressings and compete favourably with and suppress growth of any surviving weeds. Although in cereal-fields it is usually possible to find seedling weeds in the later stages of crop growth, these rarely have time to seed before harvest and cultivation for the next crop destroys them. Thus pre-emergents largely eliminate the capacity for any weeds present to set seed. Campbell *et al.* (1996) made a similar point, suggesting that the decline of weeds had been significantly underestimated because most extensive studies were based on distribution surveys made when plants were still small. The historic seed bank of the soil could ensure a supply of such plants to be sprayed out annually before growing much beyond the seedling stage. The plants certainly continue to exist in the soil's seed bank, as their re-emergence with set-aside shows.

These points were demonstrated to me by a brief comparison I made in 1984 of the weeds found in stubbles on our family farm, intensively managed for cereals, with the use of autumn pre-emergents as standard practice, and those in stubbles on a nearby farm where weed control still depended on cultivations and occasional applications of MCPA. In each sample all dicotyledonous weeds were counted in a series of randomly selected square metre quadrats; headlands were excluded. The results are shown in Table 7.3. Although each farm had similar numbers of species, only one-third were common to both but of those found only on Halsey's *Sinapsis, Raphanus, Cerastium* and *Polygonum* were all once common at Oakhurst. Numbers of weeds overall were also the same on both sites. The outstanding difference was the nature of the plants present. At Oakhurst all were seedlings, carrying neither flower nor seed and about to be destroyed by autumn cultivations. At Halsey's all plants were carrying abundant seed and were left to stand over winter.

Thus the change wrought by pre-emergent herbicides especially is not extermination of the plant species but their elimination in the field as maturer plants, particularly at the flowering and seeding stage. As the work of Roberts and his colleagues showed (see above, also Roberts & Dawkins 1967), the elimination of seeding plants erodes the soil's seed bank, a pattern that Robinson & Sutherland (2002) showed has continued to the present. The process has removed not just seeds as a food source for birds but whole swathes of the invertebrate fauna of fields as well, which were dependent, not on the crop, but on associated plants. To anyone brought up to farming before 1960 the totality and scale of the disappearance of arable weeds within crops is extraordinarily visible and it has always struck me very forcibly as the key ecological change in modern arable farming, not only for its direct impact on the flora and fauna of

TABLE 7.3. *The numbers of weeds per square metre found in stubbles in two Sussex farms under different forms of weed control in 1984*

Oakhurst		Halsey's	
Species[a]	Plants/m²	Species[a]	Plants/m²
Ranunculus	9.7	Ranunculus	0.1
Capsella	0.03	Sinapsis arvensis	1.4
Viola	0.2	S. alba	0.1
Stellaria	2	Raphanus	2
Spergula	1.03	Viola	1.6
Geranium	0.8	Cerastium	1
Vicia	0.05	Stellaria	3.3
Rumex	0.6	Geranium	1.2
Anagallis	0.15	Epilobium	0.1
Veronica	3.5	Euphorbia	0.1
Lamium	0.35	Polygonum	3.3
Plantago	0.18	Anagallis	2
Anthemis	1.85	Convolvulus	0.7
Carduus/Cirsium	0.25	Lamium	1.6
Sonchus	0.03	Galium	0.2
		Anthemis	0.9
		Carduus/Cirsium	1.1
Total	20.7		20.7

[a] Because many were seedlings, plants were usually identified only to genus.

farms but also because the chemicals are basic to many changes in farming methods that are themselves unfavourable to wildlife.

The use of pre-emergents has been particularly important in root crops, most of which are brassicas, vegetables and legumes. These plants are especially susceptible to many contact herbicides used against dicotyledonous weeds. Root crops therefore remained a major source of weed seeds for birds until the development of pre-emergent and pre-planting techniques of weed control, using soil-acting herbicides. An important limitation is that many soil-acting herbicides may not work in peat soils or light sands. So the Fens, an area of most intensive arable cropping, paradoxically provides a significant refuge for declining breeding species such as Turtle Dove, Skylark and Corn Bunting (Gibbons et al. 1993) and winter flocks of seed-eaters such as Tree Sparrows and Linnets (Lack 1986).

The availability of these functions has had a fundamental impact on the ecology of arable farmland, underlying changes in the principal season of cultivation (Chapter 8) and radically altering the flora. An extensive literature has been built up in the last 15 years or so, with papers chronicling the decline of farmland birds generally, of farmland specialists, of farmland seed-eaters, of buntings and of particular species such as Skylarks and Song Thrushes. The reasons elucidated for these declines vary (see Table 7.4 for a selection) but they virtually all date from the decade 1965–75 or just after, the decade when these chemical functions became widely applied. There cannot be much doubt that the decline in the birds of arable farmland ultimately stems from the decline in the diversity of the habitat brought about by these highly efficient herbicides, of which decline in birds is simply just a part. This idea is sharply at variance with that of Murton & Westwood (1974), who concluded from their work at Carlton, Cambridgeshire, that 'the structure and, more important, the diversity of the farmland avifauna is not affected by relatively drastic changes in land use' and that 'the conservationist should be less concerned with what happens on the farm *per se* [than with individual species], provided agricultural practice does not result in widescale hazards to wildlife through indiscriminate use of toxic chemicals or the pollution of waterways'. Their work at Carlton, however, was largely completed by 1972 and was mainly carried out before the wide-scale introduction of the herbicide capacity discussed here and their conclusion was well supported by the contemporary evidence of the CBC (Chapter 8).

Other pesticides

In the long term herbicides are the most significant pesticides in their impact on farmland ecology, because they are fundamental to modern farming systems and rotations; other pesticides are not. Nevertheless insecticides and fungicides have a much longer history of use. Grigg (1989) noted that, by the late nineteenth century, plant derivatives – nicotine, derris and pyrethrum – were being used as insecticides in high-value crops such as fruit and hops. Even earlier Peter Hawker was complaining that chemical seed dressings were poisoning partridges in the 1840s. The chemical of which he complained was blue vitriol, used to control smut, a seed-borne disease of cereals. It was replaced by organomercury for this purpose in the early 1950s. There is little evidence that, in Britain, the use of mercury, now banned, caused problems to wildlife (e.g. Cooke *et al.* 1982), but its use led to the virtual elimination of smut in cereals. By the 1920s, despite the depression in farming, the chemical industry was taking greater interest in the potential profitability of pesticide manufacture. The major expansion in use, however, dates since the 1940s.

TABLE 7.4. *Recent trends in some farmland birds and causes of change elucidated*

Species	Recent trend	Cause	References
Grey Partridge	Long-term decline with rapid acceleration since 1960s and 1970s	Loss of arable in the west; loss of mixed farming with leys; decline of invertebrate food of chicks with pesticide use	Potts 1986
Lapwing	Long-term decline with sharp acceleration since 1980	Declining breeding success with increased grass stocking rates; loss of mixed farmland	Galbraith 1988, Shrubb 1990, Shrubb & Lack 1991 Tucker et al. 1994
Stockdove	Decline in eastern England from mid 1950s, recovery from 1962	Organochlorine pesticides; loss of spring sowing; perhaps changes in cereals harvesting	O'Connor & Mead 1984
Turtle Dove	70% decline 1970–98, mainly since 1979, after long-term increase	Declining breeding success from fewer attempts; loss of nesting sites	Browne & Aebischer 2001
Skylark	Sharp decline since 1970s	Changes in cereal management; probably more intensive grass management	Wilson et al. 1997, Donald & Vickery 2000
Song Thrush	Marked decline since mid 1970s	Reduced breeding success due to fewer attempts in intensive arable. Poor post-fledging survival	Thomson & Cotton 2000
Red-backed Shrike	Long-term decline, particularly since 1940s	Loss of old grasslands	Peakall 1962, Vanhinsbergh 2000, this study
Rook	Increase 1930s to 1960, then decline to mid 1970s then recovery	Changes in area of cereals and methods of harvesting	Sage & Vernon 1978, Marchant & Gregory 1999, this study.

TABLE 7.4. *(Cont.)*

Species	Recent trend	Cause	References
Linnet	Decline of 68% 1968–95, some recovery since	Reduced breeding success and post-fledging survival with decline of arable weeds; some recovery with spread of oil-seed rape	Siriwardena 1999, Moorcroft & Wilson 2000.
Yellowhammer	Decline from mid 1980s	Loss of cereals in pastoral landscapes; degradation and loss of hedges and field margins; cereal management	Bradbury & Stoate 2000, Robinson *et al.* 2001
Cirl Bunting	Decline since 1940s with rapid acceleration after 1968	Changes in cereal management particularly loss of winter stubbles; decline mixed farms; loss of old grasslands	Evans 1997
Reed Bunting	Increase 1963–75, then decline of 58% to 1983, since stable	Changes in winter survival caused loss of grass and other weed seeds	Peach *et al.* 1999
Corn Bunting	Long-term decline with rapid acceleration from, mid 1970s	Decline of mixed farming	Donald *et al.* 1994, Shrubb 1997, Brickle & Harper 2000

Fungicides

In crop management fungicides are used as seed dressings to control seed- and soil-borne diseases and as foliar fungicides to control diseases such as mildews and rusts in the growing crop. The use of foliar fungicides has been one of the main growth areas in pesticide use since 1980, particularly in cereals (Figure 7.8). Their use has had some marked adverse effects on some invertebrates in cereal crops, particularly polyphagous predators of the genus *Tachyporus* (Aebischer 1991) but whether this affects bird populations is unclear. But they have important indirect effects because their use has encouraged changes in cropping patterns. In cereals, for example, they have supported the switch to winter wheat, which is more susceptible than

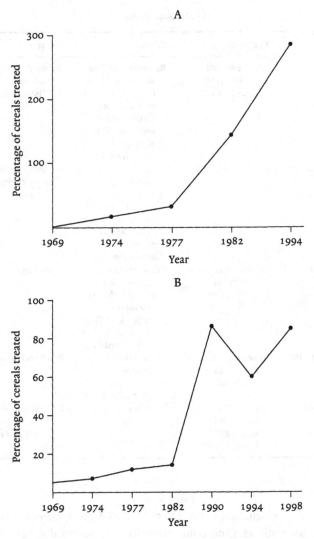

Figure 7.8. The percentage of the cereals area treated with (A) foliar fungicides and (B) insecticides since 1969. Seed treatments are excluded. Values of more 100% show more than one treatment made per field.

barley to the disease problems of continuous cropping, and to winter cereals generally. They also have some indirect herbicidal effects in winter barley, where autumn use promotes tillering and thus the competitiveness of the crop over weeds; spring applications can have similar effects (J. Evans personal communication).

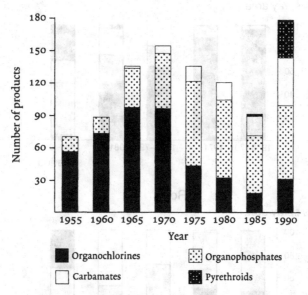

Figure 7.9. The number of products based on the four main chemical groups approved by ACAS for use as agricultural insecticides at intervals from 1955. (Data from ACAS lists and O'Connor & Shrubb (1986).)

Insecticides

Insecticides may be used as seed dressings against soil-living pests or as sprays in standing crops. They also have important uses in veterinary medicine, where the use of insecticidal dips to control sheep scab, for example, was mandatory until the 1990s. O'Connor & Shrubb (1986) observed that the major pests and risks in the principal crops were fairly well defined and that comparatively few chemicals were in really widespread use. Nevertheless these chemicals have spawned a great many competing products and the scale of use and demand is shown by the number of these which were approved for use under the Agricultural Chemicals Approval Scheme (ACAS; see below). Figure 7.9 summarises these since 1955 for each main group of chemical compounds. It shows not only the scale of the market but also the progressive changes between basic chemical groups that have taken place. Figure 7.10 takes this further and shows the proportion of use for the main crops for each chemical group.

Insecticides have been the pesticides most implicated in direct poisoning incidents in birds, particularly the cyclodiene group of organochlorines, dieldrin, aldrin and heptachlor, used largely as seed dressings, dieldrin also as a sheep-dip. They were responsible for very large mortality

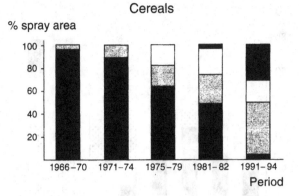
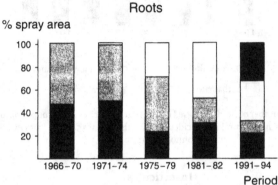
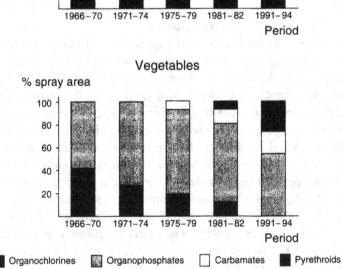

Figure 7.10. Percentage of the total sprayed area in selected crops which were sprayed with each main chemical group of insecticides at intervals since 1966. (Data from Pesticide Usage Surveys.)

of adult or full-grown birds and their predators during 1955–62. DDT (dichlorodiphenyltrichloroethane), also an organochlorine used as a crop spray, caused severe egg mortality, leading to collapsing breeding performance in several raptor species, such as Sparrowhawk (Newton 1986) and Peregrine (Ratcliffe 1980). It was used extensively in orchards, where it particularly affected Sparrowhawks in Kent and Sussex (Shrubb 1985). Mortality and deteriorating breeding success from the same causes led to marked declines in Stockdoves in arable areas of England in the 1950s and early 1960s (O'Connor & Mead 1984) and seriously affected Barn Owls (Newton *et al.* 1991). Other species for which particular studies linked population declines with reduced breeding success caused by organochlorines were Woodpigeon and Yellowhammer (O'Connor & Pearman 1987), Greenfinch (Marchant *et al.* 1990) and Corn Bunting (Crick 1997). These non-predatory species were all partly granivorous, exploiting seed grain. The history of these events is reported extensively elsewhere. Readers are recommended to consult Prestt (1965), Newton (1979, 1986), Newton *et al.* (1982a), Ratcliffe (1980) and Sheail (1985) for excellent accounts.

The problem with the organochlorine group was that, even if of comparatively low toxicity, they were highly persistent and accumulated in fat deposits; predators acquired increasingly concentrated doses from eating contaminated prey and this affected mammalian predators as well as raptorial birds. The main result of this story in Britain was that the Pesticide Safety Precautions Scheme and ACAS came into being, to examine and approve chemicals before their introduction commercially, which gave a far better opportunity to examine problems such as toxicity to wildlife and persistence in the environment before general use (see Sheail 1985). Persistence is probably more important than toxicity. Potts (1986) noted that the most widely used insecticide in cereals, the organophosphate demeton-*s*-methyl, was 40 times more toxic than DDT but caused no direct poisoning because it had low persistence, did not accumulate in the body and was applied at very low rates. He also observed that the biodegradability of insecticides had increased and toxicity declined and this pattern is clear in Figure 7.9, with the progressive switch from organochlorines through to the pyrethroids. Direct toxicity problems are not, at present, of great concern. But nasty surprises can still emerge, as the poisoning of wild geese by the organophosphate seed dressings chlorvenfinphos and carbophenothion in the 1980s showed. Agricultural insecticides are also abused to deliberately poison birds and animals and problems may yet emerge from veterinary medicines. Continued vigilance is always necessary.

The main purpose of insecticides, however, is to kill insect pests, although the most widely used chemicals tend strongly to act against a much wider spectrum of insects than just target species. Campbell *et al.* (1996) concluded that there was good evidence of both long-term and short-term declines in many types of invertebrates in farmland upon which birds feed, and that these declines were at least partly caused by pesticides. The most significant long-term study has been that of The Game Conservancy's partridge survival project (Aebischer 1991). What has proved difficult is to clearly link these declines to changes in bird populations. Campbell *et al.* frequently point to other changes in farming practice as obscuring attempts to isolate indirect pesticide effects. Nevertheless for at least one bird species, Grey Partridge, an indirect effect of pesticides, through reduction of invertebrate food for chicks, has been clearly demonstrated. The sharp recent increase in the use of insecticides in standing cereals (Figure 7.8) may well be partly responsible for the recent decline of the Yellowhammer, which often collects invertebrate food for its young from such crops, particularly wheat (personal observation).

There is a sense, however, in which trying to establish such cause and effect is irrelevant. Table 7.4 lists a number of declining farmland birds and the causes of decline that detailed studies have elucidated. These mechanisms often differ between species. Indeed there is a strong tendency for causes to be species-specific. But the causes have a common link in that they are all aspects of the declining diversity of farmland. Pesticides have made an important contribution to this by reducing populations of weeds and invertebrates upon which birds feed, both in the long term and at crucial points in birds' annual cycles. That contribution has increased over time with increasing sophistication and use of pesticides. But the most important indirect effect of pesticides for wildlife is that they are central to the forms of farming now practised. Without the facility pesticides provide to control competitive weeds, insect pests and disease, more traditional forms of rotational farming must have been retained and with them much greater diversity in animals and plants in farmland, of which bird populations are just a part.

8
Arable farming systems: after 1945

The three-year ley

The three-year ley system emerged as the most widespread system of arable farming in England and Wales between 1945 and about 1970. The system comprised three years of ley followed by three of cereals, managed around a dairy herd. It was encouraged by the State's decision, in 1952, to pay grants for every acre of permanent grass ploughed and reseeded to cultivated strains of grass, a scheme later extended to the uplands, where the levels of grant were raised from £7 to £12 per acre (Bowers & Cheshire 1983). Other major changes under the three-year ley system were the decline of fodder roots and the increase of barley, nearly all spring-sown. Barley occupied $c.$ 33% of arable in the 1960s, at the peak of three-year ley system, compared to $c.$ 10% pre-war and 15% in the 1870s. Fodder roots fell from 9% of arable in the 1930s to 3.5% in the 1960s (of which kale, not recorded in the statistics pre-war, comprised nearly half) and just over 1% in the 1980s. These changes reflected the declining importance of sheep in mixed farming systems, in favour of the dairy cow fed principally on grass and cereals. Roots had also always been the cleaning crop in arable rotations but the development of herbicides from the 1940s rendered them increasingly obsolete for such a purpose and, no longer necessary as stock feed, they steadily declined. Spring barley was the preferred cereal because barley can be most readily grown in succession without disease problems and spring sowing maximised grass weed control through cultivations. A further major development in this period was the use of insecticidal seed dressings, particularly of the organochlorine group, to control seed-borne diseases and soil pests such as wheat-bulb fly.

Changes in cropping systems in Scotland in this period were less marked, as longer leys had always been a feature of the systems developed there in the high farming period. Roots also still occupied 5% of arable at the end of the 1970s and remain regionally significant. Otherwise methods changed along the same lines as in England and Wales.

Two major surveys, Alexander & Lack (1944) and Parslow (1973) give us much general information about changes in farmland bird populations during this period. These surveys were mainly based on county avifaunas and reports and similar sources. The Common Birds Census (CBC) had started in 1961 but the severe winter of 1962-3 distorted the patterns revealed by the census up to the late 1960s. Superficially the pattern revealed for the three-year ley period by comparing Alexander & Lack and Parslow continued to be one of surprising stability in farmland bird populations. Alexander & Lack recorded declines for 18 species in the twentieth century up to 1940, to which I have added 3 more extracted from subsequent publications, compared to 30 increases and 48 species which showed no marked change. After 1940 Parslow found that 30 species declined, some after an initial increase, 31 increased and 38 showed no change. The differences in the two periods are not statistically significant. However it is more interesting to look at the species whose status was considered to have changed in each period and the reasons suggested why. These data are set out in Table 8.1.

Agricultural factors were not particularly important causes of change in Alexander & Lack's survey. The spread of arable weeds and the decline in grassland management and stocking densities contributed to increases in ten species, two declined because of the decline of arable farming and two from changes in grassland management. Persecution or protection or continued range expansions, mainly to the north and perhaps still based in climatic amelioration, were by far the most frequent causes of change quoted. The early stages of afforestation were important for a few species and excluded from the table are changes for some common woodland birds which Alexander & Lack noted as still increasing with woodland area in Scotland. Such change was outside farmland and these species are recorded as stable (s).

With the post-war period patterns began to change. The decline of persecution, increased protection and continued range expansion remained the most frequently quoted causes of increase, coupled with a spread into new habitats, such as suburbia. Afforestation encouraged a few species, such as Black Grouse and Short-eared Owl, but the advantages were

TABLE 8.1. *Changes in farmland bird populations described by Alexander & Lack (1944) and Parslow (1973) for 1900–40 and 1940–67 with suggested reasons*

Species	Alexander & Lack (1944): 1900–40		Parslow (1973): 1940–67	
	+/−	Reason	+/−	Reason
Mute Swan	+	Northward expansion	+	Unknown
Garganey	+	Northward expansion in Europe	+	Increase ended 1950s
Shoveler	+	Decline of decoying?	s	
Red Kite	−	Continued persecution	+	Continued protection
Marsh Harrier	−	Continued persecution	−*	Organochlorines
Hen Harrier	−	Continued persecution	+	Protection and decline of persecution
Montagu's Harrier	s		−*	Organochlorines, + to 1950s
Sparrowhawk	−	Continued persecution	−*	Organochlorines
Buzzard	+	Decline of persecution	+	Decline of persecution
Kestrel	s		−	Organochlorines
Merlin	−?	Continued persecution?	−	Afforestation, ?organochlorines in winter habitats
Red Grouse	s		−	Habitat deterioration and decline
Black Grouse	−	Habitat loss	+	Spread of new forestry
Grey Partridge	−	Decline in arable and keepering	−	Pesticides and decline of undersowing
Quail	−	Persecution on migration	+	Some increase after 1942
Corncrake	−	Grassland management	−	Grassland management
Stone Curlew	+	Decline in agriculture in some areas	−	Loss of old grass downland, afforestation
Golden Plover	−	Unknown	−	?Changes in moorland management
Lapwing	+	Lapwing Act 1926[a]	−	General agriculture change
Snipe	+	Expansion of grass and decline in drainage and management	−	Drainage and reseeding
Curlew	+	As Snipe	+	Expansion checked 1960s
Redshank	+	As Snipe and Curlew	−	Ploughing of grassland and as Snipe
Stockdove	+	?Spread of arable weeds	−*	Organochlorines
Turtle Dove	+	As Stockdove; ?climatic	+	Continued range expansion
Collared Dove			+	First colonised in 1960s
Barn Owl	s		−	Hard winters, loss rough grazing, organochlorines
Little Owl	+	Spread after introduction	−	Hard winters, organochlorines
Tawny Owl	+	Decline in persecution	+	Continued to 1950s
Long-eared Owl	+	Spread into plantations in north	−	?Competition with Tawny Owl, ?organochlorines
Short-eared Owl	s		+	Afforestation
Wryneck	−	Unknown	−	
Green Woodpecker	s		+	Continued northward expansion
Great Spotted Woodpecker	+	Continued northward expansion and spread of woodland	+	Continued northward expansion

(cont.)

TABLE 8.1. (Cont.)

Species	Alexander & Lack (1944): 1900–40		Panslow (1973): 1940–67	
	+/−	Reason	+/−	Reason
Woodlark	+	Declining stock densities[b]	−*	Increase to 1960s then collapse ?Increased stock densities
Pied Wagtail	−	Decrease in Scotland; ?reason	s	
Yellow Wagtail	−	Decline of arable[b]	−	Continuing trend
Whinchat	s		−	Grassland management
Stonechat	−	Habitat loss, cold winters	−	Habitat loss, cold winters
Wheatear	−	Decline in grassland management and stocking rates, loss of sheep	−	Ploughing of grassland, afforestation
Ring Ouzel	−	Unknown	−	Unknown but afforestation contributed
Blackbird	+	Northward expansion and spread of woodland	+	Continued expansion into cities
Mistle Thrush	+	Northward expansion	+	Spread into gardens and suburbs
Song Thrush	s		−	Cold winters
Sedge Warbler	+	Northward expansion	s	
Reed Warbler	s		+	Range expansion
Lesser Whitethroat	+	Range expansion	s	
Chiffchaff	s		+	Expansion in Scotland
Goldcrest	+	Spread of conifers	+	Coniferous afforestation
Coal Tit	s		+	Coniferous afforestation
Red-backed Shrike	−	Some decrease northern and western range	−	Loss of old grassland
Jay	+	Decline of persecution	+	Decline of persecution
Magpie	+	Decline of persecution	+	Decline of persecution
Chough	−	Unknown, ?persecution	s	
Jackdaw	+	Continued northward expansion	+	?Expansion of cultivation
Rook	+	Continued northward expansion	−*	Changes in cereal harvesting method
Crow	+	Decline of persecution	+	Decline of persecution
Raven	+	Decline of persecution	+	Decline of persecution
Starling	+	Continued range expansion	+	Continued range expansion
House Sparrow	−	Decline of urban horses	s	
Tree Sparrow	s		+	Unknown
Chaffinch	s		−	Organochlorines
Greenfinch	s		+	Spread into suburbs
Goldfinch	+	Agricultural decline and cessation of trapping	+	Spread into suburbs and uplands
Linnet	+	As Goldfinch	s?	
Twite	−	Unknown	−	Unknown
Bullfinch	s		+	?Decline of Sparrowhawk
Yellowhammer	s		−	Organochlorines
Reed Bunting	s		+	Expansion into dry habitats
Corn Bunting	−	Decline of arable farming	−*	Increase in some areas with arable

Notes: + indicates an increase, − a decrease. Those marked s were unchanged in one period but not the other. Under Panslow birds marked * showed an initial increase after 1940 before an overall decline for the period.
[a] But see Chapter 6.
[b] These are my interpretations; see Chapter 3 and Chapter 9.

ephemeral and, as forests grew, they affected more species disadvantageously. Moorland waders, for example, avoid the vicinity of blocks of trees and afforestation of moorland involved direct habitat loss for at least seven species listed.

Population changes deriving from agricultural causes, nearly all declines, became much more marked. The most significant factor was the development and use of new chemical seed dressings, particularly the organochlorines of the cyclodiene group, dieldrin, aldrin and heptachlor. These were considered to be the main or a contributory cause of decline for 12 species, virtually all raptors or seed-eaters. Nevertheless, although many of the species involved declined on a dramatic scale in eastern England in the late 1950s, withdrawal of the chemicals allowed extensive recovery. Potts (1970b), for example, noted that these chemicals had considerable short-term effects on partridges but no long-term effects on their density. Murton (1965) made similar observations for Woodpigeons and O'Connor & Mead (1984) traced the recovery of the Stockdove. Seed-eaters such as Stockdoves, Chaffinches and Yellowhammers were primarily affected by direct poisoning by eating dressed seed from new sowings, particularly in spring. Their recovery was rapid once a voluntary ban on spring use of such seed dressings was introduced in 1961, thus also neatly pointing up the importance of spring sowings as a feeding site for seed-eaters. Overall recovery by raptors from the effects of both the cyclodienes and from DDT was much slower but can now be regarded as complete. It must be stressed that the overall lack of long-term impact from these chemicals was the result of the strong and insistent objections to their use by conservation organisations on the grounds of evident and serious environmental damage. Without that action the damage done would have been immense and perhaps irreparable.

Otherwise rather few marked changes stemmed from any change in the arable system in the three-year ley period. The increased tendency for pastoral farming to concentrate in the west and arable in the east began to emerge more strongly in this period and was certainly affecting Corn Buntings and Grey Partridges in the west. Throughout the southeast and Midlands areas of England, however, Corn Bunting populations recovered with three-year ley farming (Shrubb 1997). Grey Partridge populations declined sharply from the early 1960s, a decline which Potts (1970a) noted as restricted to them among common birds of cereal fields. It was the first species to be adversely affected by the widespread and routine application of herbicides to cereals, through their effects on invertebrates

(Chapter 7). Rooks benefited from the expansion of cereals from 1940 but declined sharply from the early 1960s (e.g. Dobbs 1964). Poisoning by organochlorine seed dressings may have contributed, as Dobbs suggested, but O'Connor & Shrubb (1986) linked the decline of Rooks at this period to the mechanisation of the cereal harvest (Chapter 11).

Otherwise the strongest impression is one of stability and this is supported for many species by the farmland indices of the CBC up to the early 1970s. Examining the trends shown by Marchant *et al.* (1990) shows that, of 59 species indexed from the early 1960s, 27 increased up to 1975, 21 were broadly stable and 10 or 11 declined. Of the declining species Whitethroat, Garden Warbler and Spotted Flycatcher, being summer visitors, may have been more influenced by factors in the wintering area (e.g. for Whitethroat; Winstanley *et al.* 1974). Of the species that increased the trends shown for many passerines, for example Wren, Goldcrest and tits, suggest continued recovery from the exceptional winter of 1962-3 rather than any long-term population change. Whilst this analysis only examines a segment of a longer run of data, it is striking just how many of the index trends illustrated by Marchant *et al.* show a rise to a shallow peak or broad stability up to the mid 1970s. All the evidence available suggests that, outside the impact of organochlorines, arable farming during the period up to the early 1970s had little impact on bird populations. A similar conclusion was reached by Murton & Westwood (1974) from their studies in Cambridgeshire. Siriwardena *et al.* (1998) found that more species had declined in abundance in farmland since 1976 than before and Chamberlain *et al.* (2000) also found stable populations for 29 farmland species covered by CBC up to the mid 1970s despite marked agricultural change.

I suggest that the main reason for this stability was that, despite major changes in cropping patterns, the three-year ley system retained the most important features which enabled birds to adapt to high farming in the first place. Undersown stubbles continued to provide them with important winter-feeding sites widely dispersed through farmland (one field in ten) as well as essential overwintering/breeding sites for important invertebrate groups. Spring tillage still comprised 53% of arable land, providing important feeding opportunities for birds in spring and largely guaranteeing a further significant area of cereal stubble. Although stackyards and their attendant chaffheaps had disappeared as combines replaced the binder and threshing drum, grain-feeding of cattle had partly replaced them as a winter food source for birds, particularly with the wide diffusion of dairy herds in the three-year ley system. First-generation herbicides had a remarkable impact on a few arable weeds, perhaps particularly charlock (*Sinapsis*) but were otherwise comparatively inefficient (Chapter 7). Finally mixed farms, combining arable cropping and grass and stock, remained the rule.

Modern arable systems

The chemical-based revolution in British arable farming which started around 1965–70 is described in some detail in Chapter 7. It has had a number of important ecological effects which have altered farmland habitats fundamentally.

The developments in herbicides and fungicides outlined there finally did away with the cultural basis of mixed arable and stock rotations. The cultivation of roots was no longer significant for weed control in arable rotations. The development of grass herbicides did away with the need for a dual season of cultivation, in autumn and spring, for longer runs of cereal crops and therefore the ley slot vanished. The capability for chemical control of disease reduced the need for rotating crops for this reason. Parallel developments in mechanisation removed the physical basis of classical rotations, the need to spread seasonal workloads (Chapter 11). The ready availability of manufactured fertilisers and better understanding of their use removed the need for animals of any sort as sources of fertiliser or any use for straw, which was promptly burnt until recently; their significance as sources of power had long since vanished.

This whole process, together with the economic encouragement of cash-cropping and the financial pressure towards specialisation (Chapter 1), has led to the virtual disappearance of old-style rotations. Arable fodder crops now account for less than 4% of tillage in Britain, whilst wheat, barley and oil-seed rape occupy 75%. These are now virtually the only tillage crops which occupy anything approaching a significant and large-scale general distribution. Of other major crops, 70% of all sugar beet is concentrated in Lincolnshire, Norfolk, Suffolk and Cambridgeshire and the same counties account for 60% of field vegetables and nearly 40% of potatoes grown in England. Furthermore 67% of all tillage crops in Britain are now sown in autumn. The number of holdings with cattle in England and Wales has declined from 76% of farms in 1958 to 50% in 1997 but to only 29% in the arable regions of the east and southeast. Sheep populations, surprisingly, have been more stable, reflecting a recent increase in arable regions. But these are kept on grass, not in rotations in the old way. In pastoral regions the same process of specialisation has occurred, with tillage crops now absent from many parts of Wales, Northern Ireland and Scotland and occupying only 20% of improved farmland in total in these regions. Scotland perhaps retains the most mixed farming system remaining in the UK, although half of all vegetable and potato crops there are now grown in Tayside and Fife. Cattle and sheep have tended to polarise far less than they have in Wales.

This has not happened overnight. It has been a process of steadily encroaching uniformity. Uniformity has also been influenced by other changes. The area of permanent grassland has declined by *c.* 35% in arable regions, as stock and arable farming enterprises have increasingly separated, further reducing diversity, although the strong recent tendency to take pigs back outside has brought a fresh element of livestock back into arable management; fields full of arks are now a frequent sight in eastern counties. Field size has increased, reducing the patchwork effect as crops move into bigger blocks. Diversity also suffers as hedges, a major element of farmland's permanent habitat skeleton, are removed (Chapter 5). The number of holdings and thus farmers has declined, by between 45% and 55% between 1958 and 1997. With increasing farm size larger areas become devoted to a particular form of management. Fewer farmers undermines a second source of diversity attached to holding size, that of the varying interests and competence of individuals. That is also being eroded by the training and advice offered to the young entry, which tends strongly to uniformity. Generational differences also contribute a random element to the patterns of change. Older generations of farmers tend to greater conservatism. It is their successors who are the main innovators. That is a generalisation of course, but time and again one sees sweeping management changes on farms delayed until generations change.

Besides its impact on the flora, the chemical revolution, through the development of grass herbicides (Chapter 7) has had a major impact on the timing of cultivation. Because grass weeds must otherwise be controlled by cultivations, these herbicides have been the key to the decline of spring tillage since the late 1960s, illustrated in Figure 8.1. The change is one which has no historical precedent or counterpart. The figure shows only the overall pattern for England and Wales but on many individual farms the change has been much greater. Some now grow only autumn-sown crops and many grow only a small percentage of spring-sown crops. Thus on Oakhurst the proportion of tillage occupied by spring-sown crops fell from 56% in the early 1960s to 15% in the early 1980s; in common with the

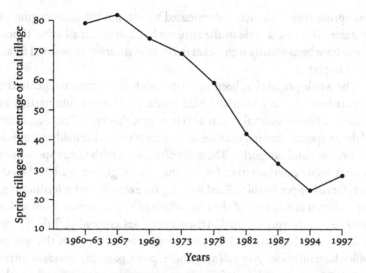

Figure 8.1. Changes in the proportion of spring tillage in England and Wales since 1960. (Data from June Census Statistics and Surveys of Fertiliser Practice.)

general pattern it rose a little in the 1990s to 19%. These patterns seem fairly typical. The change has particularly affected southeast England, where spring tillage fell from 43% of farmland in the early 1960s to c. 13% in the early 1990s, involving a total area of c. 1.4 million ha. However spring tillage has shown some recent recovery in the 1990s, with diversification into peas, spring oil-seed rape, linseed and other crops as breaks in cereal rotations. The change has also affected Scotland much less, because harvests are later and autumn sowing more uncertain. Nevertheless the proportion of autumn tillage has advanced significantly there since 1980 and now occupies 40% of all tillage.

The dominant trend in virtually all recent change in arable systems has thus been to uniformity and simplification of farmland habitats, not only on a district or farm basis but in nearly all individual fields, where the loss of the weed flora has, as Potts (1997) pointed out, brought about a catastrophic collapse in biodiversity. This has recently been exacerbated further by the increasing trend to use insecticides in standing cereal crops, the invertebrates of which are an important food source for some breeding birds, e.g. buntings. Effectively these changes have blown a huge hole in the food resources available to farmland birds, particularly in the breeding season and in the winter. Examination of Table 3.2 will show the likely impact of the loss of alternative food resources provided by pasture, leys and spring crops in a system dominated by autumn tillage. A surplus of resources is now available in the autumn but a deficit at all other times. There have been equally significant declines in diversity in pastoral farming (Chapter 9).

The whole process has been associated with significant changes in the populations of many farmland bird species, which are summarised in Table 8.2. Several general points are clear from this table. Three-quarters of the 44 species showing declines are concentrated in the field or ground-nesting/wetland categories. These are often birds with rather specialised habitat requirements, either for nesting (ducks, grouse, waders, Woodlark, Grasshopper Warbler, Reed Bunting for example) or for feeding (e.g. raptors such as Kestrel and the owls which need rough grass habitats for voles). Six of the species listed in these categories as showing little change had only stabilised after a long-term decline before 1972. Of the species which have increased over half are either raptors, pigeons, corvids or introduced. Raptors are still benefiting from increased protection or the withdrawal of organochlorine pesticides, pigeons from the increase in oil-seed rape (Chapter 10) and corvids are probably increasingly commensal with Man. The concentration of declines amongst field, ground-nesting and wetland species clearly implicates changes in field habitats and their management as the main driver of change in farmland bird populations.

A spread of -25 to $+33\%$ taken as showing little change is quite wide and it is instructive to examine those species in Table 8.2 more closely, by comparing the number of 10-km squares occupied by proven or probable breeding in the two *Breeding Atlases* (Gibbons *et al.* 1993, Sharrock 1976). Here I have considered that only differences of 10% or more in 1988–91 in the number of squares occupied in 1968–72 represented genuine change. Table 8.3 sets out the species under the categories used in Table 8.2. On this analysis only 15 farmland species have shown no change over this period, a measure of the steep decline in the underlying stability of farmland bird populations that modern farming systems have caused. The analysis also extends further the concentration of declines in field and ground-nesting/wetland birds, which now account for 79% of declines, whilst 61% of species in those categories have declined.

Birds that are persisting in farmland today are more generalist species, a process that has markedly accelerated since the mid 1970s. Chamberlain *et al.* (2000) analysed changes in the abundance of 29 farmland bird species as shown by CBC in relation to agriculture change in detail for the period 1962–95. They show a marked period of change in agricultural methods

TABLE 8.2. *Changes in farmland bird populations between 1972 and 1998*

Category	Major increase, >100%	Moderate increase, 33-100%	Little change, −25-+33%	Moderate decline, 25-50%	Major decline, >50%
Woodland	Sparrowhawk Hobby* Great Spotted Woodpecker Whitethroat Blackcap	Garden Warbler Chiffchaff Long-tailed Tit Coal Tit	Cuckoo Wren Robin Willow Warbler Goldcrest Blue Tit Great Tit Treecreeper	Dunnock Lesser Whitethroat Jay	Spotted Flycatcher Bullfinch
Woodland/ Field			Green Woodpecker Chaffinch Greenfinch	Tawny Owl Blackbird Mistle Thrush	Wryneck Song Thrush Red-backed Shrike
Field	Red Kite* Stockdove Collared Dove Magpie Jackdaw Crow	Buzzard Woodpigeon Rook	Pied Wagtail *Stonechat** *Chough** *Raven** Goldfinch	Kestrel Barn Owl* Little Owl Long-eared Owl* Swallow Starling House Sparrow Linnet Yellowhammer	Turtle Dove Tree Sparrow Cirl Bunting*
Ground-nesting/ wetland	Mute Swan Greylag* Canada Goose* Marsh Harrier* Reed Warbler	Hen Harrier* Pheasant	Mallard Garganey* *Merlin** *Red Grouse** Red-legged Partridge Quail Oystercatcher* *Golden Plover** Dunlin* Curlew* Black-headed Gull* *Wheatear** Sedge Warbler *Twite**	Teal* Shoveler* Black Grouse* Moorhen Stone Curlew* Snipe* Redshank* Short-eared Owl* Meadow Pipit Yellow Wagtail Whinchat*	Montagu's Harrier* Grey Partridge Corncrake* Lapwing Skylark Woodlark* Ring Ouzel* Grasshopper Warbler* Reed Bunting Corn Bunting

Note: Species in italics are not now changing greatly but are recorded as experiencing significant historic declines in the twentieth century.
Sources: Data are drawn from Marchant *et al.* (1999) or, for species marked * a comparison between the two *Breeding Atlases* (Gibbons *et al.* 1993, Sharrock 1976), with changes based on the difference in 10-km squares occupied by breeding evidence, using the same scale.

TABLE 8.3. *Species showing little change in Table 8.2 re-examined by a comparison between the two Breeding Atlases[a]. Only proven or probable breeding was considered*

Category	Increase	No change	Decrease
Woodland		Cuckoo Wren Robin Willow Warbler Blue Tit Great Tit	Goldcrest Treecreeper
Woodland/field		Chaffinch	Green Woodpecker Greenfinch
Field		Pied Wagtail Chough Goldfinch	Stonechat Raven
Ground-nesting/ wetland	Merlin Red-legged Partridge	Mallard Garganey Quail Oystercatcher Dunlin	Red Grouse Golden Plover Curlew Black-headed Gull Wheatear Sedge Warbler Twite

[a] Gibbons *et al.* (1993), Sharrock (1976).

and crops between 1970 and 1988, preceded and followed by periods of relative stability, and a marked period of decline in the farmland birds between about 1977 and 1986, preceded by a period of stability and followed by a period when declines were slowing or populations becoming more stable. These authors argue that the time-lag between the onset of agricultural change (which actually had its beginnings in the mid 1960s) and a major impact on birds was to be expected if the causal link was an indirect mechanism such as the reduction of food supplies. Thus a substantial reduction in winter food resources, such as seeds, would probably be needed before gregarious species such as finches began to be affected. This time-lag reflects the nature of agricultural change. Chamberlain *et al.* point out that it is not episodic but a continuous process. It also includes an important random element which is related to the attitudes and interests of farmers (see above) and the fact that farming is a highly fragmented industry and change the sum of thousands of individual business decisions. That it took less than 20 years for the modern agricultural revolution to spread right across farming as it has is unprecedented in farming history and must be the contribution that a barrage of standardised advice and education from colleges, the National Agricultural Advisory Service and its successor the Agricultural Development and Advisory Service and commercial companies has made to the process.

The species for which most study has been made in this field is the Grey Partridge (Potts 1986). However, this species provides an exception to the patterns described above in that population declines were evident by the early 1960s (Potts 1970a). The Grey Partridge may have been particularly vulnerable to the earlier changes in arable methods and cropping patterns because its young are precocious, nidifugous and self-feeding. Arable fields hold only four such species which are at all common, the two partridge species, Pheasant and Lapwing. Of these the populations of Red-legged Partridge and Pheasant are widely augmented by releases for shooting and Lapwings are much more dependent on grassland for rearing chicks (Galbraith 1988). The drawback of the nidifugous habit is that the foraging range of the chicks is less than of adults bringing food to the nest, which could be a marked disadvantage where food resources become more scattered. Nidicolous young do not face this problem and, in modern farmland, adults feeding young will range over considerable distances. At the greatest extreme Linnets are known to forage 1–3 km from the nest when feeding young (Moorcroft & Wilson 2000) and Greenfinches may do the same, at least when exploiting oil-seed rape (personal observation).

A further point which emerged from the analysis of Marchant *et al.* (1999) was that many woodland populations of farmland birds are changing in the same way as farmland populations (Table 8.4). The pattern is much more marked in field and woodland/field species, where only 13% of changes differ in woodland compared to farmland, than it is for true woodland species in which 44% of species show different patterns. Clearly field species which nest in woodland must feed outside and therefore do not entirely avoid the impact of farmland changes. For woodland/field species the important point may be the decline of winter resources, particularly for seed-eaters (see Chapter 10), a decline which again exports changes in farming methods into woodland bird populations.

Breeding success

There is little detailed information in the nineteenth-century literature about breeding performance in farmland birds. In theory at least, however, one would expect the marked changes in resources available to farmland birds in the modern era to be widely reflected in reduced breeding

TABLE 8.4. *Changes in woodland populations of farmland birds shown by CBC during the period 1972–97*

Species (category)[a]	Change in farmland[b]	Change in woodland[b]
Stockdove (F)	+ >100%	+ >100%
Woodpigeon (F)	+ 33–100%	+ >100%
Collared Dove (F)	+ >100%	+ 33–100%
Turtle Dove (F)	− >50%	− >50%
Cuckoo (W)	s	s
Green Woodpecker (W/F)	s	s
Great Spotted Woodpecker (W)	+ >100%	+ 33–100%
Wren (W)	s	s
Dunnock (W)	− 25–50%	− >50%
Blackbird (W/F)	− 25–50%	s
Song Thrush (W/F)	− >50%	− 25–50%
Mistle Thrush (W/F)	− 25–50%	s
Whitethroat (W)	+ >100%	s
Garden Warbler (W)	+ 33–100%	+ >100%
Blackcap (W)	+ >100%	+ 33–100%
Chiffchaff (W)	+ 33–100%	s
Willow Warbler (W)	s	−25–50%
Goldcrest (W)	s	− >50%
Spotted Flycatcher (W)	− >50%	− >50%
Long-tailed Tit (W)	+ 33–100%	s
Coal Tit (W)	+ 33–100%	s
Blue Tit (W)	s	s
Great Tit (W)	s	+ 33–100%
Treecreeper (W)	s	s
Jay (W)	−25–50%	s
Magpie (F)	+ >100%	+ >100%
Jackdaw (F)	+ >100%	+ >100%
Crow (F)	+ >100%	+ >100%
Chaffinch (W/F)	s	s
Greenfinch (W/F)	s	s
Linnet (F)	−25–50%	− >50%
Bullfinch (W)	− >50%	− >50%
Yellowhammer (F)	−25–50%	− >50%

Sources: As for Table 8.2.
[a] F, field species; W/F, woodland/field species; W, woodland species.
[b] + indicates an increase, − a decline and s little change as defined in Table 8.2.

success. Despite problems with organochlorine pesticides in the early part of the period, however, this does not appear to be generally so for nidicolous species, at least in terms of nest success (Siriwardena *et al.* 2000). However these authors noted that, for nidifugous species, population declines were correlated with declines in productivity. Rising stocking rates have limited breeding success in grassland for some waders for example (Chapter 9) and the effect of herbicides on Grey Partridges was noted above. Whilst this might have been expected to have affected other species, some potential nidicolous contenders (Tree Sparrow and Corn Bunting for example) are actually showing improved nest success.

O'Connor & Shrubb (1986, Table 9.6) showed that the percentage of nests succeeding (i.e. producing at least one young) in a range of common farmland passerines tended to rise between 1962 and 1980. This was possibly the result of the continued reduction in the use and effects of organochlorine pesticides, as suggested by Crick et al. (1995). These trends have continued and the overiding picture that has emerged from reports from the BTO's Nest Record Scheme (Crick et al. 1995, 1996, 1998, 2000) is of a general improvement in breeding performance. The Linnet has provided a marked exception to this general trend and Siriwardena (1999) linked its population decline to declining breeding performance. For other species, however, declining populations may be contributing to better nest success, because fewer pairs mean more nesting in optimum habitat. Siriwardena et al. (2000), however, pointed out that the nest record cards cannot measure the number of nesting attempts per pair, which is a significant element in breeding success. For example Wilson et al. (1997) found that Skylarks must make two to three nesting attempts per season for populations to be self-sustaining and that rapid crop growth precluded more than one in modern autumn cereals (Chapter 3); this has probably been important in the recent decline of the species (Donald & Vickery 2000). Thomson & Cotton (2000) found that Song Thrushes nesting in an arable area in Essex made only 2.5 nesting attempts per pair, compared to four in a mixed farming/pastoral area in Sussex, although clutch size and productivity per nesting attempt were similar. Overall productivity was therefore 33% lower in the arable area. They also found that post-fledging survival were markedly lower in the Essex site, whilst no young of the year were recruited into the breeding population the following year there, compared to 17% of the population in Sussex.

Siriwardena et al. (2000) suggested that changing survival rates represented the most important demographic factor so far investigated driving population change in farmland birds. Two points seem worth making here. I suspect that declining post-fledging survival, as shown by Thomson & Cotton for Song Thrush, may be characteristic of declining diversity in farmland. Examining Table 3.2 shows how little would have been available for young birds in the June–August period without leys particularly, but also without pasture. Whilst experienced adults can cope with such conditions, often by ranging widely over the farm (see above),

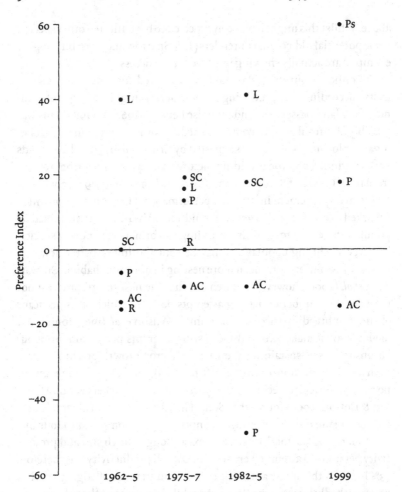

Figure 8.2. Nesting habitat preferences of Skylarks at Oakhurst at intervals since 1962. Preference Index as in Table 3.2. Positive values indicate habitat selected, negative avoided. Crop habitats available were: AC, autumn-sown cereals; SC, spring-sown cereals; L, clover leys; R, roots, not grown after 1976; Ps, peas, only available in 1999; P, permanent grass. Only peas, autumn cereals and permanent grass were available in 1999. Field beans were grown in the 1970s but not used by Skylarks.

inexperienced fledglings would be at a marked disadvantage, either because food was difficult to find or because the circumstances increased their vulnerability to predation, or both. Virtually all the Song Thrushes I found on my counts in the early 1980s at that season were in clover leys, which had been mown once and then left to grow a second crop. These leys provided both food (clover leys hold good invertebrate populations) and cover. In the past root crops would have provided similar sources of shelter and food. Secondly it is essential not to lose sight of the importance of crop diversity for many farmland breeding birds; it may be much more valuable than any single feature of farming systems. For Skylark for example, Donald & Vickery (2000) show a marked correlation between the CBC Index from 1968 to 1996 and the UK area of spring cereals. The problem of such a retrospective correlation, however, is that spring cereals in this context may simply be an indicator of a more diverse farming system. The same criticism can be levelled at O'Connor & Shrubb's (1986) correlation between breeding Skylarks and short-term leys. Indeed they are perhaps the same correlation, in view of the extent to which leys were established under spring cereals, especially in the 1960s. Wilson *et al.* (1997) made the point that a single crop type rarely provides a suitable vegetation structure for nesting throughout the season, so that Skylarks need crop mosaics if productivity is to be adequate, which is likely to be a much more general need. Figure 8.2 illustrates nesting habitat selection by Skylarks at Oakhurst at four periods with different cropping regimes from 1962. Autumn cereals were uniformly avoided but increasingly dominated the cropping system; spring cereals or rotation crops were always preferred if available.

9
Grassland and stock

Pastoral farming has always occupied over half of farmland in Britain. Since 1866 grassland has been categorised in the June Census as rough grazing (either in sole right, i.e. by a single owner/tenant, or in common, i.e. shared with other rights-holders), permanent grass or ley. Rough grazings may be broadly taken as agriculturally unimproved or semi-natural grassland, today mostly open hill grazings. Traditionally permanent grass was agriculturally improved or enclosed grassland that was not ploughed in rotation and included habitats such as water meadows, hay meadows and old pastures: ley or temporary grass was defined as grass that was ploughed in arable rotations and included crops such as clover, sainfoin and lucerne. One- or two-year leys were most common in England and two- or three-year leys in Scotland. As the prosperity of high farming declined in the late nineteenth century, leys tended to be laid down longer. Today these categories have been discarded, reflecting the frequency of re-seeding, and improved grass is now recorded as that laid down for five years or more (treated as 'permanent grass') or for fewer than five years (treated as 'ley').

During the nineteenth century the area of rough grazings in England and Wales declined with the enclosure and ploughing of the wastes (Chapter 4), whilst leys nearly trebled to c. 3 million acres (1.2 million ha). By contrast, there was little change in the area of permanent grass. In 1688 Gregory King estimated 12 million acres (4.86 million ha) of permanent pasture and meadow, which he separated from the wastes (quoted in Stamp 1955). In 1870 the June Census still recorded 11.5 million acres (4.66 million ha). The lack of change over the period indicates that this was a core of grazing and meadow land, which high farming had no pressing need to alter because of the extent of leys and fodder crops for livestock the new rotations grew. During the ecological upheavals of the enclosure period, therefore, pasture and meadow provided valuable stability in habitat, which aided adaptability in birds. In Scotland the extent of rough grazing in agricultural land (Figure 4.4) has long fulfilled the same function; comparatively little improved grass was ever permanent pasture, leys being preferred.

Stocking rates and densities

The main animals of concern here are cattle and sheep. In the early nineteenth century there were also about 1 million horses on British farms. But the management of working horses differed sharply from that of grazing stock kept for commercial enterprise and profit and it is the impact of the latter on grassland habitats which is most significant to birds.

Little accurate information about cattle and sheep numbers in the eighteenth and early nineteenth century is available. For England and Wales, however, a few exact records of sheep numbers can be found for individual farms or districts, which suggest stocking densities of about 1.24–1.73 sheep per hectare of total agricultural area (e.g. Armstrong 1973, Fussell 1952, Minchington 1956). If typical such densities suggest a total population of c. 14.5–20 million sheep around 1800. Fussell & Goodman (1930) also provided a series of county totals from *An Essay on Wool* by John Luccock (1809), who estimated contemporary stocking densities. These showed an overall density of c. 1.56 sheep per hectare of total agricultural area in nine counties, which gives a total national estimate of c. 18 million sheep. These estimates average out at 17.5 million head, which seems a reasonable total to adopt for England and Wales for around 1800.

Cattle numbers are harder to estimate, partly because they were less uniformly distributed. Stocking densities of as little as 3 acres (1.2 ha) per cow appeared to be common and Minchington (1956) and Wilkinson (1861) indicated even lower densities, although it is often unclear whether young stock were included. Grigg (1989) estimated 4.5 million cattle for 1700, declining 24% to 3.4 million in 1800. Whilst some decline is probable because the fertility of many cow pastures declined (Sturgess 1966) and because of the effects of serious outbreaks of cattle plague in mid century, 24% seems high in view of the extent of farming improvements during the eighteenth century. I suggest an estimate of 3.5–4 million cattle for England and Wales at the beginning of the nineteenth century may be more realistic.

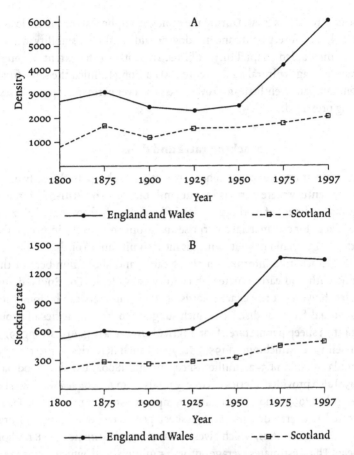

Figure 9.1. (A) Overall stocking densities and (B) stocking rates of cattle and sheep in England and Wales and Scotland per 1000 ha of all grass plus fodder crops at intervals from 1800. For definitions see text.

Numbers in Scotland were considerably lower. Symon (1959) gave 1.047 million cattle and 2.851 million sheep in 1814. Probably stocking rates in improved farmland differed rather little from England and Wales and it was in the uplands and crofting areas that numbers were low. These areas were mainly cattle-based economies and changes in upland management over the nineteenth century saw a large increase in sheep (+139%) but a fall in cattle (−8%) (Chapter 4).

These estimates allow some indication of grazing densities to be given for the last two centuries (Figure 9.1). Stocking densities are the total numbers of animals per 1000 ha of all grass plus fodder crops, stocking rates are the total number of livestock units similarly. Livestock units are a device to equalise the differences between types and ages of animals in assessing grazing pressure and I have followed Coppock (1976) in calculating these. The figure confirms for England and Wales the underlying stability of grassland management into the first half of the twentieth century, particularly with stocking densities. The sharp rise in stocking rates in 1963 reflects the increasing importance of dairy cattle in the three-year ley system after 1945. That of densities by 1987 reflects the enormous increase in sheep in the north and west from 1980. It confirms also how much less Scotland has been affected overall. This is largely because of the great area of open rough grazings there and changes in improved farmland may have been as marked as in England and Wales.

Early management systems in grass

Sheep/arable on the chalk

Throughout the chalk uplands in southern England a characteristic system of arable farming with sheep was practised. The key to the system was the flock, grazed on the down by day and folded on the arable at night, where they dropped their dung and urine to manure it (see Kerridge 1954). The system was more complex than that summary suggests but that was its essence. Downland flocks were large, rarely below 800–1000 head, and the fold was shifted daily so that the arable was evenly manured. Defoe (quoted in Furbank *et al.* 1991) suggested that this was a new method in the early 1720s but it was already well established then. Tubbs (1993) speaks of a 'rising tide of arable consuming the ancient downland turf' from the end of the sixteenth century on. Despite fluctuations, the expansion of arable continued well into the nineteenth century, particularly in the Napoleonic War period. By the 1860s, however, many farmers considered that too much down had been ploughed for the good of the sheep (Wilkinson 1861). Nevertheless sheepwalks persisted into the twentieth century. Essentially sheep in this system were manure distributors and arable farming on the chalk depended entirely on the sheepfold. It was this system that maintained the downland turf, described by Cobbett in the early nineteenth century as like 'a beautiful grey silk carpet' and by Defoe a century earlier as 'the grass and herbage of these downs is full of the sweetest and most aromatick plants, such as nourish the sheep to a strange degree... fine carpet ground, soft as velvet, and the herbage sweet as garden herbs' (Figure 9.2).

Figure 9.2. Sheep watering at a dew pond on the Wiltshire downs. Note the tightly grazed down with scattered bushes beyond: 'fine carpet ground'. (Courtesy of Rural History Centre, University of Reading.)

These grasslands had a distinctive avifauna, particularly comprising Montagu's Harrier, Grey Partridge, Quail, Great Bustard, Stone Curlew, Lapwing, larks and Wheatear. These are species typical of dry grasslands, often rather poor in nutrients. The colour described by Cobbett suggests that this was characteristic of these downland grasslands and the system involved a long-term transfer of nutrients from down to arable through the sheep. The sheepwalks of the downs themselves (the fine carpet grounds) were probably relatively unimportant as actual nesting sites for these characteristic birds, as many ground-nesting birds need some cover for concealment. Hudson (1900) remarked that numbers of nesting birds on the open downland were scant, birds using areas of scrub and torgrass which also occurred (Figure 9.2) and would have been favoured by harriers for example. The downland grass was also part of an integrated system, which included crops such as sainfoin, a favoured site for ground-nesting birds, as well as cereals and latterly roots; we should treat the system as the habitat, rather than consider the down in isolation. Thus Great Bustards used the down for feeding and actually nested on the adjacent arable, which led to their downfall as methods of growing cereals changed (Chapter 4). Smith (1887) also recorded nests in clover, a favoured site, too, for Quail (and Corncrake). Stone Curlew nested on down and arable. Walpole-Bond (1938) gave the main nest habitats as: old areas of plough on the tops long abandoned but never restored as downland turf (a habitat Cobbett (in 1822) and Hudson (1900) also described), downland with scattered scrub and molehills, and arable fallows. Most nineteenth-century authors recorded the same range of habitats, as did Gilbert White. Lapwings preferred the downland fallows (Smith 1887). Tubbs (1993) noted the presence of warren names on the Hampshire chalk, indicating another historic habitat element, places where rabbits were kept for commercial enterprise and likely to be particularly favoured by Wheatears, which nest in their burrows. Stone Curlew, Skylarks and Wheatears were probably the main species nesting on the downland proper.

Lowland heath and infield/outfield

Some heathland districts were also farmed by sheep/arable systems. Armstrong (1973) described in detail the management of the Suffolk Sandlings, where the heath replaced the down in another system centred on sheep flock and fold. The sheep grazed the heath in early summer, and in winter when heather shoots protruding above the snow provided valuable feed. They were always folded on the arable at night and leases laid down the minimum numbers to be used. Thus this system also involved long-term transfer of nutrients from grazing to arable through sheep, which maintained the low nutrient status of the heath and helped to keep it open. In the Brecks of Norfolk and Suffolk and the heathlands of some of the Midland forest districts, particularly Sherwood Forest, a rather different system, of infield/outfield, was pursued before enclosure. The infield was regularly cultivated land near the village, the outfield, or breck, was taken from the heath, cultivated for a few years and then left to revert and grazed by sheep to restore fertility until the cycle came round again, whilst another piece was taken in; parts of the Yorkshire Wolds were similarly farmed (Chambers 1955, Cunningham 1912, Dymond 1990). By the 1850s, however, the Brecks were being farmed on a similar system to the Sandlings (Read 1858). Dry sheep/arable systems were also practised on the Cotswolds.

All these dry sheep/arable systems of the chalk, limestone and heaths produced habitats that must have resembled in many ways the pseudo-steppes of Spain today. Like them, they were areas of rather low fertility and attracted many of the same birds, which included an important proportion of scarce and specialised species. The list of species discussed for the Downs was common to them all. They also had a long history of stable management, extending over some 300 years unless disrupted by enclosure.

Such systems were not practised on all lowland heathland however. In the New Forest cattle and ponies (and deer) were the principal stock and sheep were never important (Tubbs 1986). Cattle were also more important in most other lowland heathland districts, many of which also had their own local breed of ponies, known as heathcroppers (Webb 1986). Cobbett left a detailed description of the management of Ashdown Forest in Sussex in the 1820s, noting that the surrounding pastoral farms sent their young cattle to graze there for two years, before sending them to the Kent and Sussex marshes for fattening. This was a widespread system, as it was in the uplands, and was based on the principles that the sparse grazing of heath and moor promoted health and hardiness in young stock and that animals then taken to graze rich lowland pastures fatten quickly and cheaply.

Such lowland heaths lacked many of the steppe species typical of the sheep/arable. Tubbs (1986) remarked that the Dartford Warbler was the only bird species confined to the habitat although, in the past, Black Grouse were similarly restricted in lowland England and heaths were important to Montagu's Harrier. Otherwise most of the birds typical of heathland are and were more numerous in other habitats. Their comparative scarcity on heathland reflects the infertility of the habitat. Tubbs (1986) listed five species for which heathland is particularly important, Hobby, Woodlark, Stonechat, Nightjar and Dartford Warbler. The first three were undoubtedly widespread in other habitats in the eighteenth and nineteenth centuries. The valley bogs and mires of many southern heaths now support waders such as Lapwing, Snipe, Curlew and Redshank and quite significant populations are present in the New Forest. But the latter two species were absent before the late nineteenth century.

Water meadows

Although primarily associated with the streams and rivers of the southern chalk, water meadows were constructed over much of Britain, being found as far north as central Scotland and in many Midland counties of England. I have found no reference to them in the West Country or Wales, however, and in the northwest they seemed unusual, although recorded in Cumberland and Ayrshire (Dickinson 1852, MacNeilage 1906). In high-rainfall areas their construction was probably superfluous and Evershed (1869) noted that a system on the Staffordshire Dove was abandoned as

Figure 9.3. Water meadow in early spring. Note the carriers showing as dark lateral lines across the main block. Davis (1794) described the system as 'a hot-bed for grass' and the terrier gives some indication of the lush growth. (Courtesy of Rural History Centre, University of Reading.)

unprofitable. Water meadows needed a free-draining subsoil and a fall of at least 8–9 feet in the mile (1.6 m per kilometre) (Wilkinson 1861). Some were constructed on peat soils but were difficult to manage there; they did not work on impervious clays. They were first developed in the early seventeenth century, perhaps in Herefordshire or Dorset, and by the middle of that century were well established on the Wessex chalk, where they reached their most intensive development as an important part of the sheep/arable system. Davis (1794) recorded 15–20000 acres in Wiltshire alone (6–8000 ha). Their use started to decline from about 1840 (Sheail 1971). Although the construction of water meadows was complex and expensive, their principle was simple. Water from the stream was led into a network of channels and then flooded shallowly across the whole surface of the meadow, both fertilising it by the sediment the water carried and maintaining soil temperature, thus promoting early growth; water meadows rarely froze (Tubbs 1993). The water was kept flowing, an essential difference from flood meadow. The chief value of the meadows was in providing abundant nutritious feed in early spring when grazing and fodder were otherwise scarce, enabling the downland farms to keep more sheep to the benefit of the arable (Figure 9.3). The high prices realised for the early lambs produced financed the system. The annual management cycle was similar throughout the chalk regions. Irrigation started in the autumn and, after an initial soaking, meadows were watered at intervals until about mid March, when the meadows were thoroughly dried and the ewes and lambs let in to graze by day, being folded on the arable at night. Grazing was intensive, Sheail (1971) giving figures of 400 ewes and lambs per acre (= 150 livestock units per hectare) per day, and lasted for about six weeks. The meadows were then watered again for the hay crop, which was ready in about six weeks; most accounts record mowing starting in June. Further irrigation for a second crop might follow, particularly in drought years. Otherwise the meadows were grazed hard by cattle after haying until the cycle started again in October/November (Bettey 1977, Bowie 1987, Kerridge 1954, Sheail 1971, Wilkinson 1861). Information about management elsewhere is less detailed but seems not to greatly differ.

Clearly this was an intensive system and any idea that water meadows were important breeding habitat for waders and other ground-nesting birds must be discarded. Grazing was very intensive in spring in a crucial period for nesting birds. Nor was there adequate time between grazing and haying for successful breeding, even if there had been suitable sites in a period of regular irrigation and rapid growth. On the other hand they would have been valuable feeding and wintering sites, perhaps particularly for the common wildfowl, grassland plovers, Snipe, thrushes and similar species. Even in severe winter weather food would always have been available for these birds, an important factor. Mayled (1998) gives an indication of what such meadows could support. In the first winter of flooding a restored meadow system at Sherborne, Gloucestershire, significant numbers of Lapwing, Snipe and Green Sandpipers appeared and Barn Owls started hunting the area. If gamekeepers left them in peace raptors would have found these meadows excellent sites for Sheail (1971) noted that they were infested with rodents. He also observed that fish and invertebrates were frequently stranded by changes in water level during irrigation. Herons and other predators readily exploited these food supplies.

Flood meadows

Early recognition that sediment-laden flood water fertilised grass led to farmers devising water meadows. But not every valley was suitable. On the Hampshire Avon, for example, water meadows were made above Fordingbridge but there was insufficient fall downstream, where the meadows became flood meadows (Wilkinson 1861). Even in flood meadows, however,

Figure 9.4. Flood meadow on the Avon near Salisbury, cattle going to graze. This photograph and Figure 9.2 illustrate the extreme ends of the sward spectrum produced by these grazing animals – well structured swards with cattle, 'billiard tables' with sheep. (Courtesy of Rural History Centre, University of Reading.)

methods of controlling flooding were developed. In Staffordshire, for example, Evershed (1869) described works on Lord Lichfield's estate, where the River Trent was embanked and sluices constructed to regulate flooding as required. This was by no means universal and unmanaged flood meadows existed in many river valleys throughout Britain but perhaps particularly in lowland England. Their use was largely for hay and high yields were obtained in good seasons but depended largely on flooding not being of too long duration or unseasonable. It was not unusual to take two crops annually in some areas (e.g. Read 1855a). Stock grazed were cattle rather than sheep and stocking rates were low (Figure 9.4). Wilkinson (1861) noted only seven cows per 100 acres (40 ha) in the Test Valley for example. The meadows might be grazed in spring, until early March, and hay was cut in June.

Many areas of flood meadow remained in very poor agricultural condition in the mid nineteenth century. For example in Berkshire Spearing (1860) noted that the meadows in the Thames and Kennet Valleys were neglected by farmers, who concentrated on their roots and leys for their stock. Read (1855a) noted that the Thame Valley was in a particularly neglected state, as was the upper Thames where, in addition to the widespread problems of mills and locks, numerous privately owned fishing weirs impeded the river's flow. On the Severn in Shropshire Tanner (1858) noted that proprietors simply could not agree over measures to control flooding. In Huntingdonshire Murray (1868) recorded that a large extent of poor undrained grassland still remained. Such situations were widespread (Chapter 6) and plenty of wet grassland habitat clearly remained throughout the nineteenth century.

The early avifaunas show that the river valleys were important sites for breeding birds, particularly Lapwing and Snipe and, in hay crops, Quail, Corncrake and Whinchat; the waders were also common winter visitors. As breeding species, however, Curlew and Redshank had not yet colonised many inland areas and ducks such as Teal, Garganey and Shoveler were uncommon to rare generally in Britain. These sites were probably important for Barn Owls and Kestrels and, as feeding sites, also for finches and buntings (see Chapter 10). However they apparently did not hold the large concentrations of wintering wildfowl we see in the remaining sites today. Perhaps this partly reflected the much greater extent of habitat, populations were more diffused. But wildfowl populations were also heavily exploited, in spring as well as winter. Species such as Teal, Pintail and Shoveler were widely recorded as declining and the last two as uncommon or scarce (Chapter 12). Increasing wildfowl numbers against a background of declining wetland habitat has been a feature of the twentieth century (Chapter 6).

Pasture and meadow

In the early nineteenth century many grassland farmers faced increasing problems with declining fertility in their pastures and meadows, especially on clay soils. Sturgess (1966) discussed these problems in detail and particularly noted the exhausted and phosphate-deficient state of the cow pastures in Cheshire, Nottinghamshire, Derbyshire and Gloucestershire as examples; pastures in Shropshire were in a similar state (Tanner 1858). As the yields of pastures declined, fewer stock could be carried. This cycle was broken in the middle years of the century as the clays were under-drained, allowing roots and green crops for fodder to be increasingly grown, by the increasing availability and use of purchased feeds, such as brewers' grains, and by the increasing use of purchased fertilisers. Bone dust was widely advocated and used. It corrected phosphate deficiency and had a remarkable impact on productivity. Lime was important on the acid soils of the north and west. These factors reduced the demand for and increased the yield of hay, allowing more stock to be kept and a much greater return of manure to the grassland. Most nineteenth-century authors agreed that no crop benefited more from manure applications than permanent pasture. Both Sturgess (1966) and Thompson (1968) described this widespread process as the second Agricultural Revolution. One factor of major importance in the agricultural improvement of pastures at this period was that reseeding was unusual; many estates forbade it. The nineteenth-century approach was to make increased use of organic manures and improve management of grazing and mowing.

These management changes increased significantly the fertility of grass fields throughout the nineteenth century. This was widely reported and is illustrated by hay yields, which had risen by an average of 37% by the 1880s (June Census Statistics). Averages conceal much variation and my figure is based on observations of reporters such as William Marshall that a ton of hay per acre was a good crop at the beginning of the century; bog-hay meadows in Scotland yielded rather less (Smout 2000). Crops of twice this were not unusual by mid century and experimental work at Rothamsted showed by how much manuring raised hay yields (Broad 1980). This happened in farmland generally.

Increased fertility favours increased numbers (Chapter 2), so it is highly unlikely that this process had no effect on birds. Modern studies indicate that such a change would have affected food supplies for species dependent on soil invertebrates. Thus Tucker (1992) found that, for most of the birds he studied, feeding activity was consistently higher in long-established pasture where farmyard manure was applied frequently than in other field habitats. The attraction of manure was that it increased the activity of invertebrates near the surface, making them more available to birds, not that it increased invertebrate populations, which depended on the age of pastures. In the nineteenth century, however, these populations almost certainly did increase in pastures that had become nutrient deficient and exhausted, a situation similar to cultivated land in Tucker's study, where applications of manure did increase invertebrate numbers. Overall the main impact of this important and extensive change may have been to underpin the broad underlying pattern of stability in farmland bird populations in a period of profound habitat change, by providing new resources to help birds adapt.

The Starling may have provided an exception. This species became extremely rare or extinct in mainland Scotland and northern England in the late eighteenth century (Alexander & Lack 1944, Baxter & Rintoul 1953).

That pattern reversed from about 1830 and the species spread back into northern England, the whole of mainland Scotland and also west into Wales and the West Country over the next 70 years, becoming common or abundant everywhere (Alexander & Lack 1944, Harvie-Brown 1895). These changes may have been influenced by climatic change, for they followed quite closely periods of deterioration and amelioration in climate (e.g. Burton 1995). But they were also associated with the marked decline in the fertility and productivity of pastures and meadows by the late eighteenth century, followed by the equally marked recovery from about 1840, and the effect of these factors on food supplies. Feare (1984) showed that Starlings are particularly grassland feeders for which invertebrate foods are essential. He also noted that milder winters reduce migration demand in Starlings, so increasing the chance of second broods and allowing more yearlings to breed. Pasture improvement may then have improved breeding performance further by increasing food supplies generally. Whilst these ideas have now to remain speculative, this outline provides a coherent mechanism for the considerable population expansion that occurred.

Haying

When and how often hay is cut is of crucial significance to ground-nesting birds. The timing of haying depends on the growth stage of the plant. In the nineteenth century grass was cut at blossoming and clover and sainfoin just as the plant started to flower. In mixed grass/clover leys cutting was timed for the clover. When this stage was reached depended to some extent on the season and prior management but the nineteenth-century literature indicates that June was the most usual month to start mowing. For example Cobbett riding from Edgware, Middlesex, through Stanmore and Watford to St. Albans on 19 June 1822, noted that the hay crops were in ricks near London, nearly so at Watford and Stanmore and about one-third uncut towards St. Albans. The work was done by migrant gangs following in succession from tract to tract. Such migrant gangs were still working in at least parts of England into the twentieth century (e.g. Rowley 1853, Uttley 1931). Cutting started later in the hills, usually around early July (e.g. Evershed 1869, MacNeilage 1906). Smith & Jones (1991) showed that this timing still held in the Yorkshire Dales in the 1940s and 1950s. How long the haying season lasted depended on the weather and it could be protracted in wet seasons; in both the nineteenth and twentieth centuries mechanisation significantly speeded the process. Overall it seems clear that June/July constituted the haying season, with a bias towards June in the south and July in the north but in the uplands the season might extend into August, because grass was slower to mature (e.g. Dickinson 1852). A further complication in Scotland was the fairly widespread practice of feeding ley crops, such as clover, green, cutting and carrying it to the stock daily, a practice known today as zero-grazing. Dickson (1869) indicated a start date of 20 May for this system, analogous with silage today. Norris (1947), discussing the impact of haying upon nesting Corncrakes, opined that the haying season had shifted forward about 10–14 days earlier than at the end of the nineteenth century. The timing he shows, however, indicates little change in the start of the season. Probably, therefore, what his correspondents reported was the consequence of faster work with mechanisation, affecting timing in individual fields, perhaps those traditionally used by Corncrakes. In the modern era the grass harvest starts much earlier for silage crops and earlier harvesting for hay is also promoted by high nitrogen fertiliser applications.

The terms 'pasture' and 'meadow' were used to differentiate between grass that was only grazed and grass that was also mown and, until the seventeenth century, these categories probably always remained separate and distinct. But, outside water meadows and, to a lesser extent flood meadows, fertilised by irrigation, traditional meadow management was another early system which involved long-term shifts of nutrients from grass to arable. All the manure produced from feeding hay went to the arable. So always taking a hay crop from the same site annually depleted fertility, which was exacerbated by the increasing practice of taking two crops annually as yields declined. As the eighteenth century progressed, therefore, the practice grew of rotating pasture and meadow round the grass (Broad 1980) and by the late eighteenth century good practice was considered to be taking a hay crop every third year (Cadle 1867). Rotating crops in this way became regular in the nineteenth century, when estates also restricted double cropping unless meadows were properly manured. In the pastoral uplands traditional meadow tended to be retained because manure was available for it, a pattern still visible in my part of Breconshire today. In ley farming two-year leys tended to be cut in the first year and grazed the next and sainfoin, usually laid down for three years, was managed similarly. Taking two crops of hay annually off good meadows was not rare and I suspect that water meadows were double-cropped rather more often than some authorities allow; Trow-Smith (1951) noted that the mowers were busy from May to December on Bakewell's

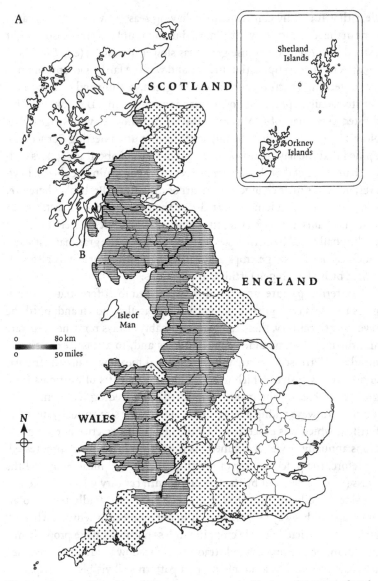

Figure 9.5. Grassland (rough grazings, permanent grass and ley) as a percentage of total farmland in (A) the early 1870s and (B) 1930s. Areas northwest of the line A–B are excluded, over 70% of this area always remained rough grazings. (Data from June Census Statistics.)

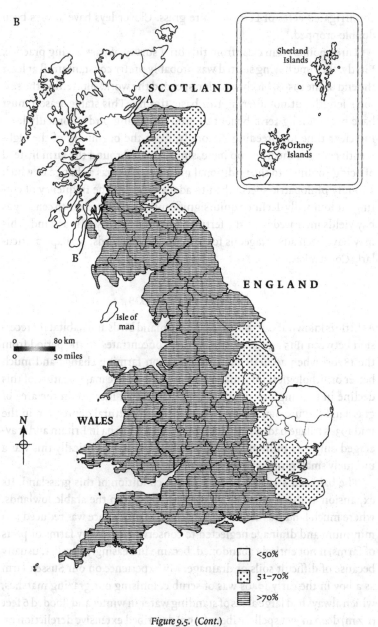

Figure 9.5. (Cont.)

200 irrigated acres of Leicestershire grass. Clover leys have always been double-cropped.

Three points seem clear from this brief survey of past haying practice. Firstly, the time haying started was probably pretty constant until at least the end of the 1950s. Mechanisation speeded the work, shortening the season's length but not altering this basic timing. This stable season must have been of long-term benefit to ground-nesting birds, normally leaving clear time for breeding. Secondly, from the beginning of the eighteenth century, the hay crop increasingly rotated round the farm instead of being confined to the traditional meadow. This was a change to which strongly philopatric species had to adapt. Thirdly, the frequency of cutting undoubtedly declined quite significantly in the nineteenth century as hay yields increased with the fertility and productivity of grassland. This may have been advantageous for ground-nesting birds, perhaps particularly Corncrakes.

Recession: 1875–1939

As little is known about the impact on farmland birds and habitats of recession between 1815 and 1840, this section concentrates on the period from the 1870s, when accurate statistical data on farming change and much better ornithological information was available. The major feature of this decline in farming prosperity was a phenomenal increase in the area of grassland (Figure 9.5). At the nadir of the agricultural depression in the mid 1930s tillage had sunk to just over 3.24 million ha in Britain and it averaged only 3.44 million throughout that decade. Historically this was a uniquely small proportion of agricultural land.

The bald statistics tell us little of the condition of this grassland. Its expansion at the expense of tillage was greatest in the arable lowlands, where much land was abandoned entirely. Maintenance was reduced to a minimum and drainage neglected to conserve cash. Many farms or parts of farms, if not entirely abandoned, became increasingly derelict, usually because of difficult soils or drainage. My experience on our Sussex farm as a boy in the early 1940s was of scrub colonising our grazing marshes, which always had large areas of standing water in winter and flooded 6 feet (1.7 m) deep in wet spells. Albery (1999) described extensive dereliction on the Hampshire chalk in the 1820s and between 1900 and 1939 and Sheail (1976) outlined the impact of the collapse of drainage systems on arable land in East Anglia and Lincolnshire. Overall it was in the arable regions of East Anglia, southeast England and the East Midlands that such factors were most important. Farming in the north and west was much less affected. Evershed (1869), for example, noted that the rising profitability of dairying in Staffordshire and other northwest counties was encouraging a return to grass and stock and rising stocking rates by the 1860s.

The expansion of grassland in this period did not simply restore farmland habitats to their pre-high farming state. Rather it produced further new habitat conditions, considered in some examples below, which were not necessarily advantageous to some of the more specialised species which enclosure and high farming had displaced. There are several elements to be considered besides the simple increase in grassland area. These included worsening drainage in many areas and thus a particular increase in damp grassland, changes in the management of hay crops and abandonment. Above all a major factor was a significant change in the patterns of stocking. During the high farming period sheep were as much animals of arable lowlands as upland farms but, in the twentieth century they have been extensively displaced by cattle. Figure 9.6 illustrates the change up to the 1930s but it has continued since. One result of this change was that stocking densities, as opposed to stocking rates, dropped sharply (Figure 9.7). Almost certainly this was beneficial for waders because fewer animals reduces the risk of nest losses to trampling and cattle produce a more tussocky sward (Figure 9.4), creating better nest sites for species such as Curlew, Redshank and Snipe.

The Corncrake provides a good example of the contrary effects of changes in management. It declined continuously in England and Wales from the 1880s (Figure 9.8), a decline which continues. Corncrakes were particularly noted as birds of hay crops in the nineteenth- and early twentieth-century avifaunas but arable crops, particularly oats in the crofting areas of west Scotland, were also frequently used. Virtually every avifauna discussing the subject linked the species' decline to the switch to mechanical mowing from the second half of the nineteenth century, pointing to the high level of nest destruction caused by machines. Here I argue in Chapter 11 that it was the change in the organisation of the hay harvest with mechanisation that was important, rather than machines *per se*, and that this affected other species. This was not the whole story. Much of the increase in grassland in Britain was grazing land and the area of hay fields stagnated at 6.5–7 million acres (2.67–2.83 million ha) throughout the period from 1880 to 1937. Grassland expansion therefore did not increase the most important Corncrake habitat, which was also

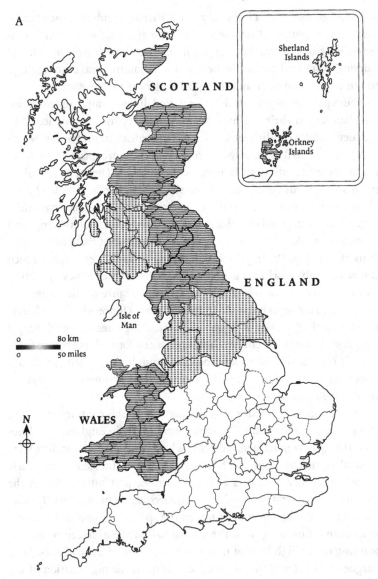

Figure 9.6. Changes in the distribution of (A) sheep and (B) cattle between the early 1870s and early 1930s. No change is taken as ±10%. (Data from June Census Statistics.)

Recession: 1875–1939

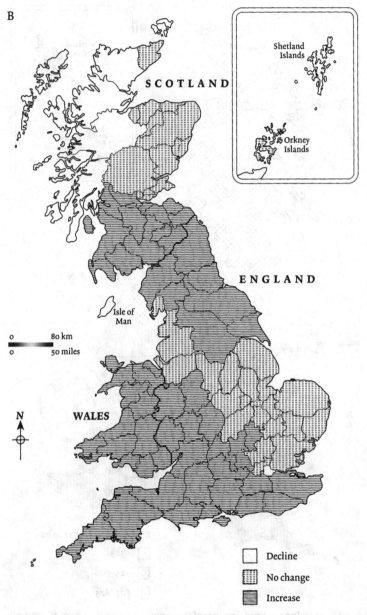

Figure 9.6. (*Cont.*)

Figure 9.7. Stocking rates and densities of sheep plus cattle per 1000 ha of all grass plus fodder crops. Early 1870s: (A) stocking rate, (B) density; 1930s: (C) (see over) stocking rate, (D) (see over) density. See text and Figure 9.1 for methods and definitions. Note that the figure of stocking densities for the 1870s overstates the utilisation of grassland because of the extent to which sheep in arable rotations were kept on roots. (Data from June Census Statistics.)

Recession: 1875–1939

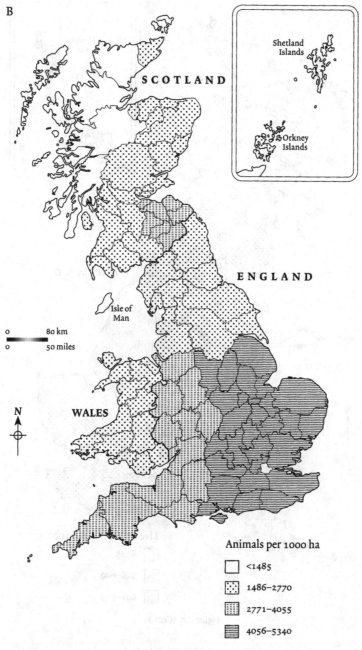

Figure 9.7. (Cont.)

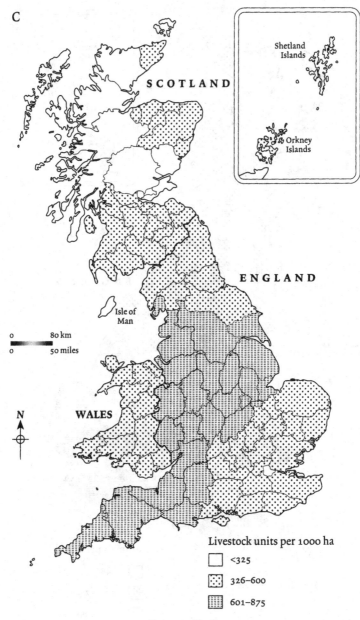

Figure 9.7. (Cont.)

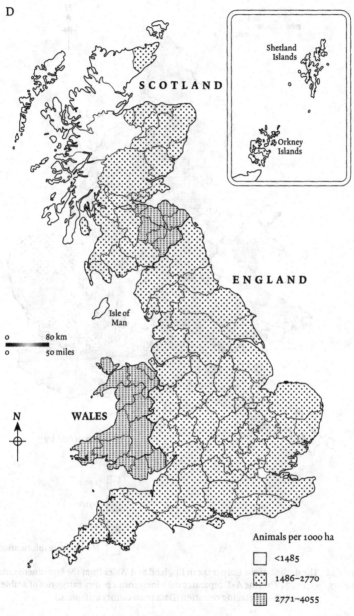

Figure 9.7. (Cont.)

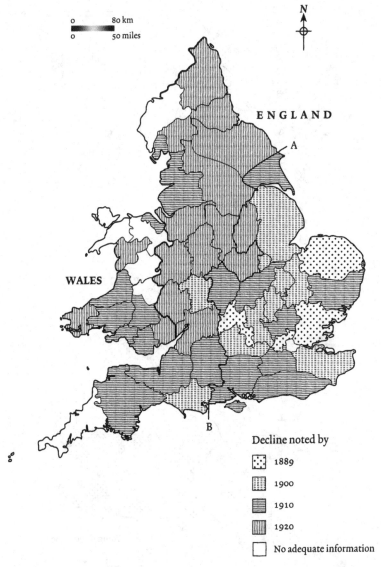

Figure 9.8. The decline of the Corncrake in England and Wales from the late nineteenth century, by county. The line A–B separates the nineteenth-century categories of arable (to east) and grazing counties. (Data from county avifaunas.)

restricted by the decline of arable rotations (Walpole-Bond 1938). Nor was the abandonment of grassland advantageous to them. Nicholson (1926) argued that the status and distribution of the Corncrake in the 1920s was similar to that of the late eighteenth century and that the early twentieth century decline therefore followed a marked expansion of range into southern and eastern England in the nineteenth century. The historical records for counties such as Sussex and Hampshire support this argument strongly. The frequency with which nineteenth-century authors linked this bird with clover crops suggests that its increase might have been encouraged by the spread of high farming rotations including them. Clover leys were not usually grazed in their first summer but taken for hay, thus providing early cover on many farms where it may otherwise have been quite scarce. The early decline of the Corncrake was largely confined to the arable districts of England, those most affected by recession, and the decline of arable rotations with clover leys was probably important (Figure 9.8). In Scotland there was a marked and fairly general decline of Corncrakes in the first 30–40 years of the twentieth century (Baxter & Rintoul 1953). Comparing earlier accounts, such as Gray (1871) or St. John (1848), with accounts for the end of the nineteenth century, however, indicates that Corncrakes were already declining, even in northwest Scotland, by the 1880s. No marked farming change seemed to account for this, although a decline of oat-growing in crofting areas may have contributed.

The expansion of grassland saw only a limited recovery in Stone Curlew populations. Their problem in the recession was often abandonment, rarely satisfactory for ground-nesting birds which have adapted to managed grassland habitats. Such birds need some degree of grazing to keep grasslands open, to create feeding areas, to detect predators easily and to allow chicks easy movement. For Stone Curlew grazing particularly by sheep was an important element on their old heathland and downland habitats. In England, however, sheep declined by 5.95 million head between the 1870s and 1930s, with 85% of that loss occurring within the old range of the Stone Curlew (Figure 9.6). As a result many of their breeding sites became overgrown from lack of grazing and were deserted, e.g. in Buckinghamshire (Lack & Ferguson 1993). The decline of sheep also meant that large areas of downland in Sussex and Hampshire and of heathland in East Anglia, no longer farmed, were afforested, resulting in total habitat loss for Stone Curlews. However in East Anglia the recession also led to some increase of heathland in parts of the Suffolk Sandlings, Brecks

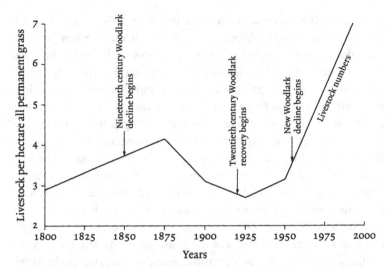

Figure 9.9. Changes in the breeding status of the Woodlark in England and Wales in relation to numbers of cattle and sheep: all permanent grass and rough grazings. When stock numbers rose above 3.5 animals/ha, Woodlark numbers fell, remaining low until the 1920s, when stock numbers declined below 3 animals/ha. With stock numbers again rising above 3.5 animals/ha in the 1950s, Woodlark again declined and, as stock numbers continued to rise, their population collapsed. The late twentieth century recovery has been outside farmland.

and central and north Norfolk heaths and some recovery in Stone Curlew numbers there (Riviere 1930, Ticehurst 1932).

Wheatears were similarly affected but the Woodlark probably benefited from declining stock densities. Yarrell (1837–43) defined its habitat as wooded parks, hedged meadows with copses, dry sheepwalks and the borders of heaths; other authors also noted grassy hillsides with scattered bushes as important. The species declined sharply in the nineteenth century but then recovered, expanding both range and population markedly from the 1920s, before a population collapse from the early 1960s (Figure 9.9). Although the species has recovered sharply in the 1980s and 1990s, this has been almost entirely in heathland and heath/forest habitats or in 'new' habitats, such as old mineworkings, not farmland or downland (Wotton & Gillings 2000). Thus, as Holloway (1996) observed, some change in habitat preferences seems to have occurred. Although habitat loss was important in the nineteenth-century decline, historic variations in habitat area do not match the observed population changes for the species in the nineteenth and twentieth centuries particularly closely, nor do climatic variables such as the 1962–3 winter. Woodlarks shrugged off the effects of previous severe winters, notably in 1916–17, 1929, the early 1940s and 1947. Burton (1995) suggested that a shift to a more maritime climate, with damp cool summers, with the climatic amelioration after 1850 may have been more significant. But this does not fit with the strong southwesterly element in distribution mapped by either *Breeding Atlas* (Gibbons et al. 1993, Sharrock 1976). What does match the historic variations in Woodlark numbers surprisingly well is change in the utilisation of grassland. Figure 9.9 sketches the timing of changes in Woodlark numbers against the ratio of total cattle and sheep to the area of permanent grass and rough grazings. Strictly these are not stocking densities, as other crops also supported stock, but the pattern illustrates the utilisation of grassland clearly. In the twentieth century the Woodlark population expanded whilst this ratio remained below the level obtaining in the early nineteenth century, but once it rose above that level it declined. An important result of reduced grazing pressure was that it allowed the spread of shrubs, such as hawthorn, effectively increasing the kind of grassland habitats Woodlarks once preferred (see Figure 4.12). In heath and downland habitats a contrary pattern operated as traditional grazing practices were abandoned, and then rabbits were lost after 1954, allowing habitats to become too overgrown. Rabbits have now returned, conservation grazing has been introduced and deer have increased, all factors that have restored heathland habitats once again. But Woodlarks are no longer farmland birds. In agricultural habitats only a rump ($c.$ 5%) of the population persists on 'low-intensity farmland' in the southwest and on set-aside (Wotton & Gillings 2000).

Abandoning grassland also involved the loss of other traditional forms of management. In Norfolk, for example, Riviere (1930) noted a decrease of both Lapwing and Yellow Wagtail in the Broads area. Both species bred on the dry marshes, formerly mown and cattle grazed annually. The coarse marsh hay went as fodder to London's bus and cab horses but, as motors replaced horses, the market died and the marshes were abandoned to rough grass, which swamped the nesting habitat. Such patterns were repeated elsewhere and Clarke (1996) noted that it also affected the few breeding pairs of Montagu's Harriers that had become established in the Broads area.

The expansion of damp grassland benefited breeding waders and Snipe and Redshank are discussed in Chapter 6. The Curlew also expanded its range rapidly in the first half of the twentieth century. In the early nineteenth century it was a common upland moorland species (MacGillivray 1837–52) but in the lowlands of England and Wales it was confined to a

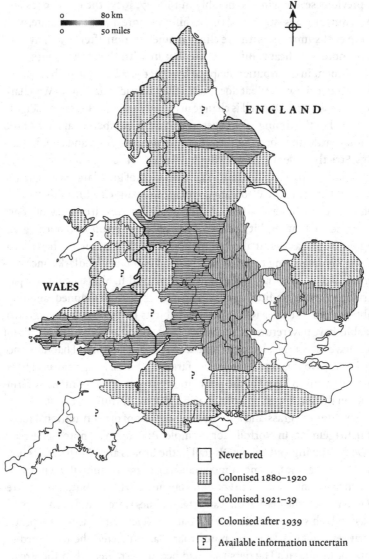

Figure 9.10. The expansion of breeding Curlew into lowland habitats in England and Wales from 1880 by county. (Data from county avifaunas, Parslow (1973).)

few extensive bogs or mosslands. A marked spread into enclosed farmland in England and Wales started around 1880 and continued broadly northwest–southeast, into the 1950s (Figure 9.10). The early spread was mainly into farmland bordering occupied hill areas but Curlew also occupied damp heathland in Dorset, Hampshire, Surrey, Sussex and Norfolk in the same period and have only rarely moved out of these habitats there. This remarkable spread coincided with the major increase of permanent grass from the 1870s, much of which in the southeast, as Orwin & Whetham (1964) noted, was under-farmed and inadequately drained, resulting in the type of damp rushy pastures that Curlew prefer. But this was not so for the northwest. Whilst the spread of Curlew apparently fits well with the pattern of changing stocking densities indicated by Figure 9.9, this is an unsatisfactory explanation for its expansion, for much apparently suitable grassland habitat actually existed in lowland England and Wales long before it was colonised by Curlews (Chapter 6). In southern counties too, the species remained a strictly heathland bird, despite increasing availability of suitable grassland habitat. Other factors were involved. Climatic considerations were possible, as Burton (1995) suggested, and this expansion may simply have been part of the general increase in Europe going on at this time.

The great plough-up

1939–45

The steady expansion of grassland farming was abruptly reversed in 1939 by the start of the wartime ploughing campaign. This saw permanent grass in England and Wales decline by 5.9 million acres (2.39 million ha, 37%) in five years and by a similar percentage, amounting to 0.54 million acres (0.22 million ha), of the limited area in Scotland. There were important regional differences in England and Wales too. Just as grass had expanded most in the arable regions of the south and east, so most was ploughed out there. The area recorded as rough grassland in England and Wales also remained stable, at $c.$ 2 million ha from the 1930s until at least 1960. The return to arable everywhere was for tillage and the area of leys stagnated or shrank during most of this short period, starting to expand in 1944. Numbers of stock also declined by about 17%, mainly owing to a further sharp reduction in sheep; cattle numbers rose by 4%.

Although it proved to be the start of a trend which has been of the greatest significance to farmland birds since, the immediate impact of this extraordinary reversal was not particularly marked. Chats continued to decline, although the immediate impact on Wheatears (and Stone Curlews) was probably lessened because important areas of downland and heathland were taken over for military training. Rooks certainly increased with the area of cereals, but most grassland waders seemed little affected. Baxter & Rintoul (1953) noted that the effect of ploughing a traditional breeding station was that Curlews simply nested in the resultant arable crops, an adaptation also recorded in Northumberland and Durham (Parslow 1973). But a likely reason for their stability was that rough grazing was little affected, providing an undisturbed core breeding habitat; stocking densities also remained low. Probably for the same reasons the impact on Redshank was also limited. They declined as breeding birds in Wiltshire, Middlesex, Hertfordshire, Essex, Suffolk and Oxfordshire but not elsewhere (Parslow 1973). Little is accurately known about Snipe populations but Lapwings, for which permanent pasture was probably as important as rough grazing, did decline (Nicholson 1951), almost certainly as a result of the sheer scale of change involved.

One other species is worth considering in detail. Red-backed Shrikes declined significantly in northern parts of their British range in the second half of the nineteenth century (Peakall 1962). They are primarily birds of grassland for foraging and distribution tends to avoid areas that are cool and rainy in summer (Lefranc & Worfolk 1997), which may explain the range contraction in the nineteenth century. Nevertheless in its core area in south and east Britain the county avifaunas, whilst recording much variation, suggest an overall stability during the late nineteenth and early twentieth centuries, which is supported by Peakall's (1962) map. The situation changed sharply in 1940, after which a progressive decline, mapped by Peakall up to 1960, continued until the bird virtually ceased to breed after 1990. This progressive disappearance coincides almost exactly with the progressive destruction of old grassland habitats in southeast Britain. This started with the great ploughing campaign of 1939–45, continued through the 1950s, encouraged by significant subsidies for the ploughing and reseeding of old grassland (Chapter 8) and has continued since with the steady intensification of grassland management (see next section). The significance of such habitats was driven home to me by looking at old grasslands in eastern Europe in 1997. There such habitats swarm with Orthoptera, Lepidoptera and shrikes. Whilst climatic considerations may limit the variety of such invertebrates in Britain, they were once abundant there in old grasslands too. They are unnaturally scarce in British grasslands today almost certainly because these are ryegrass deserts. The same factor has probably contributed to the decline of the Cirl Bunting, for which Evans's (1997) map shows similar patterns over time to those of Peakall (1962) for Red-backed Shrike.

Evans noted a dramatic increase in breeding success in Cirl Buntings once grasshoppers, typically insects of old grasslands, became available for the chicks.

After 1945

Apart from the impact of organochlorines it was in grassland that Parslow (1973) showed most agriculturally linked decline in birds after 1945. With the reversion of substantial areas of grassland to arable and the repair of drainage systems populations of Lapwing, Snipe, Curlew and Redshank all declined in lowland England. The decline of the Lapwing was probably overstated, as the increased availability of mixed spring cereals and leys must have provided satisfactory alternative habitat (Chapter 3). The loss of old pasture to drainage, improvement and ploughing was certainly affecting lowland populations of Redshank and Curlew. For example it caused breeding Curlews to desert Cambridgeshire, only recently colonised, in the early 1950s (Parslow 1973) and a major reduction in inland Redshank populations in Sussex by the mid 1960s (Shrubb 1968). However examining the maps and accounts in Sharrock (1976) shows that groups such as waders and wildfowl in fact remained widespread as breeding birds in Britain in the early 1970s and little evidence of serious decline was shown by the CBC or Waterways Birds Census before the middle of the decade (Chamberlain *et al.* 2000, Marchant *et al.* 1990). Whilst accepting that these surveys did not extend back further than 1962, this suggests strongly that breeding populations had stabilised around a rather lower level, as agriculture returned to a pattern nearer that of the Victorian era.

The maps and accounts in Sharrock (1976) in fact suggest very strongly that it was birds of dry grasslands that were most affected by agricultural change after 1945. In dry grassland habitats both Woodlarks and lowland populations of Wheatears declined sharply as breeding birds, to the plough (Wheatears), to increased stock densities (Woodlark; Figure 9.9) and to abandonment. The latter was particularly a problem on chalk downland where the decline of sheep (and rabbits after 1954) meant that many slopes too steep to plough went out of agricultural use altogether, reverting to rank grassland and then scrub woodland, a habitat change inimical to species such as Wheatears. Potts (1970b) also noted that this affected Grey Partridges because the shading effect of tall grasses and herbs progressively reduced the number and productivity of mounds of the ant *Lasius flavus*, an important food of the chicks of grassland pairs. O'Connor & Shrubb (1986) also suggested this as a cause of the decline of the Wryneck. Stone Curlews were more resilient. Ash (in Cohen 1963) noted that the Hampshire population was breeding successfully on cultivations, although at the cost of repeated layings, and recorded a density of one pair per 200 acres (81 ha). But he stressed that this population was highly vulnerable to changes in agricultural methods, particularly to changes in the timing of farm operations with increased mechanisation and the use of herbicides. Corncrakes continued to be affected by methods of haying and Whinchats vanished from many lowland areas for the same reason. Finally raptors such as Barn Owls were limited by the loss of rough grass habitats.

The modern grassland revolution

Since the mid 1960s there has been a major revolution in grassland management. It has been based on the principle that the cheapest and most efficient feed for livestock is grass. Its main features have been increased drainage, extensive reseeding, the decline of hay for forage in favour of silage, the discarding of tillage crops for forage and massive increases in inorganic fertiliser applications, particularly the use of nitrogen. These changes have supported increases in stocking rates and densities to unprecedented levels.

Drainage

This is discussed in detail in Chapter 6. But it is worth repeating here the difference in attitude between the nineteenth- and twentieth-century farmer. In the nineteenth century high water tables in grazing land were often preferred. There was a strong belief that too thorough draining of grass fields made grazing poor for fatting cattle, which was a more important enterprise than dairying; high water tables were deliberately maintained. Stock were supported in winter by arable fodder crops and the irrigation of meadows was very widely practised.

Drainage in the post-war period has reduced breeding habitat for species of wet grassland very significantly, especially for wildfowl and waders. But passerines (Meadow Pipit and Yellow Wagtail for example) are also affected and raptors such as Kestrel and Barn Owl have lost prime hunting habitat. Where drainage does not lead to total habitat loss by conversion to arable, it provides the basis of a battery of agricultural improvements in grass which may be equally inimical to breeding birds.

Reseeding

The extent to which old permanent grassland has been reseeded since 1945 differs profoundly from nineteenth-century practice. It is difficult to estimate annual changes through the agricultural statistics, as the reseeding of short-term leys and permanent grass have not usually been separated. But by 1985 the Surveys of Fertiliser Practice showed that only 13% of grass fields in improved farmland had not been reseeded in the post-war period. Fuller (1987) estimated that the area of unimproved lowland grassland in England and Wales had declined by 92% since the early 1930s and that old pastures, excluding rough grazings, in the lowlands extended to only 200 000 ha, 4% of the existing grass area; much of that was partially improved. I have little doubt that the area of such habitat has continued to decline. Fuller made the point that, by the 1970s, ploughing and reseeding had declined but intensive management and high fertiliser applications were achieving the same effect, for they particularly reduce plant diversity in favour of ryegrass-dominant swards. The loss of old grassland is one the most significant changes in farming. It is epitomised by the June Census Statistics, which now recognise only grass reseeded for five years or more, or less.

One of the most striking features of this change has been the loss of conspicuous invertebrates, obvious to anyone who walks modern grass fields. Wilson et al. (1999) noted that, in intensively managed modern grassland, the loss of grasshoppers, ants, spiders and Lepidoptera larvae removes important food sources for a wide range of bird species. Cowley et al. (2000) discussed the decline of many common butterflies to the agricultural improvement of grassland in north Wales. Species such as Small Copper *Lycaena phlaeas* and Common Blue *Polyommatus icarus*, which have no particular conservation priority (Asher et al. 2001), have still lost 91% and 75% of their distributions, as measured by flight ranges, respectively. Williams (1982) noted large reductions in the distribution and range of bumblebees since 1960, particularly in central England. In southwest England Lock (1998) linked this to large-scale loss of unimproved grassland and heathland. Wilson et al. (1999) noted that intensive cutting and grazing and the establishment of improved grass monocultures were all detrimental to grasshoppers and their disappearance with the decline of old grasslands is striking to the eye and ear. Baines (1990) found that, in his northern England study area, agricultural improvement of grassland reduced the density and biomass of spiders and carabid beetles but increased that of other Coleoptera and earthworms. Whilst not affecting the numbers of tipulids, it changed species composition. Lock (1998) had little doubt that the loss of invertebrate biomass had contributed to the declines of many bird species on pastoral farmland in southwest England. It may particularly affect the chicks of nidifugous species such as waders for which large surface-dwelling invertebrates are important in the early stages of growth (Beintema *et al.* 1991). However Baines (1990) found little evidence of this in his study of Lapwings, chicks of which fed largely on beetles, many of which are not markedly affected by the agricultural improvement of grassland in his area.

Fuller (1987) remarked that George Stapledon, who did more than anyone to influence the drive for agricultural improvement in grass, was moved to describe much old pasture as 'weed-land'. This seems a sadly narrow-minded observation. Besides their floristic and entomological interest old pastures are also a valuable winter source of seeds for finches and buntings (Chapter 10), the loss of which is particularly serious in areas where other sources such as hay or arable land, are now lacking.

The loss of invertebrates or seed is not only the result of reseeding. Intensive grazing itself increasingly prevents all plants in grass fields from flowering and seeding. This is particularly striking in sheep-farming areas such as central Wales, where pastures are now grazed tightly throughout the year and even the bottoms of the hedges are grazed out. For ground-nesting birds the limited structural diversity of such fields makes them poor breeding habitats, leading to higher predation of nests and small chicks (e.g. Baines 1990 for Lapwing). In such habitat in Wales, Lapwings simply desert such fields permanently (M.F. Peers personal communication; personal observation), which is the main reason for the massive decline there by 1998 noted by Wilson *et al.* (2001).

Fertiliser use

Patterns of fertiliser use have changed sharply. In ecological terms the most important changes involve large increases in the use of inorganic nitrogen and a decline in the frequency of applications of organic manures. In the mid 1980s 36% of permanent grass was being dressed with farmyard manure, a marked decline compared with the minimum of 50% dressed annually in the nineteenth century, when farmers made extensive use of town waste, guano, river silt and other organic substances, as well as farmyard manure. In dairying today, an increasing volume of slurry is also absorbed by the fodder maize area. Seasons of application have also changed. Victorian farmers favoured dressing grass in autumn. Today it seems more

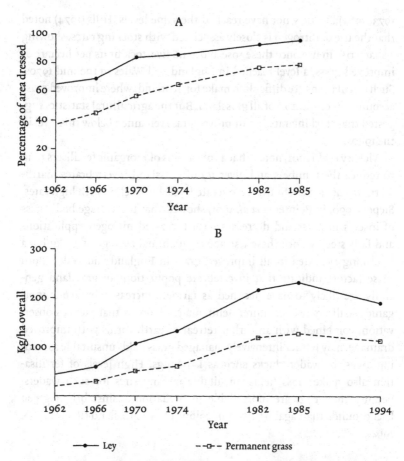

Figure 9.11. Use of nitrogen fertilisers on grassland in England and Wales at intervals from 1962. (A) Percentage area dressed, (B) volume used overall as kg/ha. (Data from Surveys of Fertiliser Practice.)

often to be done in late winter/spring; spring applications are particularly hazardous for ground-nesting birds, smothering nests.

Changes in the overall area and volume of nitrogen applications are shown in Figure 9.11. The highest levels of usage are in dairy farming, averaging 147 kg/ha, and in silage grass, averaging 200–255 kg/ha. Overall applications in livestock-rearing areas average 85 kg/ha and grass for hay receives 70–120 kg/ha. These figures are for 1985 but, as Figure 9.11 shows, the rapid increase from 1970 had levelled off by then. Patterns of use in improved grass in Scotland were similar until at least the early 1970s (Green

1973, 1974) but may not have reached the same levels. Hills (1974) noted that the use of nitrogen is closely associated with stocking rates, showing a sharp rise in use once these rose above 1.6 livestock units per hectare of improved grass, a level reached in England and Wales in the mid 1970s. Such calculations are difficult to make for Scotland, where improved grass accounts for only c. 20% of all grassland. But the agricultural statistics suggested that stocking rates on improved grass remained below this level in the 1990s.

Vickery et al. (2001) noted that applications of inorganic fertilisers tend to reduce the numbers and diversity of grassland invertebrates. Earthworm numbers benefit from moderate levels but decline with high rates. Siepel (1990, in Beintema et al. 1991) showed that the average body mass of insects in grassland decreases with increased nitrogen applications and falls steeply once these rise above an annual average of 50 kg/ha, a level long exceeded in all improved grass in England and Wales. Both these factors indicate that invertebrate populations in grassland generally are likely to have declined as farmers increasingly favour inorganic fertilisers over manure. Beintema et al. noted that Siepel's observation, combined with an earlier retreat of earthworms with improved drainage, may render intensively managed grass fields unsuitable as feeding areas for wader chicks such as Lapwings. High levels of fertilisation also make grass fields unsuitable nesting sites for such waders, because their eggs are more liable to predation against bright green backgrounds than against the dun colour of old grassland (Baines 1990, 1994).

Decline of arable

The loss of tillage in pastoral areas represents a particular loss of diversity. Tillage crops now comprise no more than 4–6% of farmland in the pastoral regions of western Britain, having declined by 36% since the early 1960s. In many pastoral districts the only tillage present today comprises reseeding leys. The loss of tillage particularly involves a loss of food resources for seed-eaters in both winter and spring, for much was spring tillage. Robinson et al. (2001) found that in grassland landscapes, increasing tillage is likely to favour granivorous birds. Their study supported the hypothesis that local extinctions of some granivorous birds in pastoral areas have resulted from the loss of tillage there, irrespective of any effects that intensifying arable management may have. It also involves the loss of safe nest sites for Lapwings (Shrubb 1990) and Skylarks.

The switch to silage

The practice of ensiling grass instead of making hay for winter fodder was first developed in the early 1870s, although problems with making silage consistently well were not properly resolved until the 1960s. Since then the area ensiled has increased from 10% of forage grass in 1962 to 66% in the mid 1980s and 75–80% in the mid 1990s, when over a quarter of the area of agriculturally improved grass was cut for silage. The development of a satisfactory method of wrapping baled silage in plastic from the late 1980s saw a rapid spread of silage in livestock-rearing areas (and an equally rapid increase of sordid plastic litter in the countryside). The change from hay to silage is virtually complete in dairying areas but hay and silage may still be interchangeable to some extent in livestock-rearing areas, depending largely on the weather; hay will always be needed for the growing horse population.

One should not be surprised at the scale of this change. Hay has always been difficult to make consistently well in our variable summer climate; silage poses far fewer problems. Nevertheless the change to silage has several important implications for birds, particularly the levels of nitrogen now used, which promote rapid and early growth, the increased frequency of cutting, mainly in dairying where there is a much greater demand for winter keep, and much earlier cutting. These factors combine to render silage fields as totally unsuitable for all ground-nesting birds. Rapid early growth makes the herbage too long and dense for adults and particularly chicks of many species to negotiate easily, reducing feeding efficiency; chicks are liable to hypothermia when grass wallows in wet spells. Lapwings reject such fields for nesting, preferring more open swards where they can readily detect the approach of predators. Although Curlews will use silage fields, early cutting, in dairying often from early May, destroys increasing numbers of nests or broods and repeated cutting simply catches any repeat breeding attempts. Wilson *et al.* (1997) found that Skylarks also lost many nests to silage cutting. This may partly result from the nature of the machinery (Chapter 11). Guest *et al.* (1992) noted in the Cheshire Plain that the pattern of multiple cutting for silage was so timed to leave little chance for Skylarks to nest successfully. Earlier cutting is not just a question of an earlier calendar date, silage is also cut at an earlier growth stage than hay, so the end product contains few seeds and wrapped silage bales least of all. This has been a significant loss of food resources for finches and buntings in winter in pastoral areas, where the seeds in hay have always been a major source of winter food.

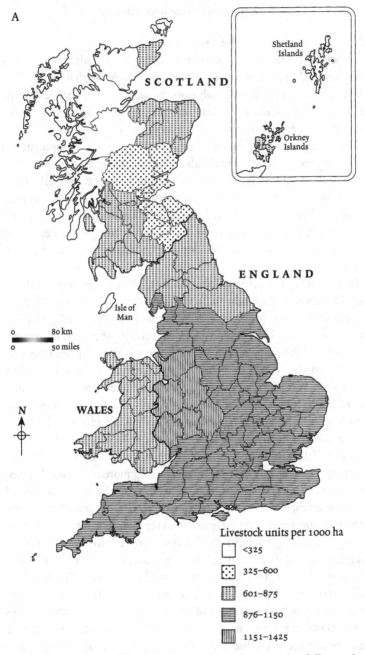

Figure 9.12. Stocking rates and densities of sheep plus cattle per 1000 ha of all grass plus fodder crops. Early 1960s: (A) stocking rate, (B) density; 1997: (C) (see over) stocking rate, (D) density. See text and Figure 9.1 for methods and definitions. Note that the figure of stocking densities for the 1960s overstates the utilisation of grassland because of the extent to which cattle in 3-year ley rotations were maintained by cereals. (Data from June Census Statistics.)

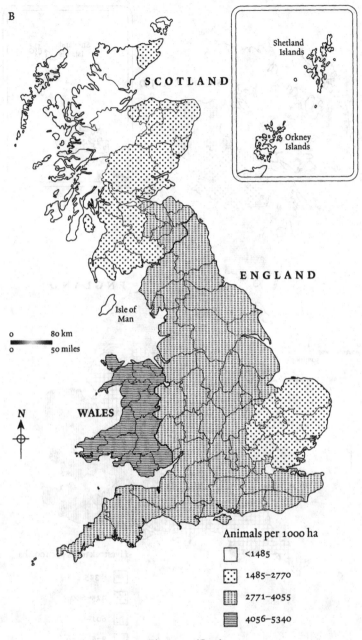

Figure 9.12. (Cont.)

Figure 9.12. (Cont.)

D

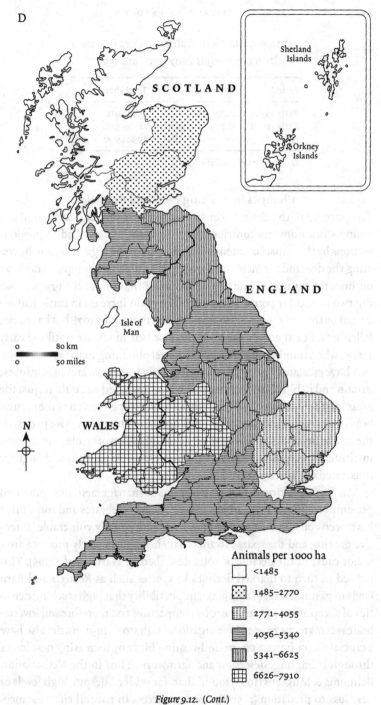

Figure 9.12. (Cont.)

TABLE 9.1. *Numbers of cattle and sheep in Britain in the early 1930s, early 1960s, and 1997*

Period	Total cattle	Total sheep
Early 1930s	7 235 358	25 262 012
Early 1960s	10 734 311 (+48%)	28 025 919 (+11%)
1997	9 901 500 (−7.8%)	39 942 900 (+43%)

Source: June Census Statistics.

Changes in stocking rates and densities

The purpose of the changes considered above is to raise stock numbers. Rising stock numbers contribute to the processes described in previous sections by the impact of intensive grazing upon the vegetation and by creating the demand for increased forage and high fertiliser inputs. Table 9.1 outlines the numbers of cattle and sheep in Britain since the 1930s, showing two particular points of interest. The main increase in cattle had occurred by the early 1960s; numbers rose a further 11% by 1978 but have since fallen 17%. For sheep by contrast, increase was moderate until the early 1960s, when numbers had reached the level obtaining in the 1870s. Numbers have escalated rapidly since, particularly from 1980 and in southwest Britain and in hill districts described as Less Favoured Areas there, just the areas least able to absorb such numbers. The result is erosion of the natural beauty and ecological interest of such uplands. This destructive pattern is the consequence of support by the crude mechanism of headage payments, instituted largely for bureaucratic convenience. The patterns of change are illustrated in Figure 9.12.

The high stocking densities and forms of management now practised pre-empt resources from all other animals. Invertebrates and some birds have been considered above but mammals are equally vulnerable. Intensive grazing and the sparse swards that sheep particularly produce are a major cause of the decline of voles described by Harris *et al.* (1995). This has led in turn to marked declines in species such as Kestrels and Barn Owls in pastoral areas (Table 9.2). The possibility that high stocking densities of sheep affect soil structure by compaction and therefore soil invertebrates seems to me to need investigation. High stocking densities also have a crucial direct impact on ground-nesting birds by increasing nest losses through trampling, desertion and farmwork. Thus in the Netherlands, Beintema & Muskens (1987) found that, for waders, despite high levels of nest loss to predation (*c.* 50%), nesting success in natural circumstances was high because of the level of renesting. In agricultural grasslands, however, trampling by stock and losses to farmwork accounted for another 30–60% of nests according to species, which led to a steep decline in nest success for all waders studied and declines in population. Increased stock numbers were associated with improved drainage and greater fertiliser use, as in Britain. Loss of nests to trampling was simply related to stock density and length of exposure to stock.

In Britain the wader most studied in this context has been the Lapwing, which breeds in a wide variety of situations in farmland and thus shows the impact of different aspects of agricultural change most starkly. In northern England Baines (1990) found that most nest losses on his study area were to predation by corvids and gulls and derived from pasture improvement reducing the cryptic defences of the clutch. Predation reduced productivity below the level of self-maintenance on agriculturally improved grass so the species declined with improvement. In an analysis of the BTO nest record cards for Lapwing for all England and Wales Shrubb (1990) found that desertion acounted for 10% of nests in grass, farmwork 9%, trampling by livestock 8% and predation 20%. Nest losses to trampling were positively correlated with the numbers of stock, and losses to farmwork particularly involved manure spreading, which smothered nests, and harrowing. Desertion was positively linked with cattle numbers which points to much nest desertion being stock-related. Thus in farmland in England and Wales agriculture had become the most potent cause of nest loss in grassland, although nest success, at 72%, was still satisfactory in the declining area of spring cereals. I calculated that nest losses in grassland in the uplands had reached a level where too few hatched to support the population. In lowland grass, where stocking rates were higher, hatching success was more satisfactory, an anomaly which probably reflected lower stocking densities with fewer sheep and increasing selection by Lapwings of the most favourable sites. As in the Netherlands, I found that nesting success in habitats outside farmland was satisfactory, with significantly larger broods hatched and more successful nests (72% vs. 54%). The difference was due to the impact of farming in the grassland sample. In grass the number of clutches started after 30 April also declined significantly from 1962, despite increased nest losses, so the declining ability to replace clutches was added to declining nesting success. But such later clutches had increased in tillage, where clutch replacement was unaffected by losses to cultivations. In the Netherlands Beintema *et al.* (1985) also found that earlier breeding had become the norm with waders breeding in agricultural grasslands. Breeding started about two weeks earlier and this coincided with earlier grazing and mowing, with drainage and with increased fertiliser applications to promote growth. They did not say whether breeding also finished earlier but Beintema & Muskens (1987) show that renesting declined with agricultural intensification.

The overall effect on birds

Modern pastoral farming emerges from this analysis as an increasingly barren environment for many breeding birds. Much attention has been paid to the effects of high stocking rates and predation on nest success. But this analysis shows that modern grass management has also amounted to a major assault on birds' food supplies, particularly of invertebrates, which may prove to be far more significant. Table 9.2 summarises an analysis I have made to compare the distribution of some common farmland birds recorded by Sharrock (1976) and Gibbons et al. (1993) in relation to three broad categories of farmland: tillage, defined as more than 66% of crops and grass in tillage crops, grass farming, defined as more than 66% of crops and grass in improved grass, and mixed farming, between those extremes. Each 100-km square of the National Grid was divided into four 50-km squares and each 50-km square was assigned to one of these categories through the county agricultural statistics from the June Census (Figure 9.13). I adopted this approach to attempt to circumvent the problem inherent in classifying counties according to broad farming type from the county statistics, particularly as mixed farmland. Today such counties are almost never composed of mixed farms but are split between districts primarily of arable farmland or of pastoral farming. The 50-km grid actually allowed a more sensitive split between types. All the species examined are widely distributed in Britain, except Oystercatcher (Scotland only), and Turtle Dove, Barn Owl, Little Owl and Yellow Wagtail (England and Wales only). Species which are supported by introductions or releases, those with restricted regional distributions which would predispose the result, or scarce or rare species were excluded. Only records of proved or probable breeding were counted.

Making such comparisons between the *Atlas* surveys involves some problems, as methods and coverage were not strictly comparable. Nevertheless a clear and very marked general trend is visible in Table 9.2. Of the 61 species included, 36 (59%) have shown some decline in distribution, of which 15 showed losses of at least 10% of their overall distribution in

TABLE 9.2. Changes in the distribution of common farmland birds in Britain expressed as the proportion of occupied 10-km squares recorded in 1968–72 (Sharrock 1976) lost or gained in 1988–91 (Gibbons et al. 1993) in different categories of farmland; the Outer Hebrides, Northern Isles, Channel Islands and Isle of Man were omitted

Category	Percentage change	Change most marked in				Little change (<2%)
		Change uniform	Tillage	Mixed	Grass	
WOOD	(16	**Sparrowhawk**[a,b]				Wren
	5					Robin
	5					Blackcap
	3			------Cuckoo[a]------		Chiffchaff[c]
	7	Dunnock				Willow Warbler
	5				**Whitethroat[a]	Goldcrest[c]
	3				**Garden Warbler[d]	Long-tailed Tit[c]
	4	Jay[a]			**Spotted Flycatcher[a]	Coal Tit
	6				*Bullfinch	Blue Tit
						Great Tit
WOOD/	10				**Tawny Owl**	Blackbird
FIELD	(5			**Green Woodpecker[c]		Song Thrush
	3	Mistle Thrush				Chaffinch
FIELD	4				**Greenfinch[a]	Pied Wagtail
	4				***Kestrel[a]	Woodpigeon
	6				***Stockdove[a]	(Magpie)
	(6	Collared Dove)				Jackdaw
	21				****Turtle Dove	Rook
	36			**Barn Owl[a]		Crow
	10		**Little Owl			
	3				**Starling	
	4				**House Sparrow[a]	
	17				**Tree Sparrow[a]	
	(23	**Goldfinch)**[f]				
	3				**Linnet[a]	

TABLE 9.2. (Cont.)

Category	Percentage change	Change uniform	Change most marked in Tillage	Change most marked in Mixed	Change most marked in Grass	Little change (<2%)
GROUND NESTING	8				****Yellowhammer[a]	Mallard
	22					Skylark[a,g]
	17				***Grey Partridge[a]	
	7				***Moorhen[a]	
	3				**Oystercatcher[h]	
	8				***Lapwing[a]	
	20		**Snipe[a]			
	8	Curlew[a]				
	18	Redshank				
	5	Meadow Pipit[a]				
	11				***Yellow Wagtail	
	18		****Whinchat[a]			
	38				****Grasshopper Warbler[a]	
	12				**Sedge Warbler[a]	
	12				**Reed Bunting[a]	
	31				**Corn Bunting[a]	

Notes: Species showing changes of more than 10% of their distribution in 1968–72 are shown in bold. Such changes are considered significant in the sense that they are very unlikely to be artefacts of differences in method in the Atlases. Values in parentheses indicate an increase, otherwise all changes were declines. Uniform indicates that change was similar across all farm categories. Asterisks indicate significant differences in the scale of change between farm categories, shown by χ^2 tests:
* $p < 0.05$, ** $p < 0.01$, *** $p < 0.001$, **** $p < 0.0001$.
[a] Indicates that a sharp decline in distribution had also occurred in Ireland.
[b] Although the overriding change for Sparrowhawk was recovery from the effects of organo/chlorine pesticides, there was some decline in grass areas and a marked contraction in Irish distribution.
[c] Three woodland species, Chiffchaff, Goldcrest and Long-tailed Tit, show marked variations but little overall change.
[d] Garden Warbler is for England and Wales only. The species expanded its Scottish range, with a net gain of 14%.
[e] Green Woodpecker showed an increase in distribution in mixed farmland, particularly in Scotland.
[f] Goldfinch showed very little change in England and Wales, the expansion being particularly notable in northern Scotland.
[g] Skylark showed a marked contraction in Ireland and, although little change is indicated for Britain, the table does not show the extent to which the species has deserted many pastoral farms in the west (personal observation).
[h] Oystercatcher is expanding its range in England and Wales.

The overall effect on birds

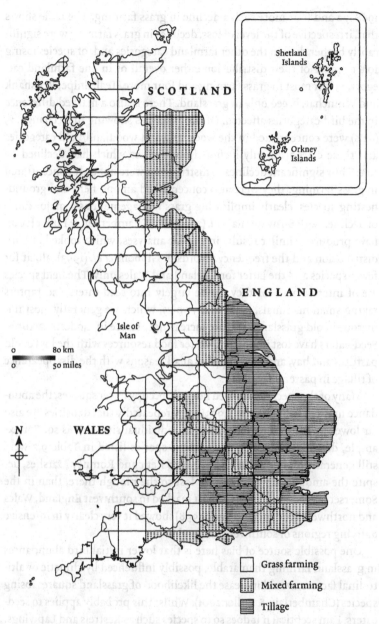

Figure 9.13. The distribution of tillage (66% or more of crops and grass in tillage crops), grass farming (66% or more of crops and grass in grass) and mixed farming (between those extremes) in Britain by 50-km squares.

1968–72 and four more such a decline in grass farming. The table shows that, irrespective of the level of loss, declines in grass farming were significantly higher than in the other farmland categories and, of species losing 10% or more of their distribution either overall or in one farmland category, 13 lost most in grass farming and four more, Teal, Snipe, Redshank and Whinchat, breed only in grassland. There is also a marked difference in the bird categories affected. Of the 21 species showing little change, 13 (62%) were concentrated in the woodland and woodland/field categories and these categories only included one species which had declined by 10%. Thus significant declines in distribution were not only concentrated in grass farming, they were also concentrated among field and ground-nesting species, clearly implicating grass management as a major cause of decline. Both Siriwardena *et al.* (2000) and Chamberlain & Fuller (2001) have produced similar results in similar analyses, which looked at both distribution and the frequency records in Gibbons *et al.* (1993), albeit for fewer species and the latter for England and Wales only. The field species are of interest here for they divide largely into seed-eaters and raptors taking small mammals, particularly voles, which are generally most numerous in old grasslands and are increasingly scarce in modern pastures. Seed-eaters have lost important winter food resources with the loss of old pastures and hay, and food supplies at all seasons with the disappearence of tillage in pastoral farmland.

Many of the species tabulated not only occupy fewer squares, the abundance maps in Gibbons *et al.* (1993) show repeatedly that densities are also far lower in pastoral regions. Even in individual sites this is so. For example, densities they show for wetland species listed in Table 9.2 were still generally higher in the Essex, North Kent and Romney Marshes, despite the amount of wet grassland lost to the plough there, than in the Somerset Levels. The pattern is most marked in southwest England, Wales and northwest Scotland and Lock (1998) showed it very clearly in intensive dairying regions of southwest England.

One possible source of bias here is that lower initial bird abundances in grassland farming than arable, possibly influenced by climatic or altitudinal factors, would increase the likelihood of grassland squares losing species (Chamberlain & Fuller 2001). Whilst this probably applies to seed-eaters, I am sceptical if it does so to species such as Kestrels and Lapwings. One might also reasonably have expected that many declining birds, for example Moorhen, Lapwing, Sedge Warbler or Reed Bunting, would have found a refuge in pastoral farmland. That they have not underlines the impact that the modern grassland revolution has had on farmland birds. The changes in grassland management cited in this chapter actually comprise the most important changes to have occurred in farmland in the second half of the twentieth century. Gibbons *et al.* (1993) also show that these patterns are even more pronounced in Ireland, where there has been a massive intensification of grassland management since 1970 and access to EU funds (e.g. Eurostat 1980, 1987). Of the species showing significant declines in Table 9.2, all that breed in Ireland have shown sharp contractions of range there since 1972, as have Corncrake, Skylark and Long-tailed Tit; so have Sparrowhawk and Goldfinch, which expanded their range in Britain.

An outstanding feature of Figure 9.12 is the much lower level of stocking in Scotland. This is associated with much larger breeding wader populations in farmland than in England and Wales. Galbraith *et al.* (1984) estimated the total population on lowland Scottish farmland, excluding machair, for Oystercatcher (20 600 pairs, density 1.03 pairs/km^2), Lapwing (60 300 pairs, 3.03 pairs/km^2) and Redshank (3000 pairs, 0.15 pairs/km^2). O'Brien (1996) recorded much higher estimates 10 years later (Oystercatcher 82 493 pairs, Lapwing 91 965 pairs, Redshank 12 076 pairs, Snipe 41 162 pairs, Curlew 35 633 pairs). Unfortunately these surveys are not comparable because of marked differences in the area and definition of 'lowlands' used. However they confirm the continuing presence of high numbers of breeding waders in Scottish agricultural land. Only for the Lapwing can a comparison be made with England and Wales. Wilson *et al.* (2001) recorded an overall density in farmland of 0.59 pairs/km^2, which is about one-third of the density recorded by O'Brien for Scottish farmland of 1.75 pairs/km^2. As already noted, one reason for this is lower stocking levels generally in Scotland. A second reason is probably that cattle tend to be more important than sheep in the Scottish lowlands, where the highest densities of waders are found. This results in fewer animals per hectare and cattle produce better-structured swards for all nesting waders. Lapwing chicks also feed avidly around their droppings and high densities of grazing young stock are advantageous to them (Galbraith 1988). Thirdly, for Lapwing and Oystercatcher, Scottish arable farmland still consists very largely of a mix of spring tillage (40% of all tillage is spring barley) and ley (41% of arable), which is the most favourable and favoured nesting situation for both. Oystercatcher is interesting as it is the only farmland wader at present significantly expanding its range, mainly in arable fields; grassland distribution actually declined in Scotland (Table 9.2). The primary reason may be that, although the chicks are nidifugous, they are fed by the parents for a considerable period. So they benefit from the parents' experience in finding food and the greater range they can cover searching for it.

How much is increasing nest predation a factor in declining numbers of ground-nesting birds, particularly waders? Their main potential nest predators are corvids and foxes, populations of which have increased markedly since about 1960 (Marchant *et al.* 1990, Tapper 1992). It is frequently argued, by landowners and farmers especially, that population declines stem from this rather than habitat change. Nest predation can certainly reduce populations, as shown by Baines (1990) for Lapwing and Grant *et al.* (1999) for Curlew in Northern Ireland. These results were obtained in grassland habitats, where these species are declining most, but they do not necessarily apply to other habitats. Thus Lapwings achieve high rates of nest success in spring tillage and non-agricultural land (Shrubb 1990), as do Oystercatchers in tillage (Heppleston 1972), although there is no reason to suppose that these species are not also exposed to nest predators in these habitats. Predation is operating selectively. Most ground-nesting species have evolved defences against their nest predators, either by laying cryptic-coloured eggs coupled with active defence or by hiding the nest and sitting tight or by nesting in wetland sites more difficult of access for ground predators. The extent to which these work is now affected by the extent to which their habitats are manipulated, altered and degraded by Man. Wet grassland is drained, making nests more accessible, old pastures are reseeded and highly fertilised, making eggs more obvious and are intensively grazed, removing cover for nest concealment. Overall the agricultural improvement of grassland has greatly increased the vulnerability of nests quite apart from any increase in predators. Nor is it just a question of these factors increasing predation risk. Population declines also reflect the extent to which such altered habitats are now rejected completely.

The Water Vole (*Arvicola terrestis*) provides another excellent example of the same process. Barreto *et al.* (1998) found an 'irrefutable association' between increasing American Mink (*Mustela vison*) and rapidly declining Water Vole numbers in the Thames catchment during the 1990s. However they argued that the decline of Water Voles there was not merely the result of an increasing and alien predator but of drainage and agricultural intensification of riverside land reducing the voles' habitat to highly fragmented linear strips, where they became increasingly vulnerable both to predation and other factors such as flood years. Barreto *et al.* observed that in more extensive wetlands Water Voles and American Mink coexisted successfully, highlighting the interaction between increased predation and habitat degradation.

Finally there is no historic evidence of a broad relationship linking the numbers of ground-nesting birds and the numbers of their predators. Both birds such as Lapwings and their predators were most abundant in the late eighteenth and early nineteenth centuries, when habitats were far less despoiled. But Lapwings have been in almost continuous decline since irrespective of the massive, usually human-induced, fluctuations in their predators, which have varied from abundance to extinction and back. We have no difficulty in ascribing population recovery of such birds to habitat recovery, especially if the latter is the result of our conservation management. Why then is the opposite difficult to accept, except for the self-interest of land managers?

10
Winter food resources

Tucker (1997) observed that ignoring non-breeding components of farmland bird populations underestimates the importance of the habitat. These components include non-breeders among populations of breeding species, as well as wintering populations of breeding species and those of species which breed elsewhere; at least 15 species that do not breed in farmland winter commonly there. This chapter considers wintering birds.

The main winter food resources exploited in farmland by birds comprise crops, crop residues, arable weed seeds and soil- and surface-dwelling invertebrates; hedgerow fruits are important for some species. All these sources tend to be most abundant or most accessible in the winter period. Crops are harvested, leaving stubbles and crop waste and accessible seeds; seeds and fruits are most abundant in autumn and decline in quantity through the winter with exploitation; young cereals offer nutritious grazing and the great extent of bare ground makes invertebrate prey easier to find (Figure 10.1). In grassland declining growth also increases accessibilty of such prey, whilst weeds in old grassland are significant for seed-eaters. In general wintering populations of birds in farmland differ in several important ways from breeding populations. They are more numerous and gregarious, a higher proportion are ground-dwelling or field species, winter visitors are more abundant than summer visitors and virtually all winter visitors come into the ground-dwelling or field categories (Chapter 2). These characteristics reflect the nature of the food resources available.

Altogether I have treated 82 species as wintering birds of farmland (Table 2.1), of which nine primarily woodland species are omitted here and raptorial birds are considered separately. This leaves 62 species, all

Figure 10.1. A good winter feeding site, Berkshire Downs, with newly turned ground, stubble, young crops and corn ricks awaiting threshing. (Courtesy of Rural History Centre, University of Reading.)

ground-dwelling, field or woodland/field species, which are fairly evenly divided between residents and winter visitors plus partial winter visitors. Table 10.1 lists these species under primary winter feeding categories, although these are not sharply distinct because many farmland birds take a wide variety of foods in winter and feeding habits also change over time, seasonally and regionally. Geese provide excellent examples of this and Table 10.2 summarises the seasonal patterns for five species, one, Bean Goose, named for the habit of feeding on crops. But, assuming the basic patterns in Table 10.1 are valid, an interesting difference emerges. Residents are more often primarily gleaners of seeds and waste grain, whereas winter visitors or partial winter visitors are more often grazers, omnivorous or primarily invertebrate feeders. For seed-eaters this may represent a significant historic change in behaviour for, in the nineteenth century, major winter immigrations of Skylarks, Tree Sparrows, Chaffinches, Bramblings, Greenfinches, Linnets and Yellowhammers into Britain were regarded as commonplace (county avifaunas, Eagle-Clark 1912, Witherby *et al.* 1940). Such major immigrations do not occur today, except with Skylark, Chaffinch and Brambling (species accounts in Lack 1986 and see below). Climatic variations may have contributed to this for Gatke (1895) noted that the migration patterns of species such as Skylarks at

TABLE 10.1. *Wintering birds in farmland according to primary feeding behaviour*

Grazers: grass, young cereals and green crops, crop residues	Feeders on soil and surface invertebrates	Gleaners of seeds/ waste grain	Omnivores
Mute Swan	**Golden Plover**	Red-legged Partridge	**Teal**
Bewick's Swan	**Lapwing**	Grey Partridge	Pheasant
Whooper Swan	**Ruff**	Stockdove	**Moorhen**
Bean Goose	**Jack Snipe**	Collared Dove	Jay
Pink-footed Goose	**Snipe**	**Skylark**	**Jackdaw**
White-fronted Goose	**Curlew**	House Sparrow	**Rook**
Greylag Goose	**Black-headed Gull**	Tree Sparrow	**Starling**
Canada Goose	**Common Gull**	**Chaffinch**	
Brent Goose	Little Owl	**Brambling**	
(dark-breasted)	Green Woodpecker	Greenfinch	
Wigeon	**Meadow Pipit**	Goldfinch	
Mallard	Pied Wagtail	Linnet	
Pintail	Wren	Yellowhammer	
Red Grouse	Dunnock	Cirl Bunting	
Black Grouse	Robin	Reed Bunting	
Woodpigeon	**Stonechat**	Corn Bunting	
	Blackbird		
	Fieldfare		
	Song Thrush		
	Redwing		
	Mistle Thrush		
	Magpie		
	Chough		
	Crow		
	Raven		

Notes: Species in bold are winter visitors or partial winter visitors. Significantly more such species are omnivorous, feed on invertebrates or are grazers ($\chi^2 = 8.45$, df 1, $p < 0.01$).

TABLE 10.2. *Seasonal patterns in the feeding behaviour of wintering geese in Britain*

Species	Autumn	Winter	Spring
Bean Goose[a]	stubble grain	roots/grass	winter cereals/grass
Pink-footed Goose	stubble grain	roots/grass	winter cereals/grass
White-fronted Goose	—— almost exclusively pasture——		grass/winter cereals
Greylag Goose	stubble grain	roots	ley/pasture/cereals
Brent Goose[b]	—— estuarine ——	——grass/cereals/estuarine——	

[a] Data for Bean Goose is for northwest Europe.
[b] Data for Brent Goose is for race *bernicla*.
Source: Owen *et al.* (1986).

Heligoland changed after the mid nineteenth century, as climatic amelioration altered wind patterns.

Grazers

The exploitation of arable crops by geese, once species of wet grasslands and saltmarshes, is now well established. Although White-fronted Geese remain almost entirely grassland birds in Britain and The Netherlands, they too are starting to exploit arable crops in parts of northwest Europe (Owen et al. 1986), and both this species and Barnacle Goose now also increasingly exploit ley grass in western Britain. When geese first exploited cereals is unrecorded, although agricultural expansion from the mid eighteenth century possibly encouraged it. But the exploitation of other arable crops is quite recent. Kear (1990) noted that it often started because severe winter weather caused individuals and flocks to try new foods, which were then absorbed into mainstream feeding habits. Pink-footed Geese were first recorded taking potatoes in Lancashire in the 1890s, when potato-growing was an expanding enterprise (Figure 1.5); when potato-growing started extensively in Lincolnshire after the 1914–18 War, the Pinkfeet there turned to them at once. They probably developed this habit by exploiting waste tubers in fields of young winter wheat planted after potatoes, a common rotation. Potatoes are now a major food throughout their range. Greylag Geese began feeding on potatoes regularly after 1945 and turned to taking turnips and swedes regularly in some areas of Scotland from the 1950s, a habit linked to changes in the management of these crops, which were increasingly left standing in fields to be grazed or lifted as required instead of being harvested and clamped for winter use (account based on Kear 1963a).

Since the early 1970s Pinkfeet have also fed on carrots in northwest England (Owen et al. 1986). Sugar-beet tops are now their principal diet between October and January/February in Norfolk (Taylor et al. 1999), a development not mentioned by Owen et al. (1986), although first recorded in 1966. Whooper Swans are now also feeding on sugar-beet tops in Norfolk and have increasingly taken stubble grain, potatoes and turnips in Scotland since 1940 (Kear 1963b). Bewick's Swans have behaved similarly since the early 1970s (Cramp & Simmons 1977) and Brent Geese started grazing leys and winter cereals regularly at about the same time. For the latter the habit is now well established and reflects the fact that estuarine habitats are no longer able to support the increased numbers locally (Vickery et al. 1995).

Although crops are important to geese and their numbers have increased markedly since the 1950s (Atkinson-Willes 1963, Owen et al. 1986, Lack 1986), these factors are not necessarily related. Potatoes, for example, have declined by 67% in area in Britain since the late 1940s, whilst the sugar-beet area has remained more or less constant since the 1960s. However the availability of these crop wastes has changed. Mechanised harvesting wastes more potatoes than handwork (Kear 1963a) and the exploitation of sugar-beet tops probably reflects the declining importance of livestock on many East Anglian farms, where the crop is now concentrated. The tops were once widely harvested for stock feed. Geese have mainly increased because persecution has declined, perhaps particularly spring shooting (Owen et al. 1986). Their success in opportunistically exploiting nutritious crops and crop residues, however, must have underpinned that increase by maintaining winter survival. In addition the application of nitrogen fertilisers to grass and young cereals increases their food value to geese (higher protein levels) and has probably contributed to population increases (Vickery & Gill 1999). Distribution is still broadly governed by the existence of secure traditional roosts, in major estuaries or large areas of open fresh water. Within that framework distribution has also changed with changing cropping patterns, for example the increase of barley in Scotland changed the comparative distribution of Pinkfeet in Scotland and eastern England in the 1950s (Owen et al. 1986). The creation of new waters, such as reservoirs, and curling ponds in Scotland, also provided new roosts for geese, enabling them to exploit new arable feeding areas (Kear 1990).

The common surface-feeding ducks listed in Table 10.1 also exploit such food sources. Mallard were recorded feeding on potatoes before geese, with the first definite record in 1863. The habit may have been encouraged by the mid-nineteenth-century outbreaks of potato blight, previously unknown, which would have left large quantities of decaying tubers in fields, perhaps especially in low-lying areas occupied by ducks (Kear 1963b). Ducks' use of such food sources today is probably partly related to the increasing extent of inland waters, such as gravel pits, reservoirs and nature reserves, available as secure roosts, which has encouraged feeding in adjacent farmland by what were much more estuarine species in the nineteenth century. In the Fens, for example, Thomas (1976) found that more than half the Mallard and smaller numbers of other dabbling ducks using the Ouse Washes fed in the surrounding arable fenland. Stubbles were exploited until ploughed in October, Mallard and Pintail exploited waste potatoes from November and up to half the Wigeon present grazed germinating winter wheat. More ducks used these resources when the Washes were heavily flooded.

The other important grazer in farmland is the Woodpigeon, a species long adapted to the availability of various crops as food sources. Murton *et al.* (1964) found that population regulation in Cambridgeshire farmland primarily rested on the availability of stubble grain in October–December and particularly clover during December–March. Their study area lacked fodder roots such as turnips, however. In my Sussex area turnip leaves were the main winter food until turnips ceased to be grown in 1968, the birds switching to clover, both cultivated crops and clovers in pastures, in early spring. The size and number of winter flocks declined as turnip-growing declined but increased rapidly again with the later expansion of oil-seed rape (O'Connor & Shrubb 1986). Inglis *et al.* (1990) continued the Cambridgeshire study into the 1980s and also found a changed situation. Clover had declined but a large area of oil-seed rape was grown, which had become the main winter food resource. Its availability had caused a sharp decline in winter mortality from food shortage, removing a major check on the population. The national expansion of oil-seed rape, now comprising nearly 10% of tillage crops, has repeated the nineteenth-century situation of massive winter food supplies becoming available from the leaves of a standing brassica crop and promoting and supporting a marked population increase (Figure 10.2). It offers a good example of the importance of winter food supplies in regulating farmland bird populations and Inglis *et al.* remarked that limiting factors in the breeding season now regulate Woodpigeon numbers.

The winter feeding habits described for grazers above involve major adaptations in response to changes and availability of crops and crop residues. These have often been accompanied by population increases or have supported population increases rooted in other causes. They seem to provide good examples of the idea propounded by Newton (1998) that some species may be able to occupy more habitats than they currently use, making it difficult to define their full potential habitat range. This seems particularly so for wildfowl.

Feeders on invertebrates

Important winter visitors under this heading are Golden Plover and Lapwing. These plovers detect prey visually and winter feeding habitats

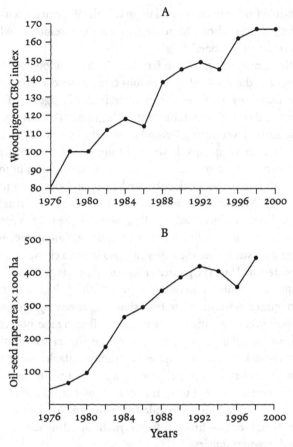

Figure 10.2. (A) The CBC index for Woodpigeon from 1976 (the first year it was calculated) to 2000 compared to (B) the area of oil-seed rape grown in Britain over the same period.

balance the need for bare ground to detect prey easily and the impact of frost, which seals the ground, making prey unobtainable; snow compounds such problems. Coastal populations often feed on arable land, particularly winter cereals, but in midland England grassland is preferred (e.g. Fuller 1986, Fuller & Youngman 1979, Mason & MacDonald 1999, Shrubb 1988). This is probably related to soil temperatures and the incidence of frost, the former being consistently higher and the latter less frequent in coastal areas (Coppock 1976). In Nottinghamshire Barnard & Thompson (1985) found a relationship between temperature and feeding on leys rather than old pastures. Leys had less dense swards, thus insulating the soil less, and were mainly used in mild weather. In Essex

Burton (1974, in Fuller & Youngman 1979) found that Golden Plover preferred grass leys to permanent grass and and switched to winter cereals in early spring. The latter was more widely recorded by Fuller & Lloyd (1981) in south and east England, where soil temperatures rise quickly in early spring. In southwest Sussex Lapwings preferred cereals but feeding on grass fields increased as the winter progressed and soil temperatures declined, particularly in frost (Shrubb 1988).

Fuller & Lloyd (1981) suggested that the Golden Plover's preference for feeding in permanent grass reflected the greater volume of soil invertebrates there. For Lapwing in North Yorkshire, however, Gregory (1987) found similar rates of capture of large prey items in both winter cereals and permanent grass. I obtained similar results in Sussex but earthworms were taken less frequently in cereal fields there as cereal sequences lengthened. In these fields Lapwings apparently sought surface invertebrates, as they did in cereals after rape (Shrubb 1988). All these case studies suggest temperatures as an important determinant in winter habitat use in plovers. Another marked pattern everywhere was that plovers used different habitats for roosting than feeding. The main characteristics of roosting sites were well-structured ground habitats, which provided shelter and cryptic protection from predators. Larger fields were also preferred.

Gregory (1987) pointed out that young cereals are now an important winter feeding habitat for these plovers. This represents an historic change in habitat use, as nineteenth- and early twentieth-century avifaunas rarely mention them. The arable habitats most frequently recorded then were fallows or roots, the latter either as standing crops or harvested fields. Nicholson's (1938) survey of habitat use recorded much the same pattern, with fallow as the most important arable habitat for December flocks and arable green crops, only part of which comprised cereals, fifth in importance. This switch must partly reflect the greatly increased proportion of tillage now occupied by winter cereals. A further historic change concerns Golden Plover, which was once noted as making important use of heathland habitats in winter in southern Britain (e.g. in Surrey, Hampshire, Middlesex, Norfolk and Leicestershire). This habit declined as heathland was enclosed and grazing in remaining areas declined, a change associated with a decline of wintering Golden Plover. Old upland grazing commons remain favoured habitats in central Wales, particularly where bracken is still mown in autumn (M.F. Peers personal communication).

Further changes are now taking place. Mason & MacDonald (1999) found a consistent annual decline in the number of plovers wintering in the three years of their Essex study, where grass was an insignificant habitat, and linked this to the more rapid and dense growth of cereal crops, the main habitat used, in the warmer autumns of the later years of the study; plovers avoided cereals more than 100 mm high. Similarly, on my Sussex study area, Golden Plover no longer winter and Lapwings have become confined to major areas of old grassland. These changes are also related to the management of winter cereals but because cereals are now sown in September–October rather than October–November, again resulting in much higher and denser plant stands over winter. A second factor has been a change in drill widths. In the past we used a drill that sowed in rows 7 inches (180 mm) apart, which was then standard. Today drills commonly sow in rows 4 inches (100 mm) apart and the crops quickly grow to completely cover and conceal the soil, making feeding by plovers impossible. Mason & MacDonald suggested that such changes may have long-term implications for the conservation of wintering plovers in southern England, where grassland is scant.

Other important winter visitors in this group are Fieldfares and Redwings. Berries and fruits (these species are fond of orchards) form major winter foods but, particularly later in the winter, large numbers also feed on grassland. Tucker (1992), in examining winter feeding behaviour on farmland in Buckinghamshire, found that all the thrushes preferred permanent grass and selected small fields with tall hedges; Fieldfares and Redwings also selected fields dressed with farmyard manure. I studied winter feeding behaviour by these birds on two pastoral farms in central Wales in the early 1990s and the main factor influencing use was again the presence of large and dense boundary hedges, which acted as refuges from Sparrowhawks, Merlins and Peregrines, which were all recorded hunting these flocks (unpublished data). I found little evidence that factors such as the age of leys or sward height influenced the choice of grass fields used, although most were, in fact rather uniform. Farmyard manure applications were not made. More recently M.F. Peers and I have noted a sharp reduction in the number of these thrushes using such habitat in the area; most depart when the berry crop is exhausted.

Starlings are also winter visitors to British farmland on a large scale for which invertebrate foods are essential, although cereals are widely taken and may be an important energy source (Feare & McGinnity 1987). Feare (1994) linked a marked decline of wintering birds on his study site in

Lincolnshire between 1975 and 1992 to the loss of preferred feeding habitats, grass fields, winter fallow (plough, cultivated ground and stubble) and stockyards or feed-lots, in favour of autumn-sown cereals over the period. In 1975 grassland was the most favoured feeding habitat, accounting for 59% of birds. Feare's data, however, suggest that the overall decline of Starlings utilising the area may have been related as much to a loss of diversity in food resources as the loss of any individual feature and he suggested that the loss of such winter food resources may be contributing to the general decline in northern Europe. Certainly Tiainen et al. (1989) attributed declining breeding populations in Finland to the decline of mixed farmland in favour of specialist arable farming, which led to reduced breeding success. This has contributed to the decline of populations wintering in Britain.

Theoretically all the species in this group should benefit from the availability of permanent grass, since invertebrate populations are always highest in areas left uncultivated (Tucker 1992). Today, however, with frequent reseeding, permanent grass is an illusion. Long-established pastures are rare and fertiliser regimes are affecting populations of soil invertebrates further (Chapter 9). Large areas of pastoral farmland seem underutilised by any of these species, for example in central Wales (personal observation). Figure 10.3 illustrates this pattern for three species, showing no obvious link between areas with most grass and most birds in England and Wales. The core area of Lapwing distribution lies in the south Midlands, where there is historical evidence that pastures have been very long established (Broad 1980). The Starling map suggests they avoid uplands and Fieldfares winter primarily in mixed farming areas. All the species discussed in this section also illustrate the importance of a diversity of resources within wintering areas, either as roosts or refuges or alternative feeding sites. Such diversity is perhaps more important than any single habitat feature.

Gleaners

The major food resource for these species is the availability of waste grain and seeds of arable weeds, which are important to gamebirds, pigeons, larks, Starlings, corvids, finches, buntings and often wildfowl. The main places where birds found such foods were, or remain, stooks and stubbles, stackyards and chaffheaps (Figure 10.4), stockyards and other places where stock is fed, freshly tilled land and amongst standing fodder, root

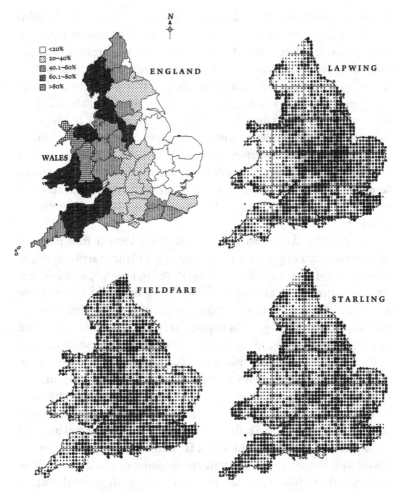

Figure 10.3. Permanent grass (see text) as a percentage of improved farmland compared to the distribution of three grassland species in winter as shown by the *Atlas of Wintering Birds in Britain and Ireland* (Lack 1986). (Maps reproduced by permission of the BTO.)

and vegetable crops. The availability of such feeding sites has fluctuated markedly over the past 200 years, their peak coinciding with the apogee of high farming in the mid nineteenth century, a decline with the arable recession until the 1940s, some recovery with three-year ley farming in the 1960s and a collapse since the 1970s (Figure 10.5). The decline between 1870 and 1940 simply reflected the contraction of an unchanging farming system but that since 1970 reflects a revolution in farming systems and

Figure 10.4. Threshing, here at harvest, Somerset 1953. Note the monumental chaffheap at right. This one wouldn't be left but in winter these were important feeding sites for birds. (Courtesy of Rural History Centre, University of Reading.)

methods. Appendix 1 shows how these figures were obtained and gives additional information for other important features discussed below.

The importance of these food supplies and sources, particularly in winter, was never studied in any detail in the past. But one cannot read the nineteenth-century literature without getting an overwhelming impression of the sheer profusion of seed-eating birds in winter on farmland and around farmsteads. All early avifaunas comment upon it. Larks occurred in flocks of thousands and finch and bunting flocks swarmed on stubbles and at every stackyard, often in sufficiently large numbers at the latter to be regarded as pests, as they stripped the thatch off ricks. Storing grain in ricks meant a constant supply of fresh seeds becoming available as winter threshing progressed. The bulk of that profusion is now history. It seems quite likely that the large winter immigrations of seed-eaters in the nineteenth century were, in fact, exploiting a considerable surplus of waste grain and seeds in British farmland. Although this can now only be supposition, such birds would clearly find it difficult to invade and live in winter farmland in Britain in great numbers today. Thus in the nineteenth century Skylarks were universally regarded as seed-eaters in winter, hence I have classified it as a gleaner, but Green (1978) noted that weed seeds were

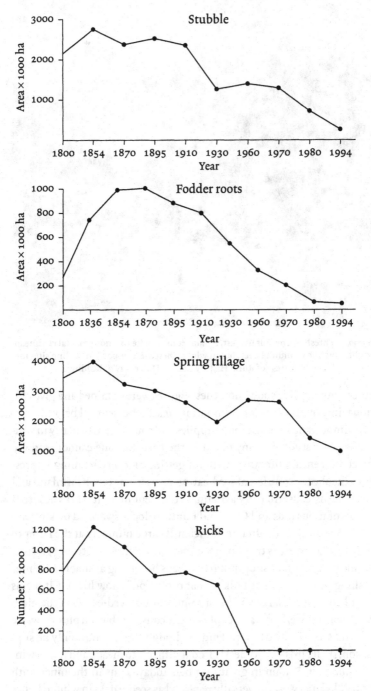

Figure 10.5. Changes in extent of important food sources for seed-eaters in farmland since 1800. Data and methods in Appendix 1.

often too scarce to be a profitable food source in mid-winter and grazing, particularly of winter cereals, had become the main foraging technique. Birds feeding primarily on seeds in winter are not only more often residents, they have also almost certainly shown more declines than the other categories in Table 10.1. Although accurate national population figures are scant, this is supported by the analysis of flock sizes below. Grazers, particularly wildfowl (see Pollitt et al. 2000), have shown more increases. This fairly reflects the present availability of winter food resources and suggests that resident seed-eaters are facing particular difficulties.

Undersown stubbles were the most important section of the stubble area because they stood unploughed for the whole winter and because of the difficulty of weeding them, either by hoe or, until about 1970, by herbicides. In high farming one arable field in five was an undersown stubble in winter and the proportion was still one in ten with the three-year ley system and stood at around a half or a third of all stubbles from the 1870s until about 1970 (Appendix 1). Undersown stubbles were also major overwintering sites for many invertebrates of cereal ecosystems which are important to farmland birds. This has been studied in detail for Grey Partridge (Potts 1986) and Corn Bunting (Aebischer & Ward 1997) but it cannot be doubted that all species with broadly similar ecological requirements benefited.

Hay ricks, straw ricks and stockyards were also important sources of seeds in winter and were also distributed throughout farmland until the mid twentieth century. Hay ricks were particularly noted as favoured feeding sites of Cirl Buntings. Not only seed-eaters exploited such food sources. At Oakhurst our stockyards were favoured winter haunts of species such as Dunnocks, Wrens, Robins, Blackbirds and tits, feeding in hay, straw, on invertebrates in the dung and scavenging the troughs. The Yellowhammer is the species which is perhaps most associated with this habit in county avifaunas and O'Connor & Shrubb (1986) found a general link between its winter survival and cattle-farming. Probably the population decline now evident in southeast Britain at least partly reflects the steady decline of holdings with cattle there. The switch to bales from handling hay loose made little difference to the availability of seeds to birds from this source. Until recently a favoured winter feeding site for finches and buntings in pastoral areas was where hay was spread on fields for animals to pick up. Baled silage today offers no such resources. For buntings, particularly Yellowhammer, this has been as significant a loss as that of arable crops in pastoral areas. Root crops grown for stock feed provided other sources of seeds for they were often infested with weeds. Thus

TABLE 10.3. *The most important feeding habitats used by wintering seed-eaters in England during 1965–1995 (species as for Figures 10.6 and 10.7)*

Species	Farmland habitats[a]	Non-farmland habitats[b]
Skylark	Stubble	Saltmarshes
Tree Sparrow	Kale/roots, stubble	Sewage farms, reservoirs, gravel pits
Greenfinch	Kale/roots, stubble, linseed	Sewage farms, reservoirs, saltmarsh, gardens
Goldfinch	Grassland, linseed	A wide range of habitats but none dominant
Linnet	Kale/roots, oil-seed rape, stubble, grass	Reservoirs, saltmarshes
Yellowhammer	Stubble, farmyards/cattle-feeding stations	Reservoirs, saltmarshes
Corn Bunting	Stubble, grassland	Reservoirs, saltmarshes

[a] Stubble includes set-aside.
[b] Saltmarsh includes beaches.

the increase of crops such as kale, widely grown for dairy cows in southern England, where it formed 6% of tillage in the early 1960s, provided major new feeding sites for finches. County bird reports for the 1960s and early 1970s show, for example, that winter flocks of 1000–5000 Linnets were not rare in weedy kale crops in counties such as Sussex, Hampshire, Wiltshire and Dorset. The frequent habit of using such crops as holding cover and feeding sites for game enhanced their attractions for other birds and remains a significant food source for passerines today. Old grassland is also an underestimated source of food for seed-eaters but bird reports record them as significant sites for Goldfinches, Linnets and Corn Buntings (Table 10.3).

The other great source of seeds was cultivation. This continued throughout the year in high farming systems, with progressive ploughing of stubbles and fallows in winter, followed by spring planting and the summer fallowing of root ground, until the hay and corn harvest opened new sources of supply. Every time the soil is stirred, fresh seed from its seed bank is exposed and this was a particularly valuable food source in spring, when seeding plants, indeed all the winter food resources considered here, were scarcest.

Some general points

The results of modern studies of farmland birds leave little room to doubt that the effects of nineteenth-century changes in cropping systems and rotations and farming methods were overwhelmingly beneficial to large numbers of farmland birds. I suggest that resources on the scale indicated by Figure 10.5 underpinned the broad stability of farmland bird populations shown by Figure 2.3. The extensive new habitat resources which emerged with high farming also enabled many birds to adapt to the losses of semi-natural habitat that occurred in the nineteenth century, except for some ground-nesting or wetland specialists.

The most important point here may have been that these new resources improved winter survival for many species. A number of the species shown as increasing in Table 4.2 undoubtedly benefited but the nineteenth-century literature largely confines itself to generalities in this area. It is easier to find evidence that the later disappearance of these resources led to the loss or decline of several important farmland birds. This has been demonstrated for Cirl Bunting (Evans & Smith 1994) and the importance of stubbles as winter feeding sites for the species has been amply confirmed by its population recovery in its core areas in the southwest following conservation measures to ensure that such sites are available in winter (Evans 1997). Donald & Evans (1994) showed that weedy stubbles were also the most favoured winter habitat of Corn Buntings, and Donald (1997) and Shrubb (1997) both suggested that the loss of such feeding sites in winter was an important contributory factor to the species' decline. Historically it has shown little flexibility in winter habitat use. Stubbles and stackyards always remained the primary winter habitats despite rapidly disappearing in the core area of the Corn Bunting's distribution. The conclusion that this major loss of winter feeding sites has contributed to population decline is supported by the findings of Donald & Forrest (1995) who found little relationship between breeding population decline and changes in farm structure and cropping practice on a sample of 29 CBC farms. They concluded instead that such decline was based on the loss of winter food supplies. Peach *et al.* (1999) also found that the loss of winter food supplies was the main cause of the recent decline of Reed Buntings. This is likely to have been true of many other species, as argued by Chamberlain *et al.* (2000). Nor is just the loss of stubble or of weed seeds involved. Modern short-strawed cereal varieties and the use of growth regulators to shorten and stiffen straw further have largely eliminated lodging in cereal crops and so significantly curtailed the amount of waste grain available to birds.

The decline of stackyards and stockyards has probably particularly reduced the capacity of seed-eating passerines to survive severe winters. A point that every nineteenth-century avifauna I have consulted stressed in this area was the value of stackyards, stockyards, farmsteads and flail threshing floors as major refuges for these species in severe winter weather. *British Birds* has published assessments of the effect of severe winters in Britain since 1918 and, since 1962, the CBC has provided more exact data. The reports for the winters of 1916–17 (Jourdain & Witherby 1918), 1929 (Witherby & Jourdain 1930), 1939–40 (Ticehurst & Witherby 1941), early 1947 (Ticehurst & Hartley 1948) and 1962–3 (Dobinson & Richards 1964) showed a consistent pattern of only limited impact on the common sparrows, finches and buntings compared to much more severe impacts on other small farmland passerines. The pattern was most clearly expressed by Ticehurst & Hartley for 1947 and Dobinson & Richards for 1962–3, who showed that only 17% and 28% respectively of reports for common finches and buntings in their surveys were for severe decline or extinction compared to 46% and 52% for other small farmland passerines. The seed-eaters most severely affected were consistently reported as Goldfinch and Reed Bunting and to a lesser extent Linnet. These are species which are rarely recorded in the nineteenth-century literature as exploiting stockyards and stackyards and my own records agree. The pattern for 1962–3 shown by Dobinson and Richards is supported by the CBC (Table 2.3), with only Linnet and Reed Bunting significantly reduced among common finches and buntings when abundant stockyards were still available, compared to eight species in this group significantly affected in later severe winters when such food resources were much scarcer. It seems reasonable to conclude that the greater vulnerability suggested to severe winters is the direct result of declining winter food supplies and, accepting that, it is likely that this has altered the overall pattern of winter survival.

As the availability of traditional winter feeding habitats has declined habitat use by wintering seed-eaters has changed. Figure 10.6 examines this for seven species, which are common components of winter flocks in farmland, for three periods. The figure is compiled in the same way as Table 3.1, with each record of habitat use in each species account in the county and regional avifaunas scoring one. Records for farmland and non-farmland habitats were then totalled and percentage use in each category plotted. The figure shows an increased tendency for these species to exploit habitats outside farmland in winter as the twentieth century progressed. The records are subject to two possible biases. The non-farm habitats involved are principally saltmarshes or gravel pits, reservoirs, sewage works and rubbish tips, all of which tend to be associated with significant areas of uncultivated land. Except for the first (i.e. saltmarsh) the extent of

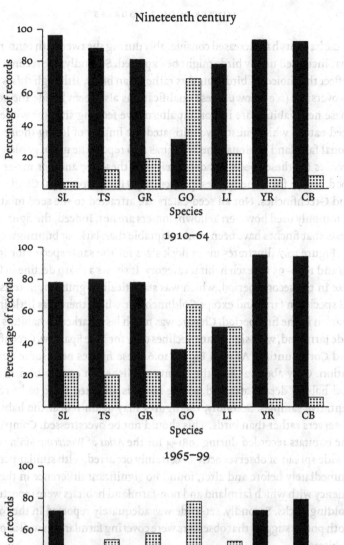

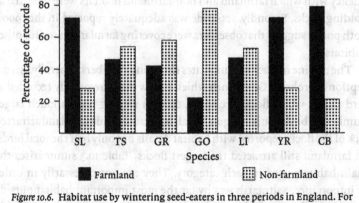

Figure 10.6. Habitat use by wintering seed-eaters in three periods in England. For methods see text. Species are: (SL), Skylark; (TS), Tree Sparrow; (GR), Greenfinch; (GO), Goldfinch; (LI), Linnet; (YR), Yellowhammer; (CB), Corn Bunting. (Data from county avifaunas.)

these habitats has increased considerably during the twentieth century, so that increased use by birds might be expected. Secondly, the records may reflect the choices of bird-watchers rather than birds, although this is easy to overstate (see below). These qualifications also seem beside the point: these new habitats are important alternative feeding sites for wintering seed-eaters, which must have mitigated the impact of losing their traditional farmland habitats. The avifaunas also report the use of garden bird feeders for these species inadequately, but these are another major new food source for some species, particularly House Sparrows, Chaffinches and Greenfinches. Not all seed-eaters are attracted to the seed mixtures commonly used however; Yellowhammers are not. Indeed, the figure suggests that finches have been more adaptable than larks or buntings.

Figure 10.7 illustrates mean flock sizes for the same species for 1965–79 and 1980–95 for each habitat category. It shows a sharp decline in flock size in the second period, which was statistically significant (Z tests) for all species in farmland except Goldfinch, for which there was little information in the first period. Change was much less marked in habitats outside farmland, with significant declines only for Tree Sparrow, Greenfinch and Corn Bunting. As with Figure 10.6 these figures need some qualification. Only about 25% of the counts published in county bird reports had habitat details attached, so the analysis assumed these to be representative samples. Secondly, the figures may again reflect the habits of observers rather than birds. This should not be overstressed. Comparing the habitats recorded during 1981–4 for the *Atlas of Wintering Birds*, when a wide spread of observer activity certainly occurred, with similar periods immediately before and after, found no significant difference in the frequency with which farmland and non-farmland habitats were reported as holding flocks. Secondly, set-aside was adequately reported in the 1990s. Both points suggest that observers were covering farmland as well as other habitats.

The decline in flock size indicates declining numbers generally. An exception is probably Greenfinch which is now most frequently recorded at bird tables, where flocks are small, though ringing reveals much larger numbers as birds come and go. Overall habitats outside farmland attracted 63% of the flocks reported with habitat details but only half the total birds. So farmland still attracted the largest flocks. Table 10.3 summarises the main habitats used in each category. They are not necessarily in order of importance. Saltmarsh was by far the most important habitat outside farmland, whilst stubble remains the most important in farmland. Kale

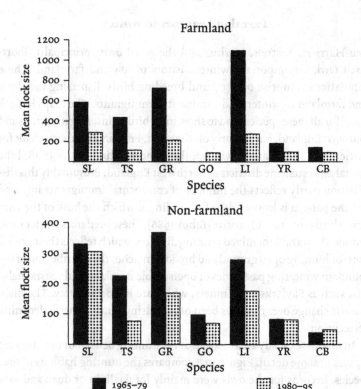

Figure 10.7. Mean flock sizes of wintering seed-eaters recorded in two periods in England since 1965 in farmland and non-farmland habitats. Records of roosts are excluded. Winter is taken as November to March. Species as for Figure 10.6. (Data from county bird reports.)

crops, often infested with fat-hen (*Chenopodium*), attracted very large flocks in the three-year ley period and its place has been taken to some extent by oil-seed rape and linseed since 1980. Buntings tend to avoid oily seeds (Cramp & Perrins 1994), so have benefited little from the expansion of these crops. The analysis suggested that buntings had been particularly badly affected by the loss of traditional habitats yielding waste grain and they may also have been affected by the decline of grass seeds available in cereals with the modern use of grass herbicides. All our common buntings feed extensively on grass seeds (Cramp & Perrins 1994). Rotational set-aside has partly restored weedy stubble to the arable system but it is unclear how permanent an arrangement this will be. The reappearence of linseed as an arable crop has also been beneficial to wintering finches, and to Stockdoves (county bird reports).

Farmland raptors in winter

Hen Harriers, Kestrels, Merlins and the eared owls, principally Short-eared Owl, are important winter visitors to lowland farmland. These populations comprise both upland breeding birds dispersing into lowland farmland to winter, and continental immigrants. The maps in Lack (1986) for all these species tend to show most birds wintering in eastern and southern England, particularly on the coast, Fens and southern chalk for harriers and Short-eared Owls. Another concentration occurs around the coastal and estuarine districts of northwest England. Presumably this distribution partly reflects the number of continental immigrants involved and the pattern is least evident for Merlin for which the bulk of the wintering birds are probably native (Bibby 1986). These predators winter most commonly in arable or mixed farming districts, which reflects their preference for hunting open ground and broadly matches the distribution of the abundant wintering passerines of open arable farmland and coastal habitats, such as Skylarks and Linnets, which are important prey. The main historic change over time has been of increasing numbers with declining persecution.

In my Sussex study area I looked at the hunting behaviour of wintering raptors in some detail. Figure 10.8 compares the hunting habitats of four species. Records for the owls were mainly for daylight or dusk and may not have been strictly representative. But Cayford (1992), using radio tracking in arable farmland in Suffolk, found a closely similar pattern for Barn Owl. Several general points emerged. Hunting frequency in each habitat differed between species but the outstanding feature of the records was the importance to all of undersown stubbles and field boundaries. These comprised only 8.5% of the total area. Only Kestrels made much use of the largest habitat available, young cereals, hunting there almost entirely for invertebrate prey, particularly earthworms. The overall importance of structured ground habitats, which provide cover so that these species can approach prey undetected, is clear.

There was also clear separation in diet. Table 10.4 shows the diet recorded for the two most numerous species, Kestrel and Barn Owl. More than half the Kestrel's diet by frequency and weight comprised birds, particularly seed-eating passerines, and invertebrates, both of which were unimportant to Barn Owls. Among mammal prey Kestrel also took many fewer shrews and mice. I had rather little information for Hen Harrier but eight batches of pellets from roosts contained two voles, a lagomorph

TABLE 10.4. *The food of Kestrels and Barn Owls in winter on 890 ha of mixed farmland in Sussex compared by percentage frequency (F) and percentage weight (W)*

Species	Shrews	Voles	Mice	Rats	Birds	Invertebrates
Kestrel %F	4	20	8	1	20	47
%W	3	31	5	8	49	4
Barn Owl %F	28	50	16	3	3	0
%W	12	55	12	14	7	0

Sources: Shrubb (1980) and unpublished data; weights for mammals from Yalden (1977), for birds from *Birds of the Western Palaearctic*.

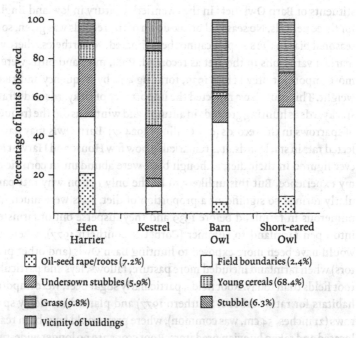

Figure 10.8. Hunting habitats used by some farmland raptors in winter on 890 ha of mixed farmland in Sussex. Hunting records were taken as kills, attempted kills or prey detected only. Only 11 records for Short-eared Owl were available. Young cereals includes new-sown ground. Percentages in the key are of total area.

and 17 small passerines. Field observations of hunting also indicated a high proportion of passerines as prey, perhaps particularly Skylarks. That suggested that Hen Harrier and Kestrel were preying off different groups of birds, probably reflecting the differences in their selection of hunting habitats. Yalden (1985) drew attention to the similarity in the diets of Barn, Long-eared and Short-eared Owls, with their high dependence on Field

Voles, which may also be important to Tawny Owls in farmland (Hardy 1992). Yalden found that dietary separation between these owl species in the Peak District was largely the result of differences in habitat and hunting behaviour. This is presumably a general pattern.

Changes in farmland habitats have had a marked impact on the ecology of these farmland predators. The main changes of significance have been the decline of old pastures and rough grazing, loss of hedgerow, the loss of root crops and leys and the disappearance of rickyards and strawed stockyards. The species for which most information on diet can be examined against these changes is the Barn Owl. Table 10.5 lists the main constituents of Barn Owl diets in the twentieth century in lowland England for three periods. No seasonal break-down of the records was given, so any seasonal bias in the samples cannot be estimated. Nevertheless there were marked variations in the diet as recorded. Rats, mice and birds were the most important prey before 1939, forming 43% by frequency and 58% by weight. This must have reflected the importance of rickyards and strawed stockyards as hunting grounds in autumn and winter, as did the frequency of sparrows in the records (e.g. Collinge 1924–7). Barn Owls may have selected rats in stackyards. It is remarkable how few House and Harvest Mice ever figured in their diet, although both were abundant in corn ricks in my experience. But this unlikely to be the only reason why rats particularly formed so significant a proportion of diet. Rats were much more numerous in farmland before 1939 and they disperse out of farmsteads into open farmland to summer (Corbet & Southern 1977), where they would have been more exposed to hunting Barn Owls (and other predators) when farmland included more pasture, fallows, leys and particularly root fields than today. Root fields, particularly sugar beet, were important habitats for rats (Corbet & Southern 1977) and planted in widely spaced rows (21 inches, 54 cm, was common), where prey would have been readily located and taken by avian predators. Root crops are no longer widespread but sugar beet remains important in eastern England. This undoubtedly underlay the importance of rats in the diets of wintering Barn Owls and Short-eared Owls there noted by Glue (1974, 1977). After 1939 shrews and Field Voles dominated Barn Owl diet and rats and birds progressively declined in importance. Love et al. (2000) convincingly attributed the sharp increase in *Apodemus* spp. in Barn Owl diets in the 1990s to the area of set-aside then available.

Several general points arise from Table 10.5. Firstly, the average size of mammal prey recorded has consistently declined, from 20.6 g before 1939,

TABLE 10.5. *Diets of the Barn Owl in lowland England at three periods in the twentieth century*

Prey	Pre 1939			1956–74			1993–97		
	F	EF	Percentage weight	F	EF	Percentage weight	F	EF	Percentage weight
Common Shrew *Sorex aranus*	845	943	6.5	7007	5998	10	6970	7881	9
Pigmy Shrew *S. minutus*	125	335	<1	1355	2129	1	3782	2798	2
Bank Vole *Clethrionomys*	261	213	4	1281	1357	4	1811	1783	4.5
Field Vole *Microtus agrestis*	1334	1817	27	14243	11551	55	12966	15176	43.5
Apodemus sp.	1043	915	18	3074	5820	10	10264	7646	29.5
Harvest Mouse *Micromys minutus*	30	90	<1	244	572	<1	1140	752	1
House Mouse *Mus musculus*	167	76	2	420	480	1	600	631	1
Brown Rat *Rattus norvegicus*	404	88	23	731	561	8	251	737	2
Birds	370	102	14	761	647	7	469	851	4

Notes: Frequency (F) is the number of items. Expected frequencies (EF) assume that no change in the proportion of prey items occurred. Weights calculated as for Table 10.4.
Sources: Collinge (1924–7), Ticehurst (1935), Glue (1974), Glue & Jordan (1989) and Love et al. (2000).

Figure 10.9. Winter threshing 1955. Note the heap of dead rats, prime Barn Owl fodder, (Courtesy of Rural History Centre, University of Reading.)

to 17.4 g during 1956–74 and to 15.8 g in 1993–7. This reflects changes in the abundance and availability of different prey, particularly the declining importance of rats and birds and the increasing importance of shrews. I am not competent to discuss the implications for the energetics of Barn Owl hunting but it seems likely that this will have affected factors such as winter survival and breeding success unfavourably. Love *et al.* (2000) discuss this subject more fully in considering the changes in diet they observed since 1974. Probably this change has affected all farmland owls to some degree.

Secondly, the present dominance of Field Voles in the Barn Owl's diet has coincided with an enormous decline in primary vole habitat, old pastures, rough grass and hedgerows, in lowland farmland. Harris *et al.* (1995) noted a long-term decline of Field Voles in the twentieth century from these causes. Voles benefited from the reduction of rabbit grazing with myxomatosis in the 1950s but decline has resumed since 1970, with further habitat loss and agricultural intensification, particularly rising stocking rates. Thus Barn Owls increasingly depend on a declining prey animal inhabiting a shrinking habitat. In arable farmland this may be particularly important in summer, when the proportion of Field Voles taken rises, because of the long-term loss to cereals of hunting habitats such as undersown stubbles and leys. These factors have probably affected all farmland owls and perhaps increased competition between them.

Thirdly, the decline of rats in Barn Owl diets is indicative of a major decline of these animals in farmland. Both Tapper (1992) and Love et al. (2000) ascribe this mainly to modern methods of control by poisoned baits. Nevertheless, although they probably remain common at the numerous feeding stations for game that now exist, rats, like many farmland species, have suffered major habitat losses in farmland. Rickyards and root fields were both major food sources and the former a prime breeding habitat. Comparing figures for breeding productivity in corn ricks given by Venables & Leslie (1942) with Appendix 1 suggests just how many rats might have been pumped into the countryside by populations breeding there (Figure 10.9). Modern grain stores are far less attractive and valuable habitats. Rats would certainly have been far less common in modern farmland as a result of these changes.

Finally it is likely that the loss of the rickyard has altered the pattern of winter mortality in Barn Owls, in severe winters particularly. I can find little evidence in the nineteenth-century literature that Barn Owls were thought to be much affected by the not infrequent severe winters, even that of 1880–1, viewed as the most severe of the century (*Zoologist* 1881). Nor does there seem any obvious reason why such weather should have greatly affected them in the presence of the readily available food supplies farmyards provided, even in periods of snow cover. The reports in *British Birds* on the effects of severe winters since 1916 on birds, also show only limited impact on Barn Owls until 1947, when snow lay unusually late into April. Since then the pattern has changed and Barn Owls suffered severe declines following the winters of 1962–3 (Dobinson & Richards 1964) and 1979–80 and 1981–2 (Bunn et al. 1982), when rickyards had disappeared.

11
Labour, machines and buildings

Modern agricultural machines and implements have a surprisingly long lineage. Mowers, combine harvesters, combine drills, manure distributors, reversible ploughs, spring-tine cultivators and so on all have Victorian progenitors. Such machines have long had an impact on the ecology of farmland, whilst mechanising crop and materials handling has influenced the design of farm buildings. Machines used in minimal cultivation techniques may prove to be important in the future.

Power sources

Until well into the twentieth century draught animals were the main source of power on farms. Although oxen were still used in some areas as late as the early 1900s (Grigg 1989, p. 150), horses were most important by the nineteenth century. Draught animals had one significant effect on farmland ecology, they required fodder, particularly cereals (usually oats) or beans for energy. This ensured a wider distribution of arable crops than has existed since their demise. The loss of these crops has been an important contributory factor in the declining diversity of pastoral farmland, particularly for seed-eating birds (Robinson et al. 2001)(Chapter 9).

Where it could be tapped water power was quite widely harnessed on farmsteads in the nineteenth century. Water wheels drove winnowing machines for cleaning grain, mills, chaff-cutters and mixers for preparing animal feeds, and threshing machines (drums) on farmsteads laid out to facilitate such mechanisation. But steam provided the main source of mechanical power. As with water wheels, its most widespread application was for barn machinery and driving threshing drums. By the 1860s

Figure 11.1. Steam ploughing with a balance plough, an unwieldy implement in such a confined space (compare Figure 11.3A). Horses were still needed for water and coal cart. (Courtesy of Rural History Centre, University of Reading.)

most farms were using steam power for threshing. The switch from flail to threshing drum and winnowing machine was ecologically important. Both machines were fitted with screens and sieves designed to clean grain of impurities such as weed seeds. Their development was a major step towards eliminating weed seeds from crop seeds (Chapter 7) and a basic first stage in the long-term decline of arable weeds.

Mechanisation of cultivations and other field work was much more limited. Steam engines were used, working in pairs and pulling implements back and forth across the field by cables. Long (1963) estimated that c. 200 000 acres (81 000 ha) were being cultivated by steam in England by 1867, but this was less than 2% of tillage. Although capable of high work rates, steam engines never became established on farms for field work. By the 1860s the pattern was set that these were contractors' machines, hired as needed, except on some large estates, for farms also had to maintain their horses for other work (Murray 1868). Field layouts were also often unsuitable for the most efficient use of such large machines (Figure 11.1). Evans (1956) quoted the example of John Goddard of Blaxhall, Suffolk, who farmed 800 ha in the late nineteenth century and equipped his farms for steam cultivation. He discarded it as uneconomic in his circumstances after two years.

Thus, although the machines were available, British farming lacked mechanical power in the field to raise productivity in an era of low prices in the last quarter of the nineteenth century, an important contributory cause of the arable recession. Not only was capital investment lacking but many estates came to be supported by industrial or commercial money and were valued for amenity or field sports rather than agricultural investments (Chapter 5). As a result tenant farmers were often prevented from re-organising field systems to make more efficient use of machinery by landlords who valued the dense pattern of hedges for game. Where this was not so there was substantial hedge clearance in the nineteenth century, as field systems were reorganised into more economic patterns (Chapter 5). The limiting effect of poor field patterns is well illustrated by comparing East Anglia with southeast Scotland. In the Lothians, on farms laid out for arable farming in the early nineteenth century, better field shapes and sizes were reckoned to almost halve demand for horse-power; on similar soils the scale of horse-power was one team (two horses) for 95 acres (38 ha) in the Lothians compared to 50 acres (20 ha) in East Anglia. Differences in rotations, mainly the use of two-year rather than one-year leys in the Lothians, contributed some of this saving in power but the rest reflected greater efficiency in the use of horse-power in more convenient working areas. It contributed to the greater resilience of Scottish farming. Mechanisation also enabled American farmers to supply Britain's wheat competitively. American farm machines were designed for speed and easy working. Much larger teams of horses were used than in Britain, the advantage of space and scale in action. Two-furrow ploughs, ploughing 16-inch (41-cm) furrows, instead of one 9-inch (23-cm) furrow (Figure 10.1) were typical (Hobhouse 1989) and Hart (1981) shows a late-nineteenth-century photograph of an American disc harrow 32 discs wide, pulled by a team of eight horses. Unlike British implements, it had a seat for the driver; American farmers favoured faster-moving types of horses and made a particular study of efficient multiple hitches for their machines. Figure 11.2 shows the apogee of horse-drawn technology, an American horse-powered combine harvester of the late nineteenth century.

The horse was displaced by the motor tractor, which has provided the farmer with a geometric increase in power (Figures 11.3 and 11.4). Essentially the application of tractor power has been a post-1945 development. The scale and flexibility of the power available in modern farming has two

Figure 11.2. Not a cavalry charge but a horse-drawn and horse-powered combine in the American West in the late nineteenth century; 27 horses hitched in tandem. (Courtesy of Rural History Centre, University of Reading.)

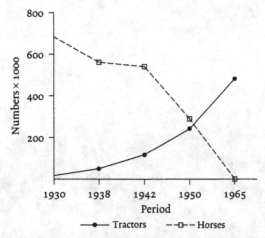

Figure 11.3. The decline in horse power and the increase in tractors in English farming. Note that horses ceased to be recorded after 1965. (Data from June Census Statistics.)

fundamental effects on farmland ecology. Firstly, crop rotations formerly spread work throughout the year, making the best economic use of labour. Without modern power sources farming would still have to include a greater element of seasonal crop rotation to achieve this. Secondly, the capital demand of modern machinery promotes specialisation in crops.

A

B

Figure 11.4. From horse to tractor. (A) The six teams of horses in Berkshire in the 1920s (four harrowing and two drilling) are doing the same as (B) the tractor and drill in Wiltshire in 1951, the latter being fitted with cultivating attachments. Note steam cultivation in the background to (A). (Courtesy of Rural History Centre, University of Reading.)

Large expensive machines need large crop areas to support the capital investment. In some cases this dilemma is met by co-operatives. But these have never been popular in Britain. One reason may be our uncertain oceanic climate, which limits the time available for most tasks; everyone needs the shared machine at the same time, perhaps the only opportunity in a poor season.

Increasing mechanisation and power on farms has been matched by a declining labour force. This peaked in the early 1850s and has declined virtually continually ever since (Grigg 1989). Farm work has always been comparatively poorly paid and labour increasingly left farms in the second half of the nineteenth century, attracted by better pay and more free time in industrial work and able to get there owing to the increased mobility offered by trains, then bicycles and eventually omnibuses and motorcars; agricultural wages and conditions were always better in areas nearer alternative employment (Mutch 1981, Orwin & Whetham 1964). Similar factors affected the supply of seasonal, casual and migrant workers upon which the hay and cereal harvests particularly and heavily depended from the late 1830s, largely because of competition for labour from railway building (Collins 1969). Although the situation eased in the later 1840s with an influx of migrant Irish labour after the potato famine, this declined after the 1850s and the availability of casual and seasonal labour always remained uncertain from this period (Collins 1969). In some areas cottage trades such as weaving were an important source of seasonal labour for farming in peak periods of demand but this source was lost with the increasing concentration of textile production in factories (Mutch 1981). By the late 1860s, the Victorian farm literature constantly refers to the need for mechanisation if farmers were to guarantee their harvests. The other great pool of casual labour was provided by women and children. This progressively dried up with factors such as laws limiting the use of child labour and introducing compulsory schooling (Orwin & Whetham 1964) and Collins (1969) also remarked on a change in attitude by women, against farm work. Labour has long left farming, given the opportunity, quite as much as it has been shed by farmers. Mechanisation has been as much a necessity as a choice.

Drills and hoes

The first new machine of major importance to farmland ecology was the seed-drill. Jethro Tull's seed-drill of the early eighteenth century was not the first seed-sowing machine developed in Britain. At least two workable machines were designed in the seventeenth century (Trow-Smith 1951) (Figure 1.1). But with his drill Tull developed a system of 'drill-husbandry', based upon sowing crops in widely spaced rows (or drills) and keeping them weed-free during growth by regular passes along the drills with either a manual or a horse-drawn hoe (an implement he also designed) or both. Until the development of herbicides such methods became commonplace in the nineteenth and twentieth centuries. Thoroughly applied and managed they were very effective in controlling weeds (Chapter 7). Nevertheless early drills were inefficient and a successful and popular machine was not developed before the nineteenth century; sowing cereals in drills by hand-dibbling was a frequent practice. County farming accounts were still noting whether or not all corn was drilled into the 1850s. The advantages of Tull's methods for establishing and managing root crops, however, were recognised and adopted much more quickly. Root crops were the basis of all high farming rotations and weed-free root crops meant a clean rotation cycle.

The ecological significance of drilling and hoeing is that they provided the first real opportunity for efficient weed control in growing crops, simply because efficient hoeing in broadcast crops is impossible and hand-weeding difficult; crops need to be spaced in rows, allowing easy systematic access (Figures 7.3 and 7.4). Salisbury (1961) noted that a number of weed species largely vanished from the farm flora as a result of the development of drill husbandry, citing darnel (*Lolium temulentum*), dodder (*Cuscuta trifolia*), red chamomile (*Adonis annua*) and annual larkspur (*Delphinium ambigua*) amongst other examples. Newton (1896) had little doubt that the efficient weed control possible in high farming with the use of seed-drill and hoes contributed significantly to the nineteenth-century decline of some seed-eating passerines such as Goldfinches (Chapters 3 and 7).

Drill technology has changed a great deal over time. One such change has some ecological significance. Modern corn-drills sow at significantly narrower row spacings which greatly reduces the suitability of winter cereal fields as feeding sites for birds such as plovers (Chapter 10). The modern technique of 'tramlining', where fertiliser distributor and sprayer cover the same span, which may be several drill widths, gives more exact use of chemicals and promotes greater uniformity in crop density. The latter can create problems for ground-nesting birds, chicks of which may find such crops difficult to negotiate. Donald & Vickery (2000) also found that 34% of Skylarks' nests in autumn cereals were next to 'tramlines', a proximity which significantly increased nest losses by facilitating access for predators.

Haymaking

In the past the hay and corn harvests were the largest users of labour and the jobs most vulnerable to and dependent on the weather in the farming year (Figure 11.5). Mechanising haying involved haymaking machines, mainly tedders to scatter it to dry and rakes to collect it up again in the nineteenth century, and mowers. By the end of the century turners, sweeps and loaders were also developed. Haymaking machines probably came into widespread use earlier than mechanical mowers. Thus Read (1855a) and Rowley (1853) both note the widespread use of the former in Oxfordshire and Derbyshire but do not even mention the latter. Within a decade mowing machines were described as widespread in many counties, for example Berkshire (Spearing 1860), Leicestershire (Moscrop 1866), Staffordshire (Evershed 1869) and Worcestershire (Cadle 1867); in southwest Lancashire over 70% of farms were cutting hay by machine by 1877 (Mutch 1981).

Mechanical mowing was perceived as having a deleterious impact on some ground-nesting birds because of the scale of nest losses caused. The evidence of the nineteenth-century avifaunas, however, was that mowing by whatever method caused numerous nest losses in species such as Quail, Corncrake, Whinchat and Corn Bunting which frequently nested in hay crops. Quail were then regarded as quite as vulnerable to mowing as Corncrakes and modern observations, though scant, hint at a change in nesting habitat, with Quail now particularly associated with cereals rather than grassland. For Corn Buntings Dennis, in his Sussex diary for 1848, remarked for that 'all the numbers of nests brought to me this mowing season' were in seeds hay (i.e. leys) and again on 22 June that 'the mowers have cut out a great many nests, some fresh, some hard set'. Corn Bunting nesting habitats seem also to have changed, with declining use of improved grass such as leys and increased use of cereals, field boundaries and rough grass (Table 11.1). Although the sample of nests up to 1950 is small, it continues the trends shown by the later periods. Whinchats have behaved similarly. Although Alexander & Lack (1944) recorded little overall change in numbers in Britain, detailed examination of county avifaunas up to the early 1950s suggested that distribution had actually become more patchy since the early 1900s and indicated a habitat change. Up to 1914 63% of

Figure 11.5. These two photographs show the scale of labour needed to harvest by hand. (A) The scythe team (Suffolk 1888) are probably working in corn, as the scythes have cradles but similar teams mowed hay. (B) There are at least 22 hands visible in the haying scene (Berkshire 1906). (Courtesy of Rural History Centre, University of Reading.)

TABLE 11.1. *Nest sites used by Corn Buntings*

Period[a]	Number (and %) of nests reported in				
	Cereals	Other tillage crops	Improved grass[b]	Field boundaries	Rough grass and scrub
1. Up to 1950	7 (13)	6 (11)	20 (37)	15 (28)	6 (11)
2. 1950–70	60 (29)	5 (2)	39 (19)	58 (28)	46 (22)
3. 1970–89	99 (36)	8 (3)	16 (6)	99 (36)	56 (20)

[a] The differences between periods were statistically significant; 1–2, $\chi^2 = 23.77$, df 4, $p < 0.01$; 2–3, $\chi^2 = 20.86$, df 4, $p < 0.01$.
[b] Virtually all nests in improved grass were in leys.
Source: Information up to 1950 is taken from county avifaunas, from 1950 from BTO nest record cards.

avifaunas recording nesting habitat for Whinchats noted a strong liking for nesting in hayfields. After 1914 only 19% of accounts of nesting habitat included them and 62% did not. That this was a genuine change in habitat use is supported by more recent studies. Gray (1974) noted many nest sites within easy reach of hay fields in Ayrshire but none in them. His study also showed how much mowing could affect nesting success, as 46% of nests in road verges were lost when amenity mowing and nesting clashed. Fuller & Glue (1977), in an analysis of nest record cards, noted that 58% of Whinchat nests were in agricultural grassland but predation was the main cause of nest loss and hay crops were not mentioned at all. Presumably changes in nesting habitat by these species, once common in hayfields, arose because populations using hay crops have tended to die out with the mechanisation of haying. Few young were produced to perpetuate the habit. They still occupy other habitats successfully.

In this they differ sharply from the Corncrake, which has always been recorded as closely tied to hay crops, although it was also once common in the fen habitats also favoured by Quail and certainly used cereal and bean crops in many areas. Virtually every early avifauna discussing the subject linked the species' decline from the late 1880s to the switch to mechanical mowing during the second half of the nineteenth century, because of the high level of nest destruction it caused. Modern research has continued to link decline to modern methods of cutting hay or silage combined with changes in the timing of this harvest. The importance of silage can, perhaps, be overstated; for example Corncrakes disappeared from most of Wales well before it became a significant factor in grassland management there.

MESSRS. H. AND G. KEARSLEY'S COMBINED MOWING AND REAPING MACHINE.

Figure 11.6. A combined reaper/mower, a common machine in the Victorian era. (Courtesy of Rural History Centre, University of Reading.)

It is difficult to construct an accurate chronology for the introduction of mowing machines. It is reasonably safe to assume that their use would have spread rapidly from around 1880 with increasing labour shortages, an assumption supported by Wilson (1902) and Speir (1906). One problem is that the Victorians did not consistently distinguish mowers (for hay) and reapers (for corn). Early machines used for cutting hay often combined both functions (Figure 11.6) but were usually discussed under reapers. Thus Wilson (1864) described and compared such machines from 11 makers and noted that one had sold 279 machines in the north of England and Scotland during 1860–2 alone. He was discussing them as reapers but noted 'the machines in the above class – in fact all machines on a clipping principle – come under the category of Grass Mowers, and are ordinarily used as such'. Many such machines came with separate cutter bars for corn and for grass. Mutch (1981), in noting the high proportion of hay cut by machine in southwest Lancashire by 1877, reported that 60% of farms cutting mechanically used combined machines during 1877–81 and 72% during 1887–91. Only when the reaper–binder, which tied the sheaves mechanically, came into widespread use for corn at the end of the nineteenth century, was the use of the mowing machine more clearly defined. Their use also varied in districts. Fields where grass lay in old ridge-and-furrow were difficult to mow by machine, so their introduction there was delayed until the conformation of the land was changed, although haymaking machines could be used (Jonas 1846). In Berkshire Spearing (1860) noted that, although grass was by then mainly mown by machine, farmers preferred to mow clover and sainfoin by scythe as it lost less leaf in the process. Hennell (1934) also remarked that more hay was then made by hand 'within a score of miles of London than in all the western counties together' suggesting that mechanisation might have been quicker in more uncertain western climates. It is also clear from Wilson's (1864) paper that the idea that mowers were later to reach the north of England and lowland Scotland than south and east England must be discarded; their introduction and spread was probably quicker in the north, although they probably reached the upland regions last. Overall the introduction and spread of mowing machines is not as obviously correlated with the decline of the Corncrake as Norris (1947) proposed. Haymaking machines were also important. The largest use of labour in the hay field was for making the hay, not mowing it. Haymaking machines speeded the work and freed hands to mow. A good team of scythemen would have mown hay quite as quickly as a horse-drawn mower.

In this context it is worth studying the fascinating painting 'View of the harvesting field of James Higford's manor, Dixton, Gloucester' painted c. 1725–35, reproduced in Furbank *et al.* (1991). Not only are there 23 mowers (each cutting a 10-foot swathe; Hennell 1934), there are altogether 104 people, five dogs and 18 horses in the field and another eight workers are coming to join them, singing as they come; at least three groups are having impromptu parties! This illustration makes clear that once work started in any field, Corncrakes' nests were unlikely to survive, a point also highlighted by Graham's (1890) comment that 'in Iona many nests are discovered when the grass is cut for hay'. Some artistic licence may have been taken but accounts of how the job was done suggest that the picture accurately displays the high level of human involvement (also Figure 11.5). The picture shows a social event, a community combining to do an essential job and having a party doing it.

It was the level of communal input that was important. I suggest that mechanisation increased the impact of haying on the nesting success of Corncrakes (and other species) because it changed the timing and organisation of the work. For example Jenkins (1971) set out in detail the organisation of haying in southwest Wales in the late nineteenth century. It was a system of co-operative handwork either by parishes or groups of farms, in which the fields were cut and the hay made in what was often a traditional order and time, field A on farm A followed by field A on farm B and so on,

Figure 11.7. Reaping, a romantic print from a painting by Richard Westall (1797), showing how the job was done but not the scale of labour involved. Jonas (1846) noted that he 'had set 300 men to reaping in the past'. (Courtesy of Rural History Centre, University of Reading.)

a system which gave all participants a share of good and bad weather. It also meant that many traditional hay fields would have been much securer nesting sites for Corncrakes. Such work patterns were fairly general. The introduction of machines did away with any necessity for such communal operations. Farms could and did harvest individually and, as haying is highly weather dependent, all at the same time. The decline of farm labour was an important factor in southwest Wales and Jenkins stressed the impact of the 1914–18 War, which decimated the workforces available in rural communities. There is little doubt that the loss of the traditional work patterns with mechanisation disrupted the Corncrake's breeding patterns and was a prime cause of its decline.

An important point advanced about the impact of mechanical mowing on Corncrakes is that machines, unlike scythemen, work round and round the field, so trapping the birds in the danger zone (Tyler *et al.* 1998). But anyone who has worked with a mowing or reaping machine knows that any ground-living bird trapped in a field being so cut behaves in the same way. So do rabbits and hares. The Corncrake's real problem has been an inability to adapt on any scale to any farmland habitat other than hay crops once it lost its ancestral fen grasslands. As farmland habitats continue to change and decline in diversity, increasing numbers of farmland species are facing similar problems of adaptability.

The way in which mowing machines actually cut may also be of significance for some birds. Older machines used finger bars, not unlike the gardener's hedgecutter, with a reciprocating blade running in pointed fingers (Figure 11.6). The cutter bar was hinged at the machine end and rode on a shoe at its outer end. Early observers claimed that they cut closer than the scythe and so were more liable to kill sitting birds. But they will also ride over nests without destroying them and I have found Skylarks nests still active in a freshly mown crop of clover hay. Interestingly Dennis's diary specifically remarked that the mowers brought him no Skylarks' nests and, in fact, no nineteenth-century avifauna that I have examined, except Lilford (1895), gave any indication that mowing or hayfield management affected Skylarks, a species which usually nests in a depression in the ground rather than within the herbage of the crop. Yarrell (1837–43) noted that 'the little hollows which the birds generally take care to make save most of the early broods'. Wilson et al. (1997), however, observed in their study that 26% of known nest losses derived from farmwork, mostly from cutting silage or set-aside grass. Modern mowers use counter-rotating drums fitted with cutters, which run at high speed under a cover, like the gardener's rotary mower. The chances of any nest escaping the action of such a machine is nil.

The corn harvest

In the early nineteenth century corn was still largely harvested by sickles (reaping; Figure 11.7). The first major technical change in this work was a switch to mowing with a scythe, often fitted with a cradle which gathered each slice and deposited it as a sheaf, or with a bagging hook (see Collins (1969) for a full discussion). Even in the early 1850s comments that farms now mowed their corn were frequent enough to indicate that it was then an ongoing change. Once cut, by either method, the corn had to be bound into sheaves and stooked as weather protection until it could be carted and ricked. Barley was often handled loose like a hay crop. Each stage in this process involved some waste to lie on the stubble, mowing more than reaping because it was untidier. Experiments with mechanical reapers started at the beginning of the century but satisfactory machines were not developed until mid century. They then spread quite quickly and Wilson (1864),

Figure 11.8. A mechanical reaper, one hand employed raking the cut corn clear, two raking the sheaves together and two tying. (Courtesy of Rural History Centre, University of Reading.)

for example, noted that 17% of cereals was cut by machine in East Lothian by 1860. Early machines only cut the corn. It still had to be tied manually into sheaves (Figure 11.8). Reaper-binders, always known as binders, fitted with mechanical knotters to tie the sheaf, were developed from the 1870s and more or less universal by the end of the century (Orwin & Whetham 1964). The further tasks of stooking and carting remained.

Such changes in methods had only marginal effects on birds. Hennell (1934) noted that the practice of gleaning (gathering by hand grain dropped during harvesting) declined steadily with the progressive introduction of machines because they were less wasteful than handwork. But this was partly a social change; gleaning, which once supplied the winter's bread, became less necessary in the cottage economy. Machines also meant quicker harvests so that grain stood in the field in stooks for a shorter period. Such factors were unlikely to affect most birds when the crop remained accessible in rickyards. One result of the decline of labour in the last quarter of the nineteenth century was that ricks were built in fields to save transport (Whetham 1970), spreading such winter food supplies more widely about the farm to the advantage of species such as partridges (Figure 10.1). Threshing tended to be a progressive job through the winter, so a continuous supply of chaff, weed seeds and small grains became available for birds to scratch and feed in.

It was the combine harvester that reaped and threshed the corn in one operation that completely changed grain harvesting and storage and the

Figure 11.9. The switch from binder to combine resulted in a very marked decline in the availability of cereals to birds at both harvest and in winter. (Courtesy of Rural History Centre, University of Reading.)

pattern of food supplies in autumn and winter that the traditional methods had made available to birds. The combine was an American invention and the first workable machine was patented in 1836 (Partridge 1973). By the 1850s combines were in regular use in California (Trow-Smith 1951) and were widely used elsewhere in the American West by the turn of the century (Figure 11.2). Combines did not come into Britain until the 1930s, and then very few. There were still only 10 000 in 1950 but by the early 1960s their use was virtually universal.

The introduction of combines affected the abundance of waste grain in stubbles, since they waste far less than binders, with the repeated handling of sheaves the latter involved. In addition the corn no longer stands in stook for a significant period in late summer, when the entire crop was readily available to birds (Figure 11.9). Combines, too, have contributed to the declining availability of winter stubbles because the rapid clearing of crops enables earlier autumn cultivations. Much more significantly, with combines the old rickyard has vanished completely (Figure 10.5). The implications of this are discussed in detail in Chapter 10 but it is worth noting here the totality of this change. Other major food sources exploited by birds have been affected by mechanisation or by changes in

TABLE 11.2. *Changes in Rook populations in Britain from the 1930s to 1996, together with changes in the broad composition of farmland*

Period	Proportion tillage : grass	Area, × 1000 ha (change%)			Change in Rook populations
		Tillage	Grass	Cereals	
1930s	30:70	3478	8488	2060	
1946	44:56	5101 (+47%)	6533 (−23%)	3332 (+62%)	+28%
1975	43:57	4793 (−6%)	6450 (−1%)	3703 (+11%)	−43%
1980	44:56	4936 (+3%)	6244 (−3%)	3942 (+6%)	+7%
1996	42:58	4701 (−5%)	6436 (+3%)[a]	3316 (−16%)	+33%

[a] Grassland area includes set-aside.
Sources: June Census Statistics, Sage & Vernon (1978), Marchant & Gregory (1999).

seasonal management or rotations but, as Figure 10.5 shows, they still exist if greatly reduced. The old rickyard vanished in less than 20 years and the speed with which this loss occurred was unprecedented.

The way in which such a change in the mechanical processes of farming can affect a species via its food supplies at a particular season is well illustrated by the effect of the use of combines on the population regulation of Rooks. O'Connor & Shrubb (1986) noted that the loss of stooks, which covered several million ha of the British countryside in the period late July to September before the advent of combines, greatly reduced the availability of grain in the late summer period when grassland feeding was most difficult and most mortality in Rooks occurs (Dunnet & Patterson 1968, Murton 1971). Mortality increased with this loss of food supply, making a major contribution to the decline noted from the late 1950s to 1975 by Sage and Vernon (1978).

Changes in Rook populations in Britain since the 1930s are summarised in Table 11.2, together with contemporary changes in the composition of farmland. There is no consistent relationship between the two. Dobbs (1964) related a continuous increase in Rooks in Nottinghamshire from the 1930s to 1958 to the expansion of the cereals area. Table 11.2 suggests strongly that this link was general. The decline recorded in 1975, which started around 1958–60 (Dobbs 1964), was not related to change in the cereals area but is well explained by the disappearance then of grain standing available in stooks in late summer with the use of combines. That deficiency has now been corrected with the expansion of winter barley from the early 1970s. These occupied 19% of cereals by 1980 and peaked at 26–27% by the early 1990s. Winter barley is harvested from the second

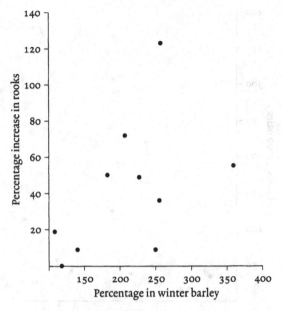

Figure 11.10. The increase in Rooks by region in Britain between 1975 and 1996 in relation to the increase in the area of winter barley. Regions as in MAFF June Census Statistics Scotland is treated as one region. The relationship is significant:

$$r_s = 0.58, n = 10, p < 0.05.$$

(*Sources:* June Census statistics and Surveys of Fertiliser Practice for barley area and Sage & Vernon (1978) and Marchant & Gregory (1999) for Rooks.)

half of July and has thus provided an extensive source of stubble grain at the crucial time in late summer. Its importance to Rooks is confirmed by Figure 11.10, which shows a significant relationship between regional changes in Rook numbers and regional changes in the area of winter barley. No similar relationship was found with changes in regional grass, tillage or cereals area. The abundance of Rooks, however, was correlated with the area of improved grass available (Figure 11.11) not that of winter barley, cereals or tillage. Thus, whilst Rooks now tend to be more numerous where there is more improved grassland, population fluctuations have been driven by the availability of stubble grain in the late summer/early autumn since at least the 1930s.

Minimal cultivations

Strictly this subject concerns systems of cultivation rather than mechanisation but it requires specialised machines, particularly drills designed

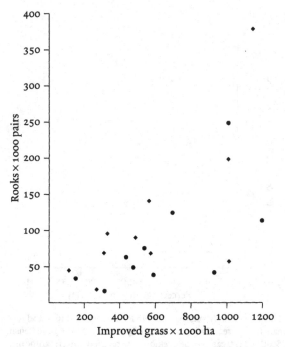

Figure 11.11. Rook populations by region in Britain in 1975 (•) and 1996 (♦) compared to the area of improved grass. The correlation was significant for both years: 1975, $r_s = 0.70$, $n = 10$, $p < 0.001$: 1996, $r_s = 0.64$, $n = 10$, $p < 0.05$. (Data as for Figure 11.10.)

to sow directly into unploughed ground such as stubble. Such systems have been the subject of extensive research and experiment since the 1950s and interest in them has accelerated, as they offer significant savings in running costs. They require, however, considerable capital investment in specialised machines and economies of scale may eventually govern how widely applied and profitable they are. Their limited application to date has probably resulted from problems in weed control, particularly grass weeds.

There are three main ideas involved in such systems, often known today as lo-till. One is to drill direct without preliminary cultivations. A second is to work the surface shallowly, using disc harrows and cultivators to create a seedbed and incorporate trash from previous crops in the top inches of soil. A third option adopts shallow and wide ploughing and conventional cultivations. An important objective is to avoid bringing weed seeds to the surface. Weed seeds in the top inches of soil are encouraged to sprout by surface cultivations and killed by glyphosate. The main store of dormant seeds lying deeper in the soil stays undisturbed. A problem with all such systems is disposing of the trash of previous crops. For cereals, now burning is banned, the preferred method is to fit an efficient straw chopper to the combine and spread the resulting chaff evenly so that it can be worked into the top soil.

Some interesting implications for the ecology of farmland birds are involved in this methodology. Its application must further reduce feeding opportunities and food supplies for seed-eating birds, simply because such shallow cultivation will no longer bring seeds to the surface from the soil's seed bank in the way that conventional plough cultivation does. This also has long-term implications for those seed banks. However, the modern predominance of autumn cultivation may mean that this is of limited immediate significance, since it is the loss of spring tillage which has been most important here, spring being the season of shortest supply from other sources. By contrast increases in populations of soil invertebrates, particularly earthworms, have been observed (Allen-Stevens 2000). But availability is also important to birds; large quantities of invertebrate food are of no value if unobtainable. Just how important the flush of such food that traditional ploughing makes available to birds is would be worth examining. Ausden (2001), examining invertebrate food supplies in flooded grassland, suggested that availability may be more important than abundance. However, availability may also be influenced by incorporating trash in the top inches of the soil profile instead of ploughing it down. This is likely to encourage earthworm activity near the surface, within range of many birds' feeding techniques (see Tucker (1992) for a parallel case with manure applications to pasture). Compaction problems may arise from the size and weight of the machines used, although this can be dealt with by subsoiling. An increase in surface invertebrates with lo-till has also been observed (Allen-Stevens 2000), including numbers of beneficial aphid predators. Such an effect will provide food for birds and reduce pesticide demand. Indeed reading recent reports on these techniques in the farming press suggests that they could lead to significant changes in pesticide and fertiliser regimes in cereal farming, which would seem to warrant close examination by ecologists. They seem particularly likely to see an increase in molluscicide use. Finally, if they become widespread, minimal cultivation techniques will complete the modern agricultural revolution by doing away with the historic basis of all cultivation, the plough.

Farmsteads

Every farm had a farmstead, with house and garden, orchard and sheltering trees, barns, stables and byres, space for building corn and hay ricks and a source of water, usually a pond, for stock, especially horses. These elements were repeated throughout farmland but the size and complexity of such steadings varied with farm size and use. House, garden, shelter, barn and pond were those most widely repeated; features such as rickyards and horse ponds are now generally redundant. Many farms had more than one set of buildings.

Change in the design and purpose of farm buildings has always continued. Much new building was done after 1875, as landlords sought to attract new tenants and encourage enterprises which might be more profitable than corn. New farmsteads were laid out particularly to facilitate the handling and mechanised feeding of stock (Wade Martins 1991) and to meet new hygiene regulations in the growing dairying industry. Old timber-framed buildings were often also dismantled and rebuilt elsewhere in a more convenient place or style. Barns on our family farm showed much evidence of this. The materials used in traditional farm buildings tended to reflect the underlying geology and resources of the area: timber in the southeast, brick and pantile in the east, stone and slate in the west and north (Darley & Toler 1981). Thatch was a common roofing material in arable areas. New buildings in the nineteenth century continued to use traditional materials but often from commercial sources. This was obvious, again, on our family farm. The older barns were built with timbers cut locally and roughly shaped, nineteenth-century buildings used sawmill timber but were constructed in the traditional way. Wade Martins' (1991) photographs often show the same point.

Since 1945 there has been a revolution in the design and materials used for farm buildings. The modern demand is for widespan buildings, with prefabricated steel or prestressed reinforced concrete frames clad with corrugated asbestos or metal sheeting or, for stock, with wooden slatting (Figure 11.12). This revolution has been largely fuelled by the mechanisation of materials handling and the need for grain stores for threshed grain in bulk behind the combine. Tractor-mounted front-end loaders or forklift trucks, fitted with a variety of attachments, now load and stack hay and straw, handle manure and palleted goods and materials and shovel bulk grain and roots, and need uncluttered spaces to work efficiently. Perhaps the greatest advantage of modern materials is the ability to construct

Figure 11.12. The old and the new. (A) A fine set of thatched buildings but awkward to adapt to modern demands. (B) Grain handling and store (with caddy rooves) at back.
(Courtesy of Rural History Centre, University of Reading.)

buildings with wider and completely clear spans, doing away with the clutter of posts, beams and kingposts common in traditional construction. Illustrations of traditional barns often incidentally make clear the point that a great many are all roof (Figure 11.12A), greatly restricting their storage space and utility. Nor could buildings of traditional construction be adapted for the bulk handling and storage of grain and other crops, which need buildings designed to cope with the stresses involved (Figure 11.12B). It is such demands that have led to many old farm buildings becoming obsolete and redundant. In the past 25 years or so new buildings have also spread rapidly into pastoral regions. Even in sheep-farming districts, traditionally a form of farming that used few buildings, most holdings now have purpose-built lambing sheds to lamb their flocks under cover.

In the past redundant buildings would probably simply have been bypassed, used occasionally but not considered worth repair and maintenance (Wade Martins 1991) and otherwise left to crumble quietly into the landscape. In much of the nineteenth and twentieth centuries dilapidated and tumbledown old farm buildings featured quite widely in the countryside, to the benefit of owls and Stockdoves and other birds that like to nest in them. Today, however, such buildings are a source of capital which, encouraged by the urge to preserve them as part of our 'heritage', results in their conversion to houses. This in fact destroys both their historic significance and ecological value within farmland. The scale of such conversions is indicated by advertisements in *Country Life* magazine. In the first six months of 2001 for example, at least 7% of all country houses offered for sale there were converted farm buildings.

Wade Martins (1991) estimated that 90% of fully working farm buildings were constructed after 1918. Effectively that means post-1945. Whilst the speed of this change owes something to the scale of grant aid available from 1947 for new buildings, arguing that such grants were a primary cause ignores the extent to which technological demands have imposed the need for buildings of different design and capacity to traditional forms.

Farmsteads are of interest for birds for three principal reasons: the storage of crops and feeding and management of livestock provides significant food sources, particularly in winter, farmsteads are significant reservoirs of habitat and the buildings themselves are of importance to some birds as nest sites.

Farmsteads as feeding areas

Farmsteads were important feeding areas in the past and the resources available there contributed significantly to winter survival (Chapter 10). That chapter emphasises the decline which has occurred with modern farming. Ever fewer arable farms retain livestock and the bulk storage of grain in silos and closed bins, increasingly bird-proofed to meet hygiene regulations, provides many fewer feeding chances. But one cannot handle grain in bulk without some spillage which birds can exploit and modern grain-cleaning plants still produce weed seeds, chaff, chippings and small grains, for disposal. This is often used to feed game but all birds can then exploit it. The spread of lambing sheds, particularly in Wales, may affect populations of the predatory scavengers – Kites, Buzzards, Ravens and Carrion Crows. For all these species the seasonal availability of carrion is important and lambing in sheds has reduced it considerably. Little sign of such impact has emerged so far but the change is relatively recent and a different pattern may yet arise.

Farmsteads as reservoirs of breeding habitat

Several authors have commented that the gardens, shelter belts, ponds and other common features of farmsteads provided a reservoir of breeding habitat that is often decreasingly available elsewhere in modern farmland. Thus I found that, on my CBC area at Oakhurst in the 1960s, farmsteads and gardens comprised c. 2% of the total area but contained 20% of the total hedge line and 46% of the total breeding population of hedge and garden birds (Shrubb 1970). In a Cambridgeshire parish Wyllie (1976) found that total bird density was seven times greater in the village (13% of total area) than in the associated farmland, with half the village territories associated with buildings and gardens. However it is easy to overestimate the importance of such proportions in considering the overall significance of farmsteads for breeding birds. Lack (1992) found that it was species such as Wren, Robin, Blackbird, tits and Chaffinch which mainly benefited, species primarily of woodland. More strictly farmland species, such as Linnet and Yellowhammer, occurred equally frequently around farmsteads and in field hedges and ground-dwelling birds avoided farmsteads. Thus although farmsteads add to the diversity of farmland for breeding birds, their importance for true farmland species is more limited.

Most farmsteads had a horse pond in the past and on some farms they may have been the only significant open water. O'Connor & Shrubb (1986) found that numbers of 33 out of 57 farmland birds they examined were enhanced by the presence of ponds but that ponds were the most important feature for five only, perhaps particularly Moorhens. Thus such ponds are again most important for the contribution they make to diversity. Many horse ponds have vanished in recent years. Relton (1972), for example, found that in the parish of Kimbolton, Huntingdonshire, the number of farm ponds declined from 152 in 1890 to only 67 in 1969. In my Sussex area at least 22% of the farm and field ponds marked on the first series of the 1:25 000 OS map (Sheet SZ 89, 1959) have also since disappeared. Field ponds are probably more valuable habitats, because they are more likely to have emergent vegetation. At Oakhurst our horse ponds never had anything except Moorhens and occasional Mallard nesting but our field ponds always also had Sedge Warblers, Reed Buntings and sometimes Reed Warblers. None was large enough to attract species such as grebes and Coot but a possibly important modern trend is the construction of sizeable farm reservoirs for crop irrigation, often in areas of low rainfall. These may be colonised by marginal vegetation and be big enough to hold a pair of Great Crested Grebes. Certainly such sites will prove attractive to Little Grebes and Coot, and Lovegrove *et al.* (1994) linked an increase of the former in Wales to the increased construction of farm ponds and reservoirs.

Buildings as nest sites

Many species use farm buildings as nest sites on occasion, for example Shelduck, gamebirds, Tawny Owl, Pied Wagtail, Wren, Robin, Blackbird, Spotted Flycatcher, Jackdaw and Starling in my own experience. For a few they provide significant numbers of sites, perhaps most importantly Barn Owl, Little Owl, Swallow and House Sparrow. Buildings are also at least locally important for Kestrel, Stockdove and possibly Jackdaw and Starling. Virtually all species regularly using buildings also nest in tree cavities and cliff sites; Swallow was presumably originally a cave nester. Birds nesting in buildings use thatch, holes in or the tops of walls, roof spaces and lofts, and hay or straw stacked inside. Whether or not farm buildings are in use seems unimportant, perhaps because use is so often seasonal.

Regional and local variations in the use of these nest sites occur. Thus abandoned farm buildings were the most frequent rural nest site for Kestrels in northwest England (Shrubb 1993a). A similar regional variation occurs with Stockdove (Spencer 1965). It arose for both species because, as new reservoirs were constructed in the late nineteenth and early twentieth centuries, all farms within the catchment areas were abandoned by order, leaving derelict houses and buildings which the birds readily occupied. Shawyer (1987) noted that Barn Owls in Britain use proportionately more buildings as nest sites in the west and tree cavities in the east; tree cavities also formed 35% of nests in England but only 7–8% in Wales and Scotland. Toms (2001) recorded the same pattern. Little Owls made far more use of farm buildings for nesting in my Sussex area than Glue & Scott's (1980) national analysis found; 83% of the nests I found were in buildings, particularly in thatch inside open-fronted sheds such as cattle hovels.

Site use also changes with availability. Kestrels in my Sussex area mainly nested in elm trees (90% of nests found) until 1974, when Dutch elm disease appeared locally. By 1979–85 56% of nests were in buildings, a high proportion of the trees having gone. Barn Owls have behaved similarly. Comparing the records Shawyer (1987) gave for nest site use from 1982 to 1985 with those he quoted from Blaker (1934) shows a decline in the importance of tree sites (from 44% to 35% of nests) and an increase in the number of agricultural buildings used (from 40% of sites to 51%); the proportion of domestic buildings and cliff sites was unchanged. The difference was statistically significant ($\chi^2 = 51.13$, df 3, $p < 0.01$). Nest boxes were excluded from these analyses but the sharp increase in nest box provision recorded in the BTO's nest record cards for Kestrel after 1974 was fuelled by the scale of tree loss to Dutch elm disease.

The concrete or steel-framed and sheeting clad construction now usual in farm buildings offers little in the way of usable nest sites – no thatch, no lofts and roof spaces, no vents ('arrow slits') in stone walls, no broad tops to such walls, little in the way of ledges and so on. The important point about the shift to using agricultural buildings by Kestrels I noted in Sussex was that the birds were invariably using a bale-stack or a nest box I supplied, not the structure of the buildings. This is also an increasing problem for Barn Owls. Shawyer (1987) recorded 62% of all nests in agricultural buildings in bale ricks (39%) or nest boxes (23%). Toms's (2001) results are not comparable because bale ricks in buildings were not shown separately but 52% of all occupied nests in agricultural buildings were in nest boxes and the presence of boxes was the most significant factor influencing occupation of a tetrad. He also examined the status of 117 of Shawyer's sites in farm buildings and at least 37 (32%) had been lost, half converted to housing and half demolished as unsafe. This is increasingly likely to happen to redundant farm buildings. The loss of established sites may lead to Barn

Owls abandoning a territory in many cases, even if apparently suitable alternative sites remain (Ramsden 1998).

Similar problems may now arise for Swallows and House Sparrows. For these species there are also other problems. Prudent farmers keep all their stores locked against thievery, grain stores and feed lots are increasingly bird-proofed to meet hygiene regulations and any store used for pesticides has to be securely locked by law. Inevitably access for birds has declined steeply. Such losses of potential nest sites through reduced access and suitability and the development or removal of buildings, may have contributed significantly to the recent decline of House Sparrows. Increased public access to the countryside may well exercerbate this as more farmers lock their buildings or remove unwanted ones, perhaps particularly to avoid legal liability in the event of damage or accident.

Swallows have been affected by another factor. Møller (1983) found that Swallows preferred cow and pig sheds for nesting to barns. I found the same on our family farm and Roberts (2001) made similar observations for two Welsh farms. Roberts convincingly linked this preference to security. Swallows used the smaller and darker buildings with more restricted access points on his study sites because they were much less likely to be predated by species such as Magpies. Dutch barns, open-sided sheds and large barns were all more vulnerable and avoided. But Møller also observed that the most marked feature of farms with the larger colonies he studied was the presence of livestock. When livestock went, as his study farms switched to arable cropping, the Swallows left also. He attributed this mainly to the consequent loss of pasture and meadows, which were the prime feeding areas for Swallows. Although Møller (2001) specifically excluded it on his site, questions of access are also likely in such situations. With animals ventilation and easy regular access are necessary. Without, buildings are more likely to be kept closed.

Buildings as winter roosts

As well as nest sites, buildings provide secure sheltered roosts for many of the same species, which may be particularly valuable in winter. Mills (1975) found for American Kestrel that the distribution of buildings and other sheltered roosts governed winter distribution in Ohio. In Britain Barn Owls and Kestrels usually have more than one regularly used roost within their ranges (Shawyer 1994, Shrubb 1993b) and even where they breed in trees Barn Owls, and Little Owls in my experience, are likely to use buildings for winter roosts. Unlike for nest sites, modern building styles provide perfectly adequate winter roosts on assorted ledges, where the need is for shelter from adverse weather. I examined the roosting behaviour of Kestrels in some detail on my Sussex area and found that, of 24 winter roosts recorded, only two were in trees and the rest in or on buildings or on bale ricks in buildings. Kestrels were perfectly happy using ledges such as sheeting rails. Barn Owls perhaps need more room and usually used beams in old buildings or bale ricks. Access to the building is not necessary for Kestrels as long as there is a perch in a sheltered site tucked under the eaves. Dutch barns, useless for nesting, provide excellent roosts under the roof. It seems likely that warm and secure roosts can contribute significantly to winter survival, especially for a species such as Barn Owl, which is susceptible to rough wet weather, a more frequent winter hazard than snow or severe cold in much of its British range. Such roosts are probably of value for other species, such as Wrens, in hard weather. Roosting behaviour appears to have been little studied for many birds but would repay detailed examination in an era when many of the most valuable sites are disappearing from the countryside.

12

Exploitation and persecution

Farmland birds in the nineteenth century were exposed to persecution as pests or by collectors, or to exploitation for food. At least 61 species in Table 2.1 were affected to some degree and several others, extinguished as breeding birds in the nineteenth century, such as Bittern, Black-tailed Godwit and Bearded Tit, were lost as much to persecution as habitat change. Early rural communities made more use of wild birds than we would consider today. Such use may have increased with enclosure which seriously damaged the economies of the rural poor. Jones (1972) noted of sparrow trapping that 'Farmers could also encourage the practice in the belief that they were using their mens' spare time to keep down a serious pest while allowing them to supplement low real wages with a protein supply.' The nineteenth-century avifaunas refer not infrequently to such things as sparrow or Corn Bunting pudding.

The exploitation and persecution of farmland birds falls under several distinct headings, exploitation for food, the cage-bird trade, pest control, game preserving and trophy collecting. How far these actually affected the size of farmland bird populations is a matter of argument. The malign influence of game preserving on predators and of trophy collecting on rare birds is undoubted. The effects of activities such as harvesting plovers' eggs are more disputable but nineteenth-century ornithologists argued that such activities interacted with habitat changes to birds' detriment. The nineteenth-century avifaunas covered such matters as bird-catching, egging and persecution thoroughly and where individual avifaunas do not record particular activities, it is probably safe to assume that they were not systematic trades, although casual exploitation probably occurred everywhere.

Food

Egging

Birds' eggs were exploited both commercially and, more casually, by the rural poor for food. The latter is illuminated by several anecdotes. MacGillivray (1837–52) recounted being regaled with tea and fried eggs at a Hebridean croft – Corn Buntings' eggs fried on a shovel – and inferred that songbirds' eggs were often cooked and eaten in this way. Pashley (1925) recorded of one of the first Sandwich Tern clutches at Blakeney Point, Norfolk, that 'G. Long tells me that he had seen them carrying sandlances and that his boys took the eggs of the first clutch and ate them on board his smack.' Ticehurst (1909) recorded that the fishing families in the Dungeness area, Kent, took every Oystercatcher's egg they could find for food, exterminating the breeding population; the same happened in north Norfolk (Lubbock 1879). Lubbock also noted that Avocet eggs were heavily exploited by country people in the Salthouse area; the species was finally lost in north Norfolk from a combination of such exploitation and draining the marshes in 1851.

Lapwings' eggs were those most widely exploited commercially but systematic commercial exploitation was recorded in rather few nineteenth-century avifaunas. Most do not mention it. It was recorded for Kent, Surrey, parts of Suffolk, particularly the Brecks, Norfolk, Yorkshire and perhaps Durham, and, in Scotland, the faunal regions of Tay and Strathmore, Deeside and the Moray Basin. Yarrell (1837–43) also included Lincolnshire, Cambridgeshire and Essex but neither Evans (1904) nor Christy (1890) refer to it for the latter two, and perhaps it had died out with a decline in Lapwing numbers, which Christy certainly recorded. All these areas are in eastern Britain. The practice seemed much rarer in the west, being recorded only for parts of Staffordshire, Cheshire and probably Dumfriesshire. The fact that Harvie-Brown and his colleagues (Harvie-Brown 1906, Harvie-Brown & Buckley 1887, 1888, 1892, 1895, Harvie-Brown & MacPherson 1904) recorded it in eastern Scotland but not in the west argues that this east–west distribution was genuine.

The scale of the trade was considerable. Stevenson (1870), for example, recorded that 600–700 eggs per week were sent by one Yarmouth, Norfolk, dealer to London and other markets during the season. Lubbock (1879) gave similar figures. Ticehurst (1909) noted that 200 dozen eggs were sent from Romney Marsh, Kent, in 1839 and that dogs were used there to find nests. Ticehurst (1932) recorded that 280 dozen per year were being sent from one Thetford, Norfolk, estate in the 1860s. The eggs were valuable. Lubbock (1879), Stevenson (1870) and Bewick (1826) all record values of between 3 and 4 shillings a dozen (£4.50–6.00 today) from the late eighteenth century until the 1870s. Not all plovers' eggs were laid by plovers, the crop from Norfolk including the eggs of Redshanks, Ruffs, Black Terns and Stone Curlews. Lubbock noted that species such as Ruff and Black Tern were less apt to re-lay than Lapwings, so intensive egging seriously affected them. Ticehurst (1932) observed that Stone Curlews' eggs were regarded as a perk of the warreners on the Brecks and excessive egging contributed to the species decline there.

All the relevant nineteenth-century authors had no doubt that systematic egging reduced Lapwing populations. Egging, however, had a long tradition in counties such as Norfolk and apparently posed few problems before the nineteenth century. In the nineteenth century, however, Southwell (in Lubbock 1879) noted that such exploitation increased with improved communications, which increased demand, particularly from London. This increase occurred when habitat was shrinking or deteriorating through drainage, enclosure and agricultural improvement. Stevenson (1870), Bucknill (1900) and Harvie-Brown (1906) all specifically ascribe decline in Lapwings to this combination, implying that the bird could withstand one but not both. This was supported by Smith (1887) who noted no historic population change in Wiltshire, as egging was not a regular trade. Several authors, for example Ticehurst (1909) and Gladstone (1910), noted that populations recovered with a decline in egging following application of the Bird Protection Acts from the 1870s.

The other farmland species which was systematically farmed for eggs was the Black-headed Gull. By the nineteenth century this bird had declined considerably in lowland England where it had lost many breeding stations to enclosure and drainage. But colonies were formerly valuable properties and farmed for their eggs, and squabs which Bewick (1826) noted sold at 5 shillings per dozen in the seventeenth century. It is likely that casual exploitation increased in the nineteenth century, the eggs then being taken as substitutes for Lapwings' eggs. Black-headed Gulls were increasing in the early twentieth century, when it had become a markedly northern and upland species (Hollom 1940), a factor which probably limited exploitation.

Wheatears and Skylarks

As with plovers' eggs, these were luxury trades, both being considered delicacies. By the second half of the nineteenth century Wheatear trapping was confined to the East Sussex Downs and the Isle of Portland, Dorset. It may have been more widespread in earlier centuries, when it was apparently practised in Wiltshire (Brentnall 1947), but White only recorded it in East Sussex (White 1789, letter to Pennant, 1768) and Montagu (1833) not at all. Wheatear trapping was a valuable secondary trade of the downland shepherds, capable of earning them £12–£14 a season, and sometimes much more, a sum equivalent to nearly half their annual wage (Borrer 1891). The birds were snared in small funnel traps (Borrer 1891, Hudson 1900, Walpole-Bond 1938). Yarrell (1837–43) noted that a shepherd and his boy could look after 500–700 traps. It is not surprising, therefore, that trappers occasionally recorded takes of up to 84 dozen per day, although the usual level was 3–4. The season for taking them extended from late July to late September. The birds were worth 6d per dozen in the eighteenth century, rising to 3s 6d (£5.25 today) by 1872, when the numbers taken had fallen markedly. The trade mainly affected migrants, usually passage birds rather than birds of British origin. The trade ended rather abruptly at the end of the century, when their employers banned the shepherds from pursuing it (Hudson 1900).

By the late nineteenth century Skylark netting appeared to be similarly limited. Sussex has a detailed literature on the subject, as has northwest England but otherwise few nineteenth-century avifaunas mention it except Stevenson (1866) for Norfolk, Harvie-Brown & Buckley (1892) for Argyll and the same authors incidentally in 1895 in describing partridge netting in Moray. It was more widespread in earlier centuries. Knox (1849) recorded it more widely in Sussex than later authors, MacPherson (1892) recorded it for Lakeland in the seventeenth century but not later, Yarrell (1837–43) recorded it in Devon in the seventeenth century but D'Urban & Mathew (1895) did not, and Brentnall (1947) records it for Wiltshire in the seventeenth century but Smith (1887) does not, although discussing the trade in London and Europe. Bewick (1826) quoted Pennant that Dunstable was famous for its larks in the eighteenth century, up to 4000 dozen a season being sent to market but Yarrell noted that the trade had ceased there in his day. Montagu (1833) says nothing about it except in Europe. Although large numbers were probably involved, assessing the scale of the trade is difficult because of the extent to which the same figures are quoted by different authors at different times. Walpole-Bond (1938) recorded that larks from Sussex were exported to France, together with other species taken at the same time, such as Song Thrushes.

Lark netting by the late nineteenth century was largely confined to extensive areas of land with open access – downland sheepwalk in east Sussex, extensive dune systems in northwest England and probably marsh, heath and dune in Norfolk, although Stevenson does not actually say. The main wintering habitat for Skylarks, however, was stubbles and Knox (1849) recorded that the principal way of taking them was to drag stubble fields at night with a long-net; partridges were similarly netted. It seems safe to assume that the trade was increasingly restricted from the mid nineteenth century by the rise in preserving game, especially partridges. The preservation of large numbers of partridges for driving and nocturnal netting for Skylarks were incompatible and neither gamekeepers nor landowners were likely to tolerate the latter. The birds taken were mainly migrants and winter visitors, occurring in hordes we can scarcely envisage today. There is little evidence that the trade ever affected native breeding populations.

Waders and wildfowl

Waders were highly prized as table birds. Accounts from gentry households in Northumberland (Bolam 1912), Yorkshire (Nelson 1907) and Norfolk (Stevenson 1870) show that a wide range of species had long been eaten – Curlew, Snipe, Woodcock, Redshank, Dotterel and other plovers, Knot, Dunlin, stints, godwits, Ruff and so on. Dotterel, Ruff and Black-tailed Godwit were among the most highly regarded of all wildfowl, particularly the last, which fetched up to 5 shillings apiece (£7.50 today), a high price which contributed to its demise as a British breeding species. Stevenson (1870) also noted that 'This species, also, as Mr. Gurney remarks, was formerly an abundant breeder in Holland, but like the Purple Heron, Spoonbill and Little Bittern has been so destroyed there of late years, that it has become comparatively rare; and this fact would also in some degree account for its scarcity on the east coast of England.' Ruffs were sold to dealers who specialised in fattening them, an extraordinary trade well described by Montagu (1833).

The nineteenth-century literature includes many descriptions of methods used for taking these birds, often apparently limited to areas where they were traditional occupations. Stevenson (1870) described the netting of plovers and other waders in the Fens and on the Lincolnshire side of the

Wash. Plovers were the principal quarry but other species were regularly taken. Netting was pursued in September–October and March–April. In the latter period breeding birds were taken and the combination of spring trapping, high prices and drainage undoubtedly exterminated native populations of Ruff and Black-tailed Godwit in the early nineteenth century. Dotterel were also sharply reduced, being tame and easy to take on spring passage, when their traditional stopping places were well known. Plover netting continued in the Lincolnshire marshes until the Lapwing received total protection in 1946 (Smith & Cornwallis 1955). In northwest England Mitchell (1885) also noted that Snipe and Woodcock were caught in snares – sprints (springes) to catch Woodcock and pantles, rows of horsehair snares attached to a line, to take Snipe; pantles were also used for larks. Although the nineteenth-century avifaunas do not record it, the snaring of these species was probably more widespread in earlier centuries. As Snipe, Woodcock and plovers were increasingly regarded as game in the nineteenth century, trapping them was probably discouraged as game preserving increased. Shore shooting of waders by professional fowlers continued on a large scale well into the twentieth century.

Wildfowl were an important article of food. Defoe, for example, recorded a major trade from the Fenland decoys to London (quoted in Furbank *et al.* 1991), whilst Mitchell (1885) and Bull (1888) both record the wildfowl carts hawking their wares in the streets of market towns as a normal activity, and Aplin (1889) noted that Oxford was served by wildfowl from Otmoor. Darby (1934) noted that the name 'decoy' was used in England long before the introduction of the Dutch type of duck decoy. It referred to tunnels or cages of netting into which wildfowl were driven when flightless in moult or as flappers. In the early eighteenth century this was recognised as damaging to native populations (it was this rather than drainage which caused the demise of native Greylag Geese) and a close season was introduced in 1739 extending from 1 June to 1 October. How effective it was is unclear, for White was recording the taking of flappers on the ponds of Woolmer Forest, Hampshire, in July 1773 (White 1789, letter 39 to Pennant, 1773). Nevertheless the draining of the Fens both reduced the opportunities for this practice and increased the operating efficiency of decoys of the Dutch type, which worked best where there was no extensive flooding to provide alternative habitat.

The design and management of these decoys was well summarised by Kear (1990). They were introduced into the Fens and southeast Yorkshire by the Dutch employed by Vermuyden for his original drainage operations in the seventeenth century. Decoys became widespread in eastern England and were found throughout the country; Defoe, for example, records two being built in Dorset in the early eighteenth century (quoted in Furbank *et al.* 1991). They never penetrated Scotland. Payne-Gallwey (1886) recorded the existence of about 250 at various times. His list was not exhaustive but more than half were sited in Essex, Norfolk, Yorkshire, Lincolnshire and the Fens. They reached their peak in the eighteenth century and were going out of use rapidly after 1850. By 1886 only 47 were operating and by 1918 only 28 (Limbert 1978). Christy (1890) gave dates for the abandoning of 17 in Essex, which were fairly evenly spread from 1790 to the 1870s. There were two reasons for the decoy's decline. One was the expansion of shooting wildfowl, particularly as breech-loading shotguns were introduced from around 1850. Decoying and shooting in the same area were incompatible and many proprietors came to prefer the latter. Secondly Kear (1990) noted that decoys became uneconomic because of the decline of native Mallard brought about by drainage.

The details of bags recorded in the county avifaunas indicate that overall numbers taken were also declining in the nineteenth century, perhaps by as much as 40% over the century. In the early eighteenth century Defoe (quoted in Furbank *et al.* 1991) recorded one decoy near Ely 'from which duckoy alone they assured me at St. Ives, (a town on the Ouse, where the fowl they took was always brought to be sent to London;) that they generally sent up three thousand couple a week'. Few such numbers are recorded for the nineteenth century, although Smith & Cornwallis reported that up to 31000 ducks were sent to London from the ten decoys in the East Fen, Lincolnshire, about 1800; that site was drained within a decade. There are quite good reasons for supposing that wildfowl were overexploited by the combination of decoys and increased shooting pressure with the improvement of guns. The wildfowling season extended well into the breeding season until late in the nineteenth century, when an Act of 1876 extended the close season back to 15 February. The taking of wildfowl in the breeding season must have reduced breeding populations, perhaps severely. The preamble of the 1876 Act stated that 'the wildfowl of the United Kingdom, forming a staple article of food and commerce, have of late years greatly decreased in numbers by reason of their being inconsiderately slaughtered during the time that they have eggs and young' (Marchant & Watkins 1897). The nineteenth-century avifaunas say little about large concentrations of wildfowl on inland wetlands by the end of the century; those present were largely confined to coasts and estuaries.

Species such as Teal, Pintail and Shoveler were widely recorded as declining and the last two as uncommon or scarce. Nelson (1907) noted of the Pintail that

> Half a century ago, the Pintail was a numerous species on the Tees, where, as Geo. Mussel tells me, it was greatly sought after by the professional gunners, who would not trouble with other fowl if they could get the Pintail and, as it was most plentiful in May, and no restrictions were at that time placed on shooting, great numbers of this delicious duck were procured and brought into market.

Such problems were European in scale. With the imposition of effective close seasons from the 1870s, many nineteenth-century avifaunas noted that these species and Garganey started to recover.

Cage-birds and pest control

Cage-birds

Trapping songbirds for the cage-bird trade was reported as widespread in the nineteenth-century avifaunas, particularly for Goldfinch and, in many parts of southern England, the Woodlark. Farmland birds were not the only ones affected of course, Nightingales were widely trapped for example and Swaine (1982) quoted a nineteenth-century Cheltenham, Gloucestershire, bird-catcher to the effect that 'Londoners will buy anything with a beak and a tail.' The trade may not have been universal, however. In parts of the east Midlands and the southwest, much of Wales and of western Scotland, the nineteenth-century avifaunas say little about it. It was primarily an urban trade, with important centres in large towns and cities – London, Brighton and the south coast towns, Glasgow, and the industrial towns of northern and northwest England.

A careful reading of the nineteenth-century avifaunas also shows that bird-catchers operated on the waste and other areas of free access, a restriction that would have become more marked as the century progressed and game increasingly closely preserved. Among farmland birds it was the Goldfinch's misfortune that it was a species of the waste far more than other popular cage-birds (Table 3.1), except perhaps Woodlark. Species such as Linnets and other finches, which were certainly taken, were protected by their much greater use of farmland habitats such as stubbles. Much of this trapping was carried out during the migrations and Gray & Hussey (1860) recorded 400–500 dozen Goldfinches being sent to London from the Worthing area annually, and up to 1050 dozen in some years. They were mainly trapped in October and November and trapping on this scale continued all along the Sussex Downs from Worthing to Rottingdean. Evidence that numbers on this scale were taken elsewhere is lacking and it may well that that part of the Sussex coast, where the Downs come right down to the sea and the towns gave immediately onto open sheepwalk, were particularly suited to the bird-trapper's pursuit. Many authors stressed that it was spring trapping which harmed local bird populations but there is rather little evidence that the trade had much impact on breeding populations of farmland birds; even the Goldfinch was as much affected by farming change (Chapter 7). A close season to protect breeding birds was imposed at the end of the nineteenth century.

Pest control

Legislation for the control of bird pests was of long standing. The species against which such ordinances were levelled were largely pests of grain crops, fruit and poultry yards – pigeons, Skylarks, Rooks, finches and sparrows, Jays and Bullfinches, and Kites, Buzzards and corvids. Control was orchestrated by paying bounties for heads and eggs of proscribed species through the Churchwardens' accounts for each parish. These accounts also record the acquisition and maintenance of guns for Rook and Crow shooting and scaring to protect new sowings (Jones 1972).

Although Churchwardens' accounts indicate that large numbers of birds were killed, it is uncertain that there was effective control, even of large birds of prey. Persecution is effective only if it adds to rather than replaces natural mortality and that effect is most likely at the start of the breeding season, when populations are at their annual low and culling breeding adults most likely (Newton 1979). Pests represented money, the weakness of all bounty systems, and I suspect that most culling would have been done when most birds were available. I have not found it possible to establish seasonal patterns in the sources I have examined. Where dates of payments for sparrows are given, for example, they are strongly biased towards April to June. But the entries leave the strong impression that they represent annual finalising and settling of accounts, rather than current payments which ought to have included many more for other seasons. Some parishes at least kept sparrow books to enter running totals. Bullfinches were persecuted for their depredations on fruit blossom, without much impact on numbers being recorded; in Kent Ticehurst (1909) recorded it as increasing despite intensive persecution.

Ticehurst (1920, 1935) showed that large numbers of Kites and Buzzards were paid for at Tenterden, Kent, in the second half of the seventeenth century. Such campaigns certainly occurred elsewhere in the seventeenth century (Tubbs 1974) but they were by no means universal. Steele-Elliot (1906) examined the Bedfordshire Churchwardens' accounts for the late seventeenth century to 1860 in full and found 'the entire absence of any mention of the birds of prey notable'. Oldham (1931) observed the same for Hertfordshire and also found that only a few Buckinghamshire parishes recorded them, whilst Tubbs (1974) found none in Hampshire. Even at Tenterden the payments for predatory birds were discontinuous, suggesting that campaigns were instituted when the birds were considered too numerous but activity declined when people found it unrewarding. Tubbs also observed that Churchwardens' accounts generally indicate that any campaign against predatory birds ceased in the mid eighteenth century, when attention became focussed on the House Sparrow. Turner (1794) noted that sparrows had very much increased in the second half of the eighteenth century. In general these accounts were often largely concerned with mammals, particularly polecats and hedgehogs, in the seventeenth and early eighteenth centuries and sparrows from the mid eighteenth (Figure 12.1). Jones (1972) noted that the Bedfordshire Churchwardens' accounts indicated that millions of sparrows and tens of thousands of their eggs must have been destroyed during the eighteenth and nineteenth centuries. Yet the nineteenth-century avifaunas are virtually unanimous that the House Sparrow was a too-abundant pest. Sparrow clubs were formed widely at the end of the century and in the early twentieth, often with official help (Jones 1972). They went bat fowling (which had nothing to do with bats; see Ennion 1960) and trapped sparrows and buntings in riddle traps in farmyards. It provided them with amusement and some protein but probably had little impact on sparrows. Rook scaring and shooting continued well into my own day but its effect on Rook numbers was immaterial.

The expansion of turnip-growing and of plantations and shelter belts from the early nineteenth century led to the emergence of another pest, the Woodpigeon. Turnips provided a certain and vastly expanded winter food supply and plantations a similarly expanded source of nest sites. Jones (1972) noted that pigeons had so increased in parts of the Cotswolds by the 1850s that it became impossible to grow vetches for stock feed. We found on our family farm that it became difficult to grow white clover mixtures in the 1970s for the same reason and abandoned them. Organised

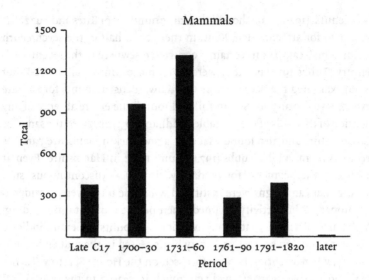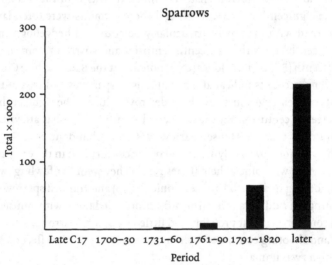

Figure 12.1. Totals of mammals and sparrows recorded as paid for in the Bedfordshire and Hertfordshire Churchwarden accounts. Note that these papers only include a selection of the entries to show the pattern of payments. Steele-Elliot published a complete analysis of the Bedfordshire accounts in 1936, which I have not seen. (Data from Steele-Elliot (1906) and Oldham (1931).)

shooting, nest-destruction campaigns and similar exercises extended from the nineteenth into the second half of the twentieth century without having much impact.

Game preserving and raptors

The nineteenth-century campaign of raptor destruction apparently started in Scotland. Harvie-Brown and his colleagues (Harvie-Brown 1906, Harvie-Brown & Buckley 1887, 1888, 1892, 1895, Harvie-Brown & MacPherson 1895) in their series of Scottish regional faunas include many lists of birds killed and rewards paid by organisations such as the United Association of the Noblemen, Gentlemen and Farmers of the counties of Sutherland and Caithness. This association paid out rewards totalling the extraordinary sum of £1115 17s $4^1/_2$d (£33 500 today) during the years 1819–26. Eagles were the most significant items, paid for at the rate of 10 shillings (£15 today) for an adult, compared to 2d (24p) for hawks. Such campaigns in Scotland were associated in time with the rise of extensive sheep-farming in the uplands in the early nineteenth century and the emphasis on eagles suggests that the two were linked. Later game preserving came to be more important and campaigns of extermination spread to the smaller raptors.

During the nineteenth century a marked change in the nature of shooting for sport occurred. Until around mid century shooting was mainly over dogs, walking up, a limitation imposed by the nature of the guns used; muzzle-loaders were slow to load and not always certain of firing. These points are clearly illustrated in numerous contemporary sporting prints. Large bags of driven game became possible with the development of the breech-loader. With them emerged the whole battery of strict preservation, intensive 'vermin' destruction, large-scale driving of reared birds, shooting with two or three guns and so on. Peter Hawker's *Diaries* underline how different game shooting was in the early nineteenth century (Hawker 1988). Vesey Fitzgerald (1946) also noted that the development of shooting as a sport attracting high rents in Scotland was strongly influenced by the spread of the railway network, greatly improving access. Sheep-farming declined as shooting became valuable. Gamekeepers appeared in large numbers. Newton (1896) indicates that the spread of high farming, by reducing holding cover, also contributed to the switch to partridge driving from the 1850s. Game preserving reached its peak in the early twentieth century, when *c.* 23 000 gamekeepers were employed

TABLE 12.1. *The status of birds of prey as breeding birds in Britain at the end of the nineteenth/early twentieth century*

Species	Status
Honey Buzzard	Probably extinct (Hudson 1923b)
Kite	20 territories with pairs in Wales (Lovegrove et al. 1994)
White-tailed Eagle	Extinct by 1916
Marsh Harrier	Extinct
Hen Harrier	Confined to Northern and Western Isles
Montagu's Harrier	Very rare breeder only
Goshawk	Extinct
Sparrowhawk	Not scarce but much reduced in numbers
Buzzard	Extinct in lowland England, scarce in uplands of north and west; Walpole-Bond (1914) estimated 250 pairs in Wales
Golden Eagle	80–100 pairs in north and west Scotland (Baxter & Rintoul 1953)
Osprey	Extinct by 1916
Kestrel	Decreased but persecution eased after the great vole plagues
Merlin	Great decrease
Hobby	Scarce breeder in lowland England; the marked increase since 1970 may be partly the result of much declined persecution
Peregrine	Nineteenth-century persecution apparently did not greatly reduce but note that the population has doubled in late twentieth century

Source: Bijleveld (1974) unless stated.

(Potts 1986), and the killing of large bags had become a matter of competition, status and fashion. The destructive effect of intensive game preserving on our breeding birds of prey is discussed in detail by Bijleveld (1974) and is summarised in Table 12.1. A similar plague was visited upon the corvids and the owls but never so complete. One might have expected that persons supposed to have some knowledge of natural history would have exempted Barn Owls for their utility in helping to control rats in rick-stored grain (Chapter 10); instead they used pole-traps.

The destruction of our birds of prey in the mindless pursuit of increasingly gross bags of game was an inexcusable act of selfishness, greed and barbarity. It is clear from early accounts that it was done on the insistence of landowners, who co-ordinated the destruction, often running private bounty schemes. The concentration of land in comparatively few hands was probably a significant factor in the scale and impact of such execution, imposing a strong element of uniformity in matters such as gamekeeping. Many landowners would dearly love to do it again today, cloaking their prejudice under a spurious concern to save our songbirds, supposedly at risk from unnaturally large populations of raptors. That particular piece of sophistry was demolished long ago by Nicholson (1926). Furthermore no nineteenth-century author that I have consulted considered that the destruction of our birds of prey benefited any songbird I have looked at, although several remarked that Woodpigeons might have been less of a pest if some natural predators existed to help control them. Authors such as Gladstone (1910) also had little doubt that the great vole plagues in the early 1890s in the southern uplands of Scotland, which caused serious economic damage to upland sheep grazings (Adair 1891, 1893, Picozzi & Hewson 1970), resulted from the extermination of vole-eating raptors by game interests. A more moderate attitude to such raptors followed. The level of ignorance displayed by game preservers offended many:

> If, however, there is little real knowledge, there is plenty of prejudice; .... Notable amongst the destroyers are the gamekeepers, who ruthlessly kill many of the most interesting and beneficial species, including even such species as the green woodpecker, which feeds largely on grubs destructive to timber, and the dusk-loving Nightjar, whose whole life is employed in doing good. (Archibald 1892)

Reading of the fate of the Hen Harrier in northern England in the twenty-first century, one wonders how much has changed.

Collecting

The curious Victorian obsession with collecting dead birds and eggs was aptly described by Nicholson (1926) as a leprosy on birds. Most country houses probably had cases of stuffed birds and many had considerable collections. Stevenson (1866–90) mentions at least 88 for Norfolk alone. Nor were just rarities collected. Most collections included examples of common birds, plumage aberrations and so on. It was E.T. Booth's declared ambition to display every British bird in its various plumages in his private museum in Brighton (Mearns & Mearns 1998). At least 76% of the Black-tailed Godwit records in the nineteenth-century avifaunas are specimens, an important reason for their demise as a British bird. Bearded Tits were virtually wiped out by collectors and Bitterns were shot on sight, which Stevenson believed long prevented their recolonisation of East Anglia. Other scarce species were similarly prevented from re-establishing themselves by collectors, Ruff and Avocet being examples. Most collectors employed or commisioned gamekeepers and other rural workers to obtain specimens or eggs and, to a shepherd, even half-a-crown for a rare bird or eggs would represent nearly one-fifth of a week's wage. So rarer birds came to have a clear monetary value, creating a market incentive which added to the mayhem. Collections of stuffed bird made significant sums at auction. Eventually, as it always does of course, monetary value led to collectors also being defrauded, most notoriously with the Hastings Rarities. Illuminating anecdotes about what went on are widely scattered in the literature. Stevenson (1870) reported that:

> Mr. Gould informs me that when he first began to collect British birds, over thirty years ago, he was in the habit of receiving weekly a basket of sandpipers and plovers supplied by Harvey, a dealer at Yarmouth, who at that time used to purchase specimens from the various gunners both on the coast and broads.

J.H. Gurney once commisioned gunners at Yarmouth to shoot every 'crest they came across for him, in the hope that one or two would be Firecrests; he wanted a Norfolk-taken specimen. The scale of trophy collecting and its essential pointlessness is well illustrated by Pashley's (1925) book on the birds of Cley, which largely consists of tedious lists of birds brought to Mr. Pashley to stuff.

Conclusion

It is difficult to avoid the conclusion that the nineteenth century was unrivalled for the sheer destruction of our bird populations that went on. I have only touched here upon certain aspects which seemed particularly likely to have affected farmland birds. Two points fairly consistently recur in considering these matters. One was the great increase in the efficiency of guns with the development of breech-loading, which allowed much more rapid and reliable firing than any muzzle-loader and flintlocks especially. The other was the great improvements in transport, both for goods and passengers, which opened up many previously remote areas. Both factors greatly increased the impact of persecution.

In the nineteenth century conservation was in its infancy, a matter of individuals more than a general movement. It was not until the 1870s that there was even a close season to protect birds in the breeding season and collectors didn't respect that. If they exterminated a bird, so what? Their collections would be more valuable for unique specimens. It is clear that, among farmland birds, wildfowl, rarer waders, raptors and owls and corvids were all seriously reduced by the level of persecution and

Conclusion 321

exploitation they suffered, as was at least partly the Goldfinch. The effect was often exacerbated by habitat change or vice versa. Evidence that many more common species, such as larks or sparrows, were affected by the scale of trapping is lacking. Whilst this may simply reflect the lack of sophisticated census information, I believe it is fairly accurate. Much trapping was an autumn and winter activity and the losses sustained may have done little to influence overall mortality, particularly in an era when farming methods guaranteed extensive winter food supplies.

13

Conclusions

Any long-term perspective shows how false the idea that farming preserves the countryside is. It is instead a catalyst for change. Farming's economic and methodological revolution during the hundred years or so after 1750 brought enormous changes in landscapes and habitats. Such revolutions habitually reach too far. Ernle's point about 'weight of metal' (i.e. money) driving arable farming into increasingly marginal areas in the nineteenth century (p. 61) is as true for the twentieth. Historically, farming expansion into such areas has long been at the taxpayers' or consumers' expense – high war-time prices in the early nineteenth century, Corn Laws over a longer period, subsidies today. With the withdrawal of such support, farming withdraws. Habitats do not revert as a result, further changes appear, as in the arable recession before 1940. Thus enclosure, high farming, recession and the farming revolution of the late twentieth century have all produced changes in the landscape and in habitats and conditions in agricultural land, often profound, which have required adaptation by wildlife.

In the nineteenth century, however, the adoption of high farming rotations virtually throughout farmland resulted in a close and varied habitat matrix emerging on a farm by farm basis, which provided greater diversity within farms as the overall diversity of the countryside declined. It provided a haven for many farmland birds displaced by the destruction of the old varied countryside that had existed before enclosure. A key feature of that haven was the extent of new food resources that emerged, particularly in winter. I suggest that this influenced winter survival in birds particularly favourably and thus maintained populations and aided adaptation. It was a factor which persisted until the early 1970s with the three-year ley system. Modern studies are making it increasingly clear that declining winter food resources are a very important determinant of decline in farmland birds. Furthermore, before enclosure birds such as Lapwings were largely immune to farming operations, they were birds of the waste. Although they adapted to high farming, that adaptation also increased their exposure and vulnerability to farming change, as it did for many birds. The farming systems of the late twentieth century have severely tested the limitations of adaptability in farmland birds.

One of the most striking points to emerge from this study is the extent to which farming methods and technology as determinants of decline in farmland bird populations are modern phenomena. These methods and technologies have led to the marked overall decline in the diversity of the farmland habitat that has become so evident since the early 1970s. The haven of mixed farming systems is being swept away and, as a result, farmland is no longer able to support the same number of species and individuals. The decline of bird populations is only part of that declining diversity, although an important symptom that has provoked attention and study because birds are conspicuous.

What does an historical perspective point to as the most important changes in modern farmland? I place first the revolution in grassland management described in Chapter 9. The retention of many old grassland habitats in the nineteenth century was an important reason why the huge habitat changes within the agricultural area in that era had a surprisingly limited impact on birds; extensive key areas of habitat were untouched. Even at the height of high farming stocking rates in grass also remained comparatively low, and sheep were as much animals of arable systems as pastoral. In the modern era birds in pastoral farmland have instead come under even greater pressure than those of arable, because of radical changes in habitat, management and stocking patterns and their impact on food supplies and breeding.

Second I would place the revolution in chemical weed control since the mid 1960s, which has led to the extensive loss of the weed flora of arable land and is steadily sapping the soil's seed bank. As noted in Chapter 7 this is the key ecological change in arable farmland. The functions of selective grass weed control and autumn use in herbicides, which are part of that revolution, have also permitted the unprecedented change in the main season of cultivation, with its concentration on autumn tillage. This has led to the loss of food supplies in winter stubbles and also of disturbed ground in spring, an important food source for many species at a difficult season.

Third I put the change in harvesting methods. Despite the scale of change through the nineteenth century, from sickle to scythe to mechanical reaper to reaper–binder, the basic methodology was unchanged – cut, sheaf, stook, cart, rick and thresh. The combine has swept all away and with it the major source of winter food for many birds. Here we should also consider grass harvesting, because the switch from hay to silage has had similar effects in pastoral farming.

Finally I would list the the loss of the undersown ley from arable systems, not only because it guaranteed overwinter stubbles but also because of the element of stability it provided for invertebrate populations in cultivation, which is otherwise an essentially unstable habitat for them. The increasing use of insecticides, particularly in cereals, has exacerbated this.

These changes are crucial because they are unique to the modern era. They have combined to cause the collapse in the diversity of modern farmland noted above and technology has become a highly significant factor of change, another marked difference from the nineteenth century. Conservation today pays too little attention to the impact on wildlife of technology in farming, in which I include developments in pesticides. But the application of technology is increasingly driving farming change. It is sharply illustrated by the series of figures in Chapter 7. It also operates independently of State agricultural policies and is little influenced by them. Policies may encourage cereal growing, for example, but say nothing about how they should be grown. But it is clear that technical aspects of cereal farming have been far more important in changing bird populations in farmland in the past 30 years than the area grown, which has changed rather little by contrast and is what policy influences. We should remember that chemical companies identified pesticides as an important new area of profit in the 1920s, at the height of the arable depression. Undoubtedly the technical changes of the modern agricultural revolution and the problems they pose would have arrived irrespective of subsidy; they would have provided a route to survival and increased profitability in farming. However, without subsidies, Ernle's point about weight of metal would have limited the scale of such change.

Conservation should therefore be looking closely at emerging technical developments and I would draw attention to three. First is the development of genetically modified crop seeds, of which I have little specialist knowledge. Common sense suggests, however, two probable reasons for making such modifications, to improve pest control and to raise yields. In improving pest control the likely advantage would be to reduce costs; pest control is otherwise more than adequate in most crops in Britain today. I find it odd that attention at present seems to be focussed upon herbicide tolerance. In the British context limitations on herbicide use seem more likely to arise from soil types than from plants' susceptibility (Chapter 7). More advantage would be gained by improving disease resistance, reducing the scale of expenditure on fungicides for example (see Figure 7.8) and improving crop yield and quality. Significant increases in yield potential are likely to bring in their train problems of surplus production, something that farmers should consider more carefully. The second development is that of lo-till, which I have suggested (Chapter 10) has important implications for birds' food supplies, as well as pesticide and fertiliser use, particularly nitrogen. These need detailed examination. In neither case do I argue that these developments are necessarily deleterious. I don't know. I am arguing that we need to understand now their likely effects. The third area of development is in veterinarian medicines, an area largely ignored in the past. The use of avermectins as worming medicines, however, gives some sign that real problems could emerge in this area. They have marked effects on the insect fauna associated with dung and therefore potential indirect effects on birds for which this is an important food source (McCracken 1993, McCracken & Foster 1993). And what environmental impacts are there with the new pyrethroid sheep-dips?

An important change in farming's direction in future will almost certainly be the large-scale adoption of agri-environment schemes, which many farmers have identified as a significant new source of income. However, general agri-environmental schemes to date have not been notably successful. In The Netherlands, where such schemes have been in place since 1981, Kleijn *et al.* (2001) found that management agreements for the benefit of waders and plants had little positive effect. Indeed, for waders, the results appeared to be perverse. These authors suggested that this arose because, for farmers, nature conservation was of secondary importance to the necessity of securing an income from farming and that the farming system itself was intensifying, whilst monitoring and supervision were inadequate. Similar problems have arisen in Britain, where Potts (1997) noted that the results of some schemes have been distinctly contradictory. The success of such schemes depends on proper monitoring and management which has proved difficult to provide. Perhaps as a result of this, some schemes also lay down rigid and impractical restrictions on what farmers may do to qualify, for example in trying to promote arable pockets in pastoral areas. Such restrictions often appear to disregard how farmland birds actually adapt to farming and simply discourage farmers from entering. The best results have emerged from schemes tightly focussed on particular species (Aebischer et al. 2000), where adequate monitoring and supervision can be applied. The recent proposal of Musters et al. (2001) that dairy farmers in The Netherlands should be paid for the breeding waders they produce, rather than participating in general environmental schemes, has similar merits – simple, clearly defined objectives, easily monitored. Experiments have shown significant increases in wader breeding success to result.

Another issue which undermines the value and efficiency of general environment schemes for wildlife in Britain has been the opposition of farming bodies to the principle of selectivity; resources are then expended on areas of little real significance. Efficient wildlife conservation often needs a strong element of selection in the area or species embraced as Murton & Westwood pointed out in 1974. Furthermore some habitats can be readily recreated, e.g. wet grasslands, but some cannot, e.g. chalk grassland on any scale without recreating the old sheep/arable systems that formed it in the first place. This should be recognised and is another argument for using environment schemes selectively. A good example of the success of a targetted scheme, here for game, has recently been described by Stoate (2001). An important point to emerge from this work was that a scheme aimed at game had major general benefits for songbirds.

Another route to improving conservation on farms that is widely favoured is a move to organic farming systems, effectively returning to high farming. Such systems have been shown to favour bird populations. However organic farming works today because its products command a scarcity premium in the market, which pays for increased production costs. If it becomes the norm that premium will disappear and organic dairy farmers already face difficulties because of this. Organic farming can therefore provide a valuable element of diversity in modern farmland but is unlikely to become generally applied.

Set-aside has proved to be valuable in some circumstances but this is an economic measure not an environmental one. Nevertheless studies of the effect of rotational set-aside in arable systems show that it offers some of the advantages to wildlife of the short-term ley, which it mimics, and it would be well worth while retaining as a general environmental measure, simple to administer. Allowing industrial oil-seed crops to be grown on set-aside has also had positive benefits for seed-eating birds. But, on our family farm, putting the old pastures into permanent set-aside has been an unmitigated disaster, ruining them as pastures and habitats. They would have been better habitats as arable.

If such schemes were widely distributed at the farm level, they would do much to restore the overall diversity of farmland. It is unnecessary for them to be universal, they never have been. Adequate monitoring is essential. The main point that general agricultural policy needs to address is the imbalance between environment and production. There is little logic to setting up environmental schemes whilst continuing to subsidise production at its present level. This is not to advocate transferring the whole battery of production subsidy to environment schemes. This will not provide value for money, merely subvert environment schemes. We need to spend less and spend it more wisely. If that means that farming contracts, that will be an ecological advantage. Furthermore conservation bodies need to recognise that the fashionable argument for redirecting subsidy from production to conservation will not necessarily influence technical change, perhaps it will accelerate it. That needs consideration.

Other trends that are emerging mirror those of the 1870s. *Country Life* (2000) noted that Savills then had 2700 potential buyers of country properties, farms and estates on their books, many of which, a survey by Knight Frank showed, were seeking such property for amenity reasons, the quality of the house, landscape, and field sports. This was a marked trend in the late nineteenth century (Orwin & Whetham 1964) but the fashion then was particularly for sporting estates in the uplands, especially in Scotland and Wales. Today the emphasis is in the lowlands, nearer the centres of commercial and industrial wealth. As in the nineteenth century, the new ownership of these properties is likely to be supported by that wealth, so there seem to be good long-term opportunities for conservation there.

The livestock industry is, at the time of writing, in a state of flux resulting from the severe epidemic of foot-and-mouth disease in 2001. This is not the place to comment upon the handling of that epidemic. In the long term, however, it may provide an opportunity to address the problems that high stocking rates pose for birds and other wildlife in pastoral farming, perhaps particularly of sheep and in the uplands. Virtually all grassland habitats in farmland would benefit from a reduction in grazing pressure. In the uplands it cannot be doubted that traditional management practices evolved by empirical observation and experience. They were discarded in pursuit of ill-designed State subsidies; a more pernicious and destructive form of support in environmental terms could hardly have been devised for upland grasslands than headage payments for sheep. Nevertheless the conservation and recovery of upland habitats is not merely a matter of reducing sheep numbers. What we see today is the final stage of a long historical process of change and deterioration. To reverse this will require a return to mixed grazing systems, with a greater use of cattle and lower stocking rates, some positive management and some regard for the nature of the habitats involved.

Finally many believe that the level of support farming receives from the public purse lacks justification. Farmers should certainly reflect that the weight of bureaucratic regulation of which they now complain is an inevitable corollary of insisting on a high level of State subsidy. There is a perfectly respectable argument for removing any sort of public subsidy from farming and leaving it to find its own level. Its proponents argue that farming will then shrink into the areas best suited for it, leaving marginal areas particularly and allowing much greater room for the conservation of wildlife. Farmers' reaction to this seems to be the curious argument that large areas of the countryside would then revert to wilderness, offered as a sort of bogeyman to frighten us. It invites the riposte – 'and what is wrong with that, it's the whole point'. Purely in terms of farmland birds, however, such a policy involves several difficulties. This study shows that farmland birds do not necessarily benefit by such a change. Species such as buntings and partridges are primarily birds of arable ecosystems, particularly cereals. They need favourable farming conditions. The waders breeding in grassland habitats need grazing in some degree to maintain their habitats in an acceptable state; their problems stem not from the presence of grazing animals, but from too many. Upland plant communities also most usually need a certain level of grazing (Jones 2001). Above all there is no guarantee that conservation would fill the gap left by farming. Land is someone's asset and it will be used. Forestry has emerged as a principal alternative in the past and deciduous woodland might be favoured in the future. Farmland birds will not benefit from a countryside that comprises intensively farmed land and some other habitat. Farmland is a distinct ecosystem in its own right, supporting even today a wide range of typical species. It needs to be considered and treated as such. Above all we need to restore its old diversity.

Appendix 1

Estimating areas of important traditional feeding sites for seed-eating birds in British farmland

In compiling the table below basic crop areas were extracted from the June Census Statistics. Additional data on the timing of sowing cereals and the age of leys was extracted from the Surveys of Fertiliser Practice. The periods covered are for the early years of the decade in each case.

The June Census Statistics for the early 1870s show that arable land was composed of 56% cereals, of which about one-third was winter wheat following a ley, 20% ley and 22% fodder roots or green crops such as vetches, fallow, or potatoes and vegetables. In this basic rotation about one-fifth of the arable area would have been cereals undersown with a ley, providing undersown stubbles throughout the winter and spring. Approximately one-fifth of arable would have been roots following cereals. Such crops were established by ploughing and working the ground at fortnightly to three-weekly intervals from March until sowing in late spring/summer. Increasingly in the mid nineteenth century, land for roots was ploughed in autumn, soon after harvest (e.g. Jonas 1846). The rest of the cereals stubble area in any autumn would have come back into cereals. Some of these crops were autumn-sown but the bulk were spring-sown and good practice would have dictated that this ground be ploughed, if possible, in autumn/winter to give extended cultivations for weed control. I estimate, therefore, that winter stubble not undersown in the 1870s was unlikely to have been more than one-fifth of the arable area, but may have been much less in a good year for cultivations. The area perhaps increased somewhat as labour declined at the end of the century.

In Scotland two-year leys were preferred and little wheat grown; 19% of arable was roots, which were often lifted and clamped in winter, and 19% was second-year ley, of which 7% (41000 ha) went to wheat. I suspect the rest would have been ploughed in autumn and left fallow over winter, coming into cereals in the spring. But I've found no confirming reference. Undersown stubble comprised 19% of arable and the remaining area of stubble at harvest (21% of arable) largely went out to roots. My guess is that much of this would have stood as stubble because of considerable pressure of work elsewhere and have estimated 210 000 ha. In the 1930s leys in Scotland tended to be extended to three years but otherwise there were few changes.

The underlying rotation in the much-declined area of arable land in England and Wales changed significantly in the 1930s, with leys extensively replaced by cash crops such as sugar beet or vegetables in eastern England but longer leys, cashed through stock, used in the west. Overall the June Census indicated that leys were laid down for an average of two years, half of which were established annually under a cereal nurse. This provides the area of undersown stubbles and the remaining stubble area was calculated as for the 1870s.

In the 1960s the pattern changed again with the introduction and dominance of the three-year ley system, three years of cereals, mainly spring barley, followed by three years of ley largely cashed through a dairy herd. The June Census for the 1960s and early 1970s recorded the area of ley established under a nurse crop annually. For the early 1960s the area recorded was one-third of the ley area, providing a reliable estimate of the area of undersown stubble and supporting the underlying importance of this system. About one-third of the ley area went out to winter wheat annually. Some 290 000 ha of fodder roots/fallow were still present and assumed to follow winter stubble and 173 000 ha were planted to winter oats or winter barley. This left $c.$ 1.57 million ha of stubbles at the end of harvest which would have come back into spring cereals or gone out to sugar beet, potatoes and vegetables. Here I suggest a maximum figure of one-third overall ($c.$ 523 000 ha) stood as stubble over winter, which assumes priority being given to ploughing spring cereal ground, but there would have been much annual variation with amounts of winter rainfall and regional variation with soil types.

The figures for the 1970s were calculated in the same way but major changes in rotations occurred from then on, with continuous cereal rotations dominated by autumn-sown crops. In such rotations I believe that the formula cereals minus autumn cereals provides a reasonable estimate of the potential area of winter stubble and I have used it for the 1980s and 1994. The area of undersown stubble was estimated from figures given in

TABLE A.1. *Estimated areas (× 1000 ha) of important traditional feeding sites for seed-eating passerines in farmland in Britain at intervals since 1870*

Period	Arable	Stubble at 1 January	Undersown stubble[a]	Holdings with cattle (000s)[b]	Corn ricks (000s)	Fodder roots	Kale[c]	Potatoes, vegetables, sugar beet	Spring tillage
1870s									
England and Wales	5930	2372	186		1031	863	Nil	156	3439
Scotland	1430	485	275		219	196	Nil	82	810
1930s									
England and Wales	3844	1266	497		652	548	Nil	344	1974
Scotland	1237	404	204		189	132	Nil	44	602?
1960s									
England and Wales	5572	1400	539	210 (77%)	Nil	326	110	550	2692
Scotland	1389	304	120?	c. 72%	Nil	99	?	68	546?
1970s									
England and Wales	5645	1293	390	121 (65%)	Nil	202	59	553	2633
Scotland	1283	235+	?		Nil	97	?	69	531
1980s[d]									
England and Wales	5520	721	<136	94 (54%)	Nil	64	30[e]	494	1436
Scotland	1686[f]	<357	?		Nil	48	11	47	489
1994[g,h]									
England and Wales	4782	<263	<80?	93 (51%)	Nil	51[i]	17[i]	438	1007
Scotland	1658[f]	<229	?	17 (51%)	Nil	27	?	37	306

Notes:
[a] The area of undersown stubble is included in the area of stubble.
[b] Figures not included in the June Census Statistics for the 1870s and 1930s. Percentages are of total holdings.
[c] The area of kale is included in the area of fodder roots.
[d] Set-aside was introduced in 1988 but not recorded until 1990.
[e] Figures include cabbage, kohl rabi and fodder rape; kale probably represents half the area.
[f] Crops and grass, the area of arable no longer being recorded separately in Scotland.
[g] 57 000 ha of linseed was grown in 1994. The crop was grown in the 1870s and 1930s but the area was not recorded separately.
[h] There were 617 000 ha of set-aside in 1994, much of which should be regarded as stubble.
[i] Figures are for 1987; the decline indicated has continued.

the Fertiliser Surveys for the 1980s and reduced in line with the decline of reseeding in 1994, when no other information was available.

The number of holdings with cattle is recorded annually but understates the total number of stockyards. Thus our farm would have been recorded as one holding with cattle but we had six cattle yards in the 1940s and 1950s and still three in 1980; today there is none. Such situations were common in the past and very probably the norm.

In calculating the number of ricks I have assumed, from personal experience, that each held the produce of 2 ha and that barley was generally stored in barns. Neither assumption may be more widely true, for ricks were not necessarily of standard size, some corn was threshed at harvest and oats as well as barley was stored in barns. But these assumptions are not unreasonable and allow an estimate to be made in a standard way for both the table and Figure 10.5. There was a considerable increase in yields in the course of the nineteenth century and the estimate for 1800 in Figure 10.5 reflects this.

The area of fodder roots up to the 1970s includes the area recorded as fallow on 1 June, the census date; this would be sown to roots later in the summer.

Appendix 2

Scientific names of birds

Avocet *Recurvirostra avosetta*
Barn Owl *Tyto alba*
Bean Goose *Anser fabalis*
Bearded Tit *Panurus biarmicus*
Bewick's Swan *Cygnus columbianus*
Bittern *Botaurus stellaris*
Blackbird *Turdus merula*
Blackcap *Sylvia atricapilla*
Black-headed Gull *Larus ridibundus*
Black Grouse *Tetrao tetrix*
Black-tailed Godwit *Limosa limosa*
Black Tern *Chlidonias niger*
Blue Tit *Parus caeruleus*
Brambling *Fringilla montefringilla*
Brent Goose *Branta bernicla*
Bullfinch *Pyrrhula pyrrhula*
Buzzard *Buteo buteo*
Canada Goose *Branta canadensis*
Chaffinch *Fringilla coelebs*
Chiffchaff *Phylloscopus collybita*
Chough *Pyrrhocorax pyrrhocorax*
Cirl Bunting *Emberiza cirlus*
Coal Tit *Parus ater*
Collared Dove *Streptopelia decaocto*
Common Gull *Larus canus*
Coot *Fulica atra*

Corn Bunting *Miliaria calandra*
Corncrake *Crex crex*
Crane *Grus grus*
Crow *Corvus corone*
Cuckoo *Cuculus canorus*
Curlew *Numenius arquata*
Dunlin *Calidris alpina*
Dunnock *Prunella modularis*
Fieldfare *Turdus pilaris*
Garden Warbler *Sylvia borin*
Garganey *Anas querquedula*
Goldcrest *Regulus regulus*
Golden Plover *Pluvialis apricaria*
Goldfinch *Carduelis carduelis*
Goshawk *Accipiter gentilis*
Grasshopper Warbler *Locustella naevia*
Great Bustard *Otis tarda*
Great Crested Grebe *Podiceps cristatus*
Great spotted Woodpecker *Dendrocopus major*
Great Tit *Parus major*
Greenfinch *Carduelis chloris*
Green Sandpiper *Tringa ochropus*
Green Woodpecker *Picus viridis*
Greylag Goose *Anser anser*

Grey Partridge *Perdix perdix*
Hen Harrier *Circus cyaneus*
Hobby *Falco subbuteo*
Honey Buzzard *Pernis apivorus*
House Sparrow *Passer domesticus*
Jackdaw *Corvus monedula*
Jack Snipe *Limnocryptes minimus*
Jay *Garrulus glandarius*
Kestrel *Falco tinnunculus*
Kite *Milvus milvus*
Lapwing *Vanellus vanellus*
Lesser Whitethroat *Sylvia curruca*
Linnet *Carduelis cannabina*
Little Grebe *Tachybaptus ruficollis*
Little Owl *Athene noctua*
Long-eared Owl *Asio otus*
Long-tailed Tit *Aegithalos caudatus*
Mallard *Anas platyrhynchos*
Magpie *Pica pica*
Marsh Harrier *Circus aeruginosus*
Meadow Pipit *Anthus pratensis*
Merlin *Falco columbarius*
Mistle Thrush *Turdus viscivorus*
Montagu's Harrier *Circus pygargus*
Moorhen *Gallinula chloropus*
Mute Swan *Cygnus olor*
Nightingale *Luscinia megarhynchos*
Osprey *Pandion haliaetus*
Oystercatcher *Haematopus ostralegus*
Peregrine Falcon *Falco peregrinus*
Pheasant *Phasianus colchicus*
Pied Wagtail *Motacilla alba*
Pink-footed Goose *Anser brachyrhynchus*
Pintail *Anas acuta*
Ptarmigan *Lagopus mutus*
Quail *Coturnix coturnix*
Raven *Corvus corax*
Red-backed Shrike *Lanius collurio*
Red Grouse *Lagopus lagopus*
Red-legged Partridge *Alectoris rufa*
Redshank *Tringa totanus*
Redwing *Turdus iliacus*
Reed Bunting *Emberiza schoeniclus*
Reed Warbler *Acrocephalus scirpaceus*
Ring Ouzel *Turdus torquatus*
Robin *Erithacus rubecula*
Rook *Corvus frugilegus*
Ruff *Philomachus pugnax*
Savi's Warbler *Locustella luscinioides*
Sedge Warbler *Acrocephalus schoenobanus*
Shelduck *Tadorna tadorna*
Short-eared Owl *Asio flammeus*
Shoveler *Anas clypeata*
Skylark *Alauda arvensis*
Snipe *Gallinago gallinago*
Song Thrush *Turdus philomelos*
Sparrowhawk *Accipter nisus*
Spoonbill *Platalea leucorodia*
Spotted Crake *Porzana porzana*
Spotted Flycatcher *Muscicapa striata*
Starling *Sturnus vulgaris*
Stockdove *Columba oenas*
Stonechat *Saxicola torquata*
Stone Curlew *Burhinus oedicnemus*
Swallow *Hirundo rustica*
Tawny Owl *Strix aluco*
Teal *Anas crecca*
Treecreeper *Certhia familiaris*
Tree Sparrow *Passer montanus*
Turtle Dove *Streptopelia turtur*
Twite *Carduelis flavirostris*
Wheatear *Oenanthe oenanthe*
Whinchat *Saxicola rubetra*
White Stork *Ciconia ciconia*
White-fronted Goose *Anser albifrons*

White-tailed Eagle *Haliaeetus albicilla*
Whitethroat *Sylvia communis*
Whooper Swan *Cygnus cygnus*
Wigeon *Anas penelope*
Willow Warbler *Phylloscopus trochilus*
Woodlark *Lullula arborea*
Woodpigeon *Columba palumbus*
Wren *Troglodytes troglodytes*
Wryneck *Jynx torquilla*
Yellowhammer *Emberiza citrinella*
Yellow Wagtail *Motacilla flava*

Bibliography of avifaunas

The following regional and county avifaunas have been used in compiling figures and tables etc. Not all necessarily appear as text references.

Aplin, O.V. 1889. *The Birds of Oxfordshire.* Oxford University Press. Oxford.
Babington, Rev. C. 1884–6. *Catalogue of the Birds of Suffolk.* Van Voorst. London.
Baxter, E.V. & Rintoul, L.J. 1953. *The Birds of Scotland.* Oliver & Boyd. Edinburgh.
Bell, T.H. 1962. *The Birds of Cheshire.* Sherrat & Son. Altrincham.
Bewick, T. 1826. *History of British Birds*, 6th edn. Bewick. Newcastle.
Bircham, P.M.M. 1989. *The Birds of Cambridgeshire.* Cambridge University Press. Cambridge.
Blezard, E. 1946. *Lakeland Natural History.* Carlisle Natural History Society. Arbroath.
Bolam, G. 1912. *The Birds of Northumberland and the Eastern Borders.* Blair. Alnwick.
Borrer, W. 1891. *The Birds of Sussex.* R.H. Porter. London.
Browne, M. 1889. *Vertebrate Animals of Leicestershire and Rutland.* Privately published.
Brucker, J.W., Gosler, A.G. & Heryet, A.R. 1992. *The Birds of Oxfordshire.* Pisces Publications. Newbury.
Buckland, S.T., Bell, M.V. & Picozzi, N. 1990. *The Birds of North-east Scotland.* North-east Scotland Bird Club. Aberdeen.
Bucknill, J.A.S. 1900. *The Birds of Surrey.* R.H. Porter. London.
Bull, H.G. 1888. *Notes on the Birds of Herefordshire.* Jakeman & Carver. London.
Buxton, J. 1981. *The Birds of Wiltshire.* Wiltshire Ornithological Society.
Cambridge Phillips, E. 1899. *The Birds of Breconshire.* Edwin Davies. Brecon.
Chislett, R. 1953. *Yorkshire Birds.* A. Brown & Sons. London.
Christy, R.M. 1890. *The Birds of Essex.* Simkin Marshall Hamilton. Chelmsford.
Clark, J.M. & Eyre, J.A. 1993. *Birds of Hampshire.* Hampshire Ornithological Society.
Clark, J.S. 1996. *Birds of Huntingdon and Peterborough.* Privately published.
Clark Kennedy, A.W.M. 1868. *The Birds of Berkshire and Buckinghamshire.* Ingalton & Drake. Eton.
Cohen, E. 1963. *The Birds of Hampshire and the Isle of Wight.* Oliver & Boyd. Edinburgh.

Cordeaux, J. 1872. Ornithological notes for Lincolnshire. *Zoologist* 7:2923, 3015, 3323.
Coward, T.A. & Oldham, C. 1900. *The Birds of Cheshire*. Sherratt & Hughes. Manchester.
Coward, T.A. & Oldham, C. 1902–5. Notes on the Birds of Anglesey. *Zoologist* (4)6:401–415, (4)8:7–29, (4)9:213–230.
Cox, S. 1984. *A New Guide to the Birds of Essex*. Essex Birdwatching and Preservation Society.
Day, J.C., Hodgson, M.S. & Rossiter, B.N. 1995. *Atlas of Breeding Birds of Northumbria*. Northumberland and Tyneside Bird Club.
Dazley, R.A. & Trodd, P. 1994. *The Breeding Birds of Bedfordshire 1988–1992*. Bedfordshire Natural History Society. Bedford.
Deans, P., Sankey, J., Smith, L., Tucker, J., Whittles, C. & Wright, C. 1992. *An Atlas of the Breeding Birds of Shropshire*. Shropshire Ornithological Society.
Dobbs, A. (ed.) 1975. *The Birds of Nottinghamshire*. David & Charles. Newton Abbot.
Donovan, J.W. & Rees, G.H. 1994. *Birds of Pembrokeshire*. Dyfed Wildlife Trust. Haverfordwest.
D'Urban, W.S.M. & Mathew, M.A. 1895. *The Birds of Devon*. R.H. Porter. London.
Evans, A.H. 1904. The Birds of Cambridgeshire. In: *Handbook to the Natural History of Cambridgeshire*. Cambridge University Press. Cambridge.
Evans, A.H. 1911. *A Fauna of the Tweed Area*. David Douglas. Edinburgh.
Ferns, P.N., Hamar, H.W., Humphreys, P.N., Kelsey, F.D., Sarson, E.T., Venables, W.A. & Walker, I.R. 1977. *The Birds of Gwent*. Gwent Ornithological Society.
Forrest, H.E. 1899. *The Fauna of Shropshire*. Terry & Co. London.
Forrest, H.E. 1907. *The Fauna of North Wales*. Witherby. London.
Frost, R.A. 1978. *The Birds of Derbyshire*. Moorland Publishing Co. Buxton.
Gillham, E.H. & Homes, R.C. 1950. *The Birds of the North Kent Marshes*. Collins. London.
Gladstone, H.S. 1910. *The Birds of Dumfriesshire*. Witherby. London.
Gladwin, T. & Sage, B.L. 1986. *The Birds of Hertfordshire*. Castlemead. Welwyn Garden City.
Glegg, W.E. 1929. *A History of the Birds of Essex*. Witherby. London.
Glegg, W.E. 1935. *A History of the Birds of Middlesex*. Witherby. London.
Graham, H.D. 1890. *The Birds of Iona and Mull*. David Douglas. Edinburgh.
Gray, R. 1871. *The Birds of the West of Scotland*. Murray. Glasgow.
Guest, J.P., Elphick, D., Hunter, J.S.A. & Norman, D. 1992. *Breeding Bird Atlas of Cheshire and Wirral*. Cheshire and Wirral Ornithological Society. Chester.
Hancock, J. 1874. *A Catalogue of the Birds of Northumberland and Durham*. Williams & Norgate. London.
Haines, C.R. 1907. *Notes on the Birds of Rutland*. R.H. Porter. London.
Hardy, E. 1941. *Birds of the Liverpool Area*. Buncle. Arbroath.
Harrison, G.R., Dean, A.R., Richards, A.J. & Smallshire, D. 1982. *The Birds of the West Midlands*. West Midland Bird Club. Studley.
Harrison, J.M. 1953. *The Birds of Kent*. Witherby. London.

Harrison, K. (ed.) 1995. *An Atlas of Breeding Birds of Lancaster and District*. Lancaster and District Birdwatching Society.
Harthan, A.J. 1946. *The Birds of Worcestershire*. Littlebury & Co. Worcester.
Harting, J.E. 1866. *The Birds of Middlesex*. Van Voorst. London.
Harvie-Brown, J.A. 1906. *A Fauna of the Tay Basin and Strathmore*. David Douglas. Edinburgh.
Harvie-Brown, J.A. & Buckley, T.E. 1887. *A Vertebrate Fauna of Sutherland, Caithness and West Cromarty*. David Douglas. Edinburgh.
Harvie-Brown, J.A. & Buckley, T.E. 1888. *A Vertebrate Fauna of the Outer Hebrides*. David Douglas. Edinburgh.
Harvie-Brown, J.A. & Buckley, T.E. 1892. *A Vertebrate Fauna of Argyll and the Inner Hebrides*. David Douglas. Edinburgh.
Harvie-Brown, J.A. & Buckley, T.E. 1895. *A Fauna of the Moray Basin*. David Douglas. Edinburgh.
Harvie-Brown, J.A. & MacPherson, H.A. 1904. *A Fauna of the North-west Highlands and Skye*. David Douglas. Edinburgh.
Heathcote, A., Salmon, H.M. & Griffin, D. 1967. *The Birds of Glamorgan*. Cardiff Naturalists' Society. Cardiff.
Hickling, R.A.O. 1978. *Birds in Leicestershire and Rutland*. Leicestershire and Rutland Ornithological Society.
Hudson, R. & Pyman, G.A. 1968. *A Guide to the Birds of Essex*. Essex Bird-watching and Preservation Society.
Hurford, C. & Lansdown, P. 1995. *Birds of Glamorgan*. Cardiff Naturalists' Society. Cardiff.
Ingram, G.C.S. & Salmon, H.M. 1939. The Birds of Monmouthshire, *Transactions of the Cardiff Naturalists' Society* 70:93–127.
Ingram, G.C.S. & Salmon, H.M. 1954. *A Handlist of the Birds of Carmarthenshire*. West Wales Naturalists' Trust. Haverfordwest.
Ingram, G.C.S. & Salmon, H.M. 1955. *A Handlist of the Birds of Radnorshire*. Hereford Ornithological Society. Hereford.
Ingram, G.C.S. & Salmon, H.M. 1957. The Birds of Brecknock. *Brycheiniog* 3:182–259.
Ingram, G.C.S., Salmon, H.M. & Condry, W.M. 1966. *The Birds of Cardiganshire*. West Wales Naturalists' Trust. Haverfordwest.
James, P.D. (ed.) 1996. *Birds of Sussex*. Sussex Ornithological Society.
Jenyns, Rev. L. 1835. *A Manual of Vertebrate Animals*. Privately published.
Kelsall, J.E. & Munn, P.W. 1905. *The Birds of Hampshire and the Isle of Wight*. Witherby. London.
Knox, A.E. 1849. *Ornithological Rambles in Sussex*. Van Voorst. London.
Lack, D. 1934. *The Birds of Cambridgeshire*. Cambridge Bird Club. Cambridge.
Lack, P.C. & Ferguson, D. 1993. *The Birds of Buckinghamshire*. Buckinghamshire Bird Club.
Lewis, S. 1952. *The Breeding Birds of Somerset and their Eggs*. Privately published.
Lilford, Lord. 1895. *Notes on the Birds of Northamptonshire and Neighbourhood*. R.H. Porter. London.

London Natural History Society. 1964. *Birds of the London Area*, revised edn. Rupert Hart-Davies. London.
Lord, J. & Munns, D.J. 1970. *Atlas of Breeding Birds of West Midlands*. Collins. London.
Lovegrove, R., Williams, G. & Williams, I. 1994. *Birds in Wales* T & A.D. Poyser. London.
Loyd, L.R.W. 1929. *The Birds of South East Devon*. Witherby. London.
Lubbock, R. 1879. *Observations on the Fauna of Norfolk*, 2nd edn, edited by T. Southwell. Jarrold & Sons. Norwich.
MacGillivray, W. 1837–52. *History of British Birds*. Scott, Webster & Geary. London.
MacPherson, H.A. 1892. *A Vertebrate Fauna of Lakeland*. David Douglas. Edinburgh.
Mansell-Pleydell, J.C. 1888. *The Birds of Dorsetshire*. R.H. Porter. London.
Mather, J.R. 1986. *The Birds of Yorkshire*. Croom Helm. London.
Mathew, M.A. 1894. *The Birds of Pembrokeshire and its Islands*. R.H. Porter. London.
Mellersh, W.L. 1902. *A Treatise on the Birds of Gloucestershire*. Bellows. Gloucester.
Mitchell, F.S. 1885. *The Birds of Lancashire*. Van Voorst. London.
Montagu, G. 1833. *Ornithological Dictionary of British Birds*, 2nd edn. Rennie. London.
Moore, R. 1969. *The Birds of Devon*. David & Charles. Newton Abbot.
More, A.G. 1865. On the distribution of Birds in Great Britain during the nesting season. *Ibis* 1:1–27, 119–142, 425–458.
Morris, Rev. F.O. 1851–7. *A History of British Birds*. Groombridge & Sons. London.
Muirhead, G. 1895. *The Birds of Berwickshire*. David Douglas. Edinburgh.
Murray, R., Holling, M., Dott, H. & Vandome, P. 1998. *Breeding Birds of South-east Scotland: A Tetrad Atlas 1988–1994*. Scottish Ornithologists' Club. Edinburgh.
Nash, J.K. 1935. *The Birds of Midlothian*. Witherby. London.
Nelson, T.H. 1907. *The Birds of Yorkshire*. Brown & Sons. London.
Noble, H. 1906. Birds. In *Victoria History of Berkshire*. Dawson. London.
Norris, C.A. 1947. *Notes on the Birds of Warwickshire*. Cornish Brothers Ltd. Birmingham.
Oakes, C. 1953. *The Birds of Lancashire*. Oliver & Boyd. Edinburgh.
Palmer, E.M. & Ballance, D.K. 1968. *The Birds of Somerset*. Longman. London.
Parr, D. (ed.) 1972. *Birds in Surrey, 1900–1970*. Batsford. London.
Pashley, H.N. 1925. *Notes on the Birds of Cley, Norfolk*. Witherby. London.
Paton, E.R. & Pike, O.G. 1929. *The Birds of Ayrshire*. Witherby. London.
Peers, M. & Shrubb, M. 1990. *Birds of Breconshire*. Brecknock Wildlife Trust. Brecon.
Penhallurick, R. 1978. *The Birds of Cornwall and Isles of Scilly*. Headland Publications. Penzance.
Radford, M.C. 1966. *The Birds of Berkshire and Oxfordshire*. Longman. London.
Rhodes, R.J. 1988. *Birds in the Doncaster District*. Doncaster and District Ornithological Society. Doncaster.
Riviere, B.B. 1930. *A History of the Birds of Norfolk*. Witherby. London.
Rodd, E.H. 1880. *The Birds of Cornwall and the Scilly Isles*. Trubner & Co. London.
Ryves, B.H. 1948. *Birdlife in Cornwall*. Collins. London.
Sage, B.L. 1959. *A History of the Birds of Hertfordshire*. Barrie & Rockcliffe. London.
Salter, J.H. 1895–1904. Observations on Birds in mid Wales. *Zoologist* 3(19):179, 3(20):24–26, 4(8):66–71.

Saunders, H. 1899. *Illustrated Manual of British Birds*. Van Voorst. London.
Saunders, H. & Eagle Clarke, W. 1927. *An Illustration Manual of British Birds*, 3rd edn. Gurney & Jackson. London.
Seago, M.J. 1967. *The Birds of Norfolk*. Jarrolds. Norwich.
Seebohm, H. 1885. *A History of British Birds*. R.H. Porter. London.
Shrubb, M. 1979. *The Birds of Sussex: Their Present Status*. Phillimore. Chichester.
Sim, G. 1903. *The Vertebrate Fauna of Dee*. Wyllie & Son. Aberdeen.
Sitters, H.P. 1988. *Tetrad Atlas of the Breeding Birds of Devon*. Devon Birdwatching and Preservation Society. Yelverton.
Smith, A.C. 1887. *The Birds of Wiltshire*. R.H. Porter. London.
Smith, A.E. & Cornwallis, R.K. 1955. *The Birds of Lincolnshire*. Lincolnshire Naturalists' Union.
Smith, C. 1869. *The Birds of Somersetshire*. Van Voorst. London.
Smith, K.W., Dee, C.W. & Fearnside, J. 1993. *Breeding Birds of Hertfordshire*. Hertfordshire Natural History Society.
Smith, T. 1930–8. The Birds of Staffordshire. *Transactions of the North Staffordshire Field Club* vols. 64–72.
Somerset Ornithological Society. 1988. *Birds of Somerset*. Alan Sutton Publishing. Gloucester.
Standley, P., Bucknell, N.J., Swash, A. & Collins, I.D. 1996. *Birds of Berkshire*. Berkshire Atlas Group. Reading.
Sterland, W.J. 1869. *The Birds of Sherwood Forest*. Reeve & Co. London.
Stevenson, H. 1866–90. *The Birds of Norfolk*. Van Voorst. London.
Swaine, C.M. 1982. *The Birds of Gloucestershire*. Alan Sutton Publishing. Gloucester.
Taylor, D.W., Davenport, D.L. & Flegg, J.J.M. 1981. *The Birds of Kent: A Review of their Status and Distribution*. Kent Ornithological Society.
Taylor, M., Seago, M.J., Allard, P. & Dorling, D. 1999. *The Birds of Norfolk*. Pica Press. Mountfield.
Temperley, G.W. 1946. A history of the birds of Durham. *Transactions of the Natural History Society of Northumberland, Durham and Newcastle on Tyne*, new series, 9:1–296.
Thom, V.M. 1986. *Birds in Scotland*. T. & A.D. Poyser. Calton.
Ticehurst, C.B. 1932. *A History of the Birds of Suffolk*. Gurney & Jackson. London.
Ticehurst, N.F. 1909. *A History of the Birds of Kent*. Witherby. London.
Trodd, P. & Kramer, D. 1991. *Birds of Bedfordshire*. Castlemead. Welwyn Garden City.
Turner, E.L. 1924. *Broadland Birds*. Country Life. London.
Venables, L.S.V. & Venables, U.M. 1955. *Birds and Mammals of Shetland*. Oliver & Boyd. Edinburgh.
Walpole-Bond, J. 1938. *A History of the Birds of Sussex*. Witherby. London.
Whitaker, J. 1907. *Notes on the Birds of Nottinghamshire*. Walter Black & Co. Nottingham.
Yarrell, W. 1837–43. *A History of British Birds*, 2nd edn. Van Voorst. London.
Yarrell, W. 1871–85. *A History of British Birds*, 4th edn. Van Voorst. London.

References

J.R.A.S.E. is the *Journal of the Royal Agricultural Society of England*, and *Trans. H.A.S.* is the *Transactions of the Highland and Agricultural Society*, of Scotland.

Adair, P. 1891, 1893. The Short-eared owl and the Kestrel in the vole plague districts. *Annals of Scottish Natural History* 6:219–231, 8:193–202.

Aebischer, N.J. 1991. Twenty years of monitoring invertebrates and weeds in cereal fields in Sussex. In: *The Ecology of Temperate Cereal Fields*. Blackwell Scientific. Oxford.

Aebischer, N.J. & Ward, R.S. 1997. The distribution of corn buntings *Miliaria calandra* in Sussex in relation to crop type and invertebrate abundance. In: *The Ecology and Conservation of Corn Buntings* Miliaria calandra. UK Nature Conservation No. 13. Joint Nature Conservation Committee. Peterborough.

Aebischer, N.J., Green, R.E. & Evans, A.D. 2000. From science to recovery: four case studies of how research has been translated into conservation action in the UK. In: *Ecology and Conservation of Lowland Farmland Birds*. British Ornithologists Union. Tring.

Agricultural Chemicals Approval Scheme (ACAS). *List of Approved Products and their uses for Farmers and Growers*. Ministry of Agriculture, Fisheries and Food. London.

Albery, A. 1999. Agriculture and wildlife conservation: accident or design? *British Wildlife* 11(1):10–16.

Alexander, W.B. & Lack, D. 1944. Changes in status amongst British breeding birds. *British Birds* 38:42–45, 62–69, 82–88.

Allen-Stevens, T. 2000. Thumbs up for lo-till. *Crops* w/e 23 September 2000.

Anderson, P. & Yalden, D.W. 1981. Increased sheep numbers and the loss of heather moorland in the peak district, England. *Biological Conservation* 20:195–213.

Aplin, O.V. 1890, 1891. On the distribution and period of sojourn in the British Islands of the Spotted Crake. *Zoologist* 14:401–417, 15:88–96.

Archibald, C.F. 1892. Wild birds useful and injurious. *J.R.A.S.E.*, series 3, 3:658–684.

Armstrong, P.H. 1973. Changes in the land use of the Suffolk Sandlings: a study of the disintegration of an ecosystem. *Geography* 58:1–8.

Arnold, G.W. 1983. The influence of ditch and hedgerow structure, length of hedgerows, and area of woodland and garden on bird numbers in farmland. *Journal of Applied Ecology* 20:731–750.

Asher, J., Warren, M., Fox, R., Harding, P., Jeffcoate, G. & Jeffcoate, S. 2001. *The Millennium Atlas of Butterflies in Britain and Ireland*. Oxford University Press. Oxford.

Ausden, M. 2001. The effects of flooding of grassland on food supply for breeding waders. *British Wildlife* 12(3):179–187.

Atkinson-Willes, G.L. (ed.) 1963. *Wildfowl in Great Britain*. HMSO. London.

Bailey Denton, R. 1863. The effect of under drainage on our rivers and arterial channels. *J.R.A.S.E.* 24:573–589.

Baines, D. 1988. The effects of improvement of upland marginal grasslands on the distribution and density of breeding wading birds (Charadriiformes) in northern England. *Biological Conservation* 45:221–236.

Baines, D. 1990. The roles of predation, food and agricultural practice in determining the breeding success of the Lapwing (*Vanellus vanellus*) on upland grasslands. *Journal of Animal Ecology* 59:915–929.

Baines, D. 1991. Long-term changes in the European Black Grouse population. *The Game Conservancy Review of 1990* 22:157–158.

Baines, D. 1994. Factors determining the breeding success and distribution of lapwings (*Vanellus vanellus*) on marginal farm land in northern England. In: *The Ecology and Conservation of Lapwings* Vanellus vanellus. UK Nature Conservation No. 9. Joint Nature Conservation Committee. Peterborough.

Baines, D. 1996. The implications of grazing and predator management on the habitats and breeding success of Black Grouse *Tetrao tetrix*. *Journal of Applied Ecology* 33:54–62.

Baker, R. 1845. On the farming of Essex. *J.R.A.S.E.* series 1, 5:1–43.

Barnard, C.J. & Thompson, D.B.A. 1985. *Gulls and Plovers: The Ecology and Behaviour of Mixed Species Feeding Groups*. Croom Helm. London.

Barr, C.J., Bunce, R.G.H., Clarke, R.T., Fuller, R.M., Furse, M.T., Gillespie, M.K., Groom, G.B., Hallam, C.J., Hornung, M., Howard, D.C. & Ness, M.J. 1993. *Countryside Survey 1990*. Main Report. Department of the Environment. London.

Barreto, G.R., Macdonald, D.W. & Strachan, R. 1998. The Tightrope Hypothesis: an explanation for plummeting water vole numbers in the Thames catchment. In: *United Kingdom Floodplains*. Westbury Press. Otley.

Beintema, A.J., Beintema-Hietbrink, R.J. & Muskens, G.J.D.M. 1985. A shift in the timing of breeding in meadow birds. *Ardea* 73:83–89.

Beintema, A.J. & Muskens, G.J.D.M. 1987. Nesting success of birds breeding in Dutch grasslands. *Journal of Applied Ecology* 24:743–758.

Beintema, A.J., Thissen, J.B., Tensen, D. & Visser, G.H. 1991. Feeding ecology of charadriiform chicks in agricultural grassland. *Ardea* 79:31–44.

Belcher, C. 1863. Reclaiming of wastelands as instanced in Wichwood Forest. *J.R.A.S.E.* 24:271–285.

Bettey, J.H. 1977. The development of water meadows in Dorset during the seventeenth century. *Agricultural History Review* 25:37–43.

Bibby, C.J. 1986. Merlin. In: *The Atlas of Wintering Birds in Britain and Ireland*. T. & A.D. Poyser. Calton.
Bibby, C.J. 1988. Impact of agriculture on upland birds. In: *Ecological Change in the Uplands*. Blackwell Scientific. Oxford.
Bibby, C.J. 1993. Red-backed Shrike. In:*The New Atlas of Breeding Birds in Britain and Ireland: 1988–1991*. T. & A.D. Poyser. London.
Bijleveld, M.F.I.J. 1974. *Birds of Prey in Europe*. Macmillan. London.
Birks, H.J.B. 1988. Long-term ecological change in the British uplands. In: *Ecological Change in the Uplands*. Blackwell Scientific. Oxford.
Birnie, A. 1955. *An Economic History of the British Isles*. Methuen & Co. Ltd. London.
Bishton, J. 1794. *A General View of the Agriculture of Shropshire*. Board of Agriculture. London.
Blaker, G.B. 1934. *The Barn Owl in England and Wales*. RSPB. London.
Bowers, J.K. & Cheshire, P. 1983. *Agriculture, the Countryside and Land Use*. Methuen. London.
Bowie, G.G.S. 1987. Watermeadows in Wessex: a re-evaluation for the period 1640–1850. *Agricultural History Review* 35:151–158.
Bradbury, R.B. & Stoate, C. 2000. The ecology of Yellowhammers *Emberiza citrinella* on lowland farmland. In: *Ecology and Conservation of Lowland Farmland Birds*. BOU. Tring.
Bradley, A.G. 1927. *When Squires and Farmers Thrived*. Methuen and Co. London.
Bravendar, J. 1850. Farming of Gloucestershire. *J.R.A.S.E.* 9:116–177.
Brenchley, A. 1984. The use of birds as indicators of change in agriculture. In: *Agriculture and the Environment*. Institute of Terrestrial Ecology. Monks Wood.
Brentnall, H.C. 1947. A Longford Manuscript. *The Wiltshire Magazine* No:187, Vol.52:1–56.
Brickle, N.W. & Harper, D.G.C. 2000. Habitat use by Corn Buntings *Miliaria calandra* in winter and summer. In: *Ecology and conservation of lowland farmland birds*. British Ornithologists Union. Tring.
Broad, J. 1980. Alternate husbandry and permanent pasture in the Midlands 1650–1800. *Agricultural History Review* 28(2):77–89.
Browne, S.J. & Aebischer, N.J. 2001. *The Role of Agricultural Intensification in the Decline of the Turtle Dove Streptopelia turtur*. English Nature Research Report No. 421. English Nature. Peterborough.
Bull, A.L., Mead, C.J. & Williamson, K. 1976. Bird-life on a Norfolk farm in relation to agricultural changes. *Bird Study* 23:203–218.
Bunn, D.S., Warburton, A.B. & Wilson, R.D.S. 1982. *The Barn Owl*. T. & A.D. Poyser. Calton.
Burton, J.F. 1995. *Birds and Climate Change*. Helm. London.
Cadle, C. 1867. The agriculture of Worcestershire. *J.R.A.S.E.*, series 2, 3:439–466.
Caird, J. 1852. *English Agriculture in 1850–51*. London.
Callaway, T. 1998. Restoration of lowland wet grassland at Pulborough Brooks RSPB Nature Reserve. In: *United Kingdom Floodplains*. Westbury Press. Otley.
Cambridge, W. 1845. On the advantages of reducing the size and number of hedges. *J.R.A.S.E.* 6:333–342.

Campbell, L.H., Avery, M.I., Donald, P., Evans, A.D., Green, R.E. & Wilson, J.D. 1996. *A Review of the Indirect Effect of Pesticides on Birds*. Joint Nature Conservation Committee Report No. 227. Joint Nature Conservation Committee. Peterborough.

Carter, E. 1982. Land drainage. *FWAG Advisory Group Spring/Summer Newsletter* 7–8.

Catchpole, C.K. 1974. Habitat selection and breeding success in the Reed Warbler *Acrocephalus scirpaceus*. *Journal of Animal Ecology* 43:363–380.

Cayford, J. 1992. Barn Owl ecology on East Anglian farmland. *RSPB Conservation Review* 6:45–50.

Chamberlain, D.E., Fuller, R.J., Bunce, R.G.H., Duckworth, J.C. & Shrubb, M. 2000. Changes in the abundance of farmland birds in relation to changes in agricultural practices in England and Wales. *Journal of Applied Ecology* 37:771–788.

Chamberlain, D.E. & Fuller, R.J. 2001. Contrasting patterns of change in the distribution and abundance of farmland birds in relation to farming system in lowland Britain. *Global Ecology and Biogeography* 10:399–409.

Chambers, J.D. 1955. The problem of Sherwood Forest. *Agriculture* 62:177–180.

Chew, H.C. 1953. The post-War land use pattern of the former grasslands of eastern Leicestershire. *Geography* 38:286–295.

Claridge, J. 1793. *A General View of the Agriculture of Dorset*. Board of Agriculture. London.

Clarke, J.A. 1848. On the Great Level of the Fens, including the Fens of South Lincolnshire. *J.R.A.S.E.* 8:80–133.

Clarke, J.A. 1851. Farming of Lincolnshire. *J.R.A.S.E.* 12:259–288, 289–414.

Clarke, J.A. 1854. On trunk drainage. *J.R.A.S.E.* 15:1–73.

Clarke, R. 1996. *Montagu's Harrier*. Arlequin Press. Chelmsford.

Cobbett, W. *Rural Rides*. Everyman's Library edition 1957. J.M. Dent & Sons. London.

Collinge, W.E. 1924–7. *The Food of Some British Wild Birds*. Privately published.

Collins, E.J.T. 1969. Harvest technology and labour supply in Britain, 1790–1870. *Economic History Review* 22:453–475.

Cooke, A.S., Bell, A.A. & Haas, M.B. 1982. *Predatory Birds, Pesticides and Pollution*. Institute of Terrestrial Ecology. Cambridge.

Coppock, J.T. 1958. Changes in farm and field boundaries in the 19th century. *Amateur Historian* 3:292–298.

Coppock, J.T. 1976. *An Agricultural Atlas of England and Wales*, 2nd edn. Faber & Faber. London.

Corbet, G.B. & Southern, H.N. 1977. *The Handbook of British Mammals*, 2nd edn Blackwell Scientific. Oxford.

Country Life. 2000. Editorial: What happens to farmland that nobody wants. 27 July.

Cowley, M., Thomas, C., Thomas, J. & Warren, M. 2000. Assessing butterflies' status and decline. *British Wildlife* 11(4):243–249.

Cramp, S. (ed.) 1985. *Birds of the Western Palaearctic*, vols. 4–6. Oxford University Press. Oxford.

Cramp, S. & Simmons, K.E.I. (eds.) 1977. *Birds of the Western Palaearctic*, vols. 1-3. Oxford University Press. Oxford.

Cramp, S. & Perrins, C.M. (eds.) 1994. *Birds of the Western Palaearctic*, vols. 7-9. Oxford University Press.

Creyke, R. 1845. Some account of the process of warping. *J.R.A.S.E.* 5:398-405.

Crick, H.P.Q. 1997. Long-term trends in corn bunting *Miliaria calandra* productivity in Britain. In: *The Ecology and Conservation of Corn Buntings* Miliaria calandra. UK Conservation No.13. Joint Nature Conservation Committee. Peterborough.

Crick, H.P.Q, Dudley, C. & Glue, D.E. 1995. Nesting in 1993. *BTO News* 196:14-16.

Crick, H.P.Q., Dudley, C., Glue, D.E. & Balmer, D. 1996. Nesting 1995. *BTO News* 207:5-8.

Crick, H.P.Q., Dudley, C., Glue, D.E., Balmer, D. & Beaven, P. 1998. Nesting 1996. *BTO News* 214:12-14.

Crick, H.P.Q., Raven, M., Beaven, P., Dudley, C. & Glue, D. 2000. Breeding trends from Nest Records lead to new alerts for declining species. *BTO News* 228:8-9.

Cunningham, W. 1912. *The Growth of English Industry and Commerce in Modern Times*. Cambridge University Press. Cambridge.

da Prato, S.R.D. 1985. The breeding birds of agricultural land in south-east Scotland. *Scottish Birds* 13(7):203-216.

Darby, H.C. 1934. Note on the birds of the undrained Fen. In: Lack, D. (ed.) *The Birds of Cambridgeshire*. Cambridge Bird Club. Cambridge.

Darby, H.C. 1968. *The Draining of the Fens*. Cambridge University Press. Cambridge.

Dargie, T.C. 1993. *The Distribution of Lowland Wet Grassland in England*. English Nature Research Report No. 49. English Nature. Peterborough.

Darley, G. & Toler, P. 1981. *The National Trust Book of the Farm*. Weidenfeld & Nicholson. London.

Darling, F. Fraser & Boyd, J.M. 1964. *The Highlands and Islands*. Collins. London.

Davies, E. 1984-5. Hafod and Lluest: the summering of cattle and upland settlement in Wales. *Folklife* 23:76-96.

Davis, B.N.K. 1967. Bird feeding preferences among different crops in an area near Huntingdon. *Bird Study* 14:227-237.

Davis, T. 1794. *General View of the Agriculture of the County of Wiltshire*. Board of Agriculture. London.

Dennis, Rev. R.N. 1848. *Diary*. Unpublished ms in library of British Trust for Ornithology, Thetford.

Dickinson, W. 1852. On the farming of Cumberland. *J.R.A.S.E.* 13:207-300.

Dickson, J. 1869. Report on the agriculture of Perthshire. *Trans. H.A.S.* 1869:160-190.

Dobbs, A. 1964. Rook numbers in Nottinghamshire over 35 years. *British Birds* 57:360-364.

Dobinson, H.M. & Richards, A.J. 1964. The effects of the severe winter of 1962/63 on birds in Britain. *British Birds* 57:373-434.

Donald, P.F. 1997. The corn bunting *Miliaria calandra* in Britain: a review of current status, patterns of decline and possible causes. In: *The Ecology and Conservation of*

Corn Buntings Miliaria calandra. UK Nature Conservation No.13. Joint Nature Conservation Committee. Peterborough.

Donald, P.F. & Evans, A.D. 1994. Habitat selection by Corn Buntings *Miliaria calandra* in winter. *Bird Study* 41:199–210.

Donald, P.F. & Forrest, C. 1995. The effects of agricultural change on population size of Corn Buntings *Miliaria calandra* on individual farms. *Bird Study* 42:205–215.

Donald, P.F. & Vickery, J.A. 2000. The importance of cereal fields to breeding and wintering Skylarks *Alauda arvensis* in the UK. In: *Ecology and Conservation of Lowland Farmland Birds*. British Ornithologists Union. Tring.

Donald, P.F., Wilson, J.D. & Shepherd, M. 1994. The decline of the Corn Bunting. *British Birds* 87:106–132.

Donaldson, J.G.S., Donaldson, F. & Barber, D. 1969. *Farming in Britain Today*. Penguin Books. Harmondsworth.

Driver, A. & Driver, W. 1794. *General View of the Agriculture of Hampshire*. Board of Agriculture. London.

Dunnet, G.M. & Patterson, I.J. 1968. The Rook problem in North-east Scotland. In:*The Problems of Birds as Pests*. Academic Press. London.

Dymond, D. 1990. *The Norfolk Landscape*. Alastair Press. Bury St. Edmunds.

Eagle Clarke, W. 1912. *Studies in Bird Migration*. Gurney & Jackson. London.

Easy, J.M.S. (ed.) 1962. Stone Curlew. In: *Cambridge Bird Report 1962*. Cambridge Bird Club. Cambridge.

Edwards, C.A. 1984. Changes in agricultural practice and their impact on soil organisms. In: *Agriculture and the Environment*. Institute of Terrestrial Ecology. Cambridge.

Ellis, E.A. 1965. *The Broads*. Collins. London.

Ellis, W. 1735. *The Shepherd's Sure Guide*. Privately published.

Elly, S. 1846. On the cultivation and preparation of gorse as food for cattle. *J.R.A.S.E.* 6:523–528.

Ennion, E.A.R. 1949. *Adventurers' Fen*. Herbert Jenkins. London.

Ennion, E.A.R. 1960. *The House on the Shore*. Routledge & Kegan Paul. London.

Ernle, Lord. 1922. *English Farming: Past and Present*, 3rd edn. Longman. London.

Eurostat. 1980. *Land Use and Production 1955–1979*. Office for Official Publications of the European Communities. Luxemburg.

Eurostat. 1987. *Farm Structure, 1985 Survey: Main Results*. Office for Official Publications of the European Communities. Luxemburg.

Evans, A.D. 1997. Cirl Buntings in Britain. *British Birds* 90:267–282.

Evans, A.D. & Smith, K.W. 1994. Habitat selection of Cirl Buntings *Emberiza cirlus* wintering in Britain. *Bird Study* 41:81–87.

Evans, G.E. 1956. *Ask the Fellows who Cut the Hay*. Faber & Faber. London.

Evershed, H. 1856. Farming of Warwickshire. *J.R.A.S.E.* 15:475–493.

Evershed, H. 1869. The agriculture of Staffordshire. *J.R.A.S.E.*, series 2, 5:263–317.

Feare, C.J. 1984. *The Starling*. Oxford University Press. Oxford.

Feare, C.J. 1994. Changes in numbers of Common Starlings and farming practice in Lincolnshire. *British Birds* 87:200–204.

Feare, C.J. & McGinnity, N. 1987. The relative importance of invertebrates and barley in the diet of Starlings *Sturnus vulgaris*. *Bird Study* 33:164–167.
Fletcher, T.W. 1961. Lancashire livestock farming during the Great Depression. *Agricultural History Review* 9:17–42.
Fletcher, T.W. 1962. The agrarian revolution in arable Lancashire. *Transactions of the Lancashire and Cheshire Antiquarian Society* 72:93–122.
Fraser, R. 1794. *General View of the Agriculture of Cornwall and of Devon.* Board of Agriculture. London.
Fuller, R.J. 1982. *Bird Habitats in Britain.* T. & A.D. Poyser. Calton.
Fuller, R.J. 1986. Golden Plover and Lapwing. In: *The Atlas of Wintering Birds in Britain and Ireland.* T. & A.D. Poyser. Calton.
Fuller, R.J. 1994. Changes in breeding populations of Curlews and Lapwings in central Buckinghamshire 1981–1990. *Buckinghamshire Bird Report* 1994.
Fuller, R.J. 1996. *Relationships between Grazing and Birds with Particular Reference to Sheep in the British Uplands.* BTO Research Report No. 164. British Trust for Ornithology. Thetford.
Fuller, R.J. & Crick, H.P.Q. 1992. Broad-scale patterns in geographical and habitat distribution of migrant and residents passerines in Britain and Ireland. *Ibis* 134 (supplement 1):14–20.
Fuller, R.J. & Glue, D.E. 1977. The breeding biology of the Stonechat and Whinchat. *Bird Study* 24:215–228.
Fuller, R.J. & Gough, S.J. 1999. Changes in sheep numbers in Britain: implications for bird populations. *Biological Conservation* 91:73–89.
Fuller, R.J. & Lloyd, D. 1981. The distribution and habitats of wintering Golden Plovers in Britain. *Bird Study* 28:169–185.
Fuller, R.J. & Youngman, R.E. 1979. The utilisation of farmland by Golden Plovers wintering in southern England. *Bird Study* 26:37–46.
Fuller, R.J., Baker, J.K., Morgan, R.A., Scroggs, R. & Wright, M. 1985. Breeding populations of the Hobby *Falco subbuteo* on farmland in the southern Midlands of England. *Ibis* 127:510–516.
Fuller, R.J., Chamberlain, D.E., Burton, N.H.K. & Gough, S.J. 2001. Distributions of birds in lowland agricultural landscapes of England and Wales: How distinctive are bird communities of hedgerows and woodland? *Agriculture, Ecosystems and Environment* 84:79–92.
Fuller, R.M. 1987. The changing extent and conservation interest of lowland grasslands in England and Wales: a review of grassland surveys. *Biological Conservation* 40:281–300.
Furbank, P.N., Owens, W.R. & Coulson, A.J. 1991. *Daniel Defoe: A Tour Through the Whole Island of Great Britain.* Yale University Press. London.
Fussell, G.E. 1952. Four centuries of farming systems in Dorset, 1500–1900. *Proceedings of the Dorset Natural History and Archaeological Society* 73:116–140.
Fussell, G.E. & Goodman, C. 1930. Eighteenth-century estimates of British sheep and wool production. *Agricultural History* 4:131–151.
Galbraith, H. 1988. Effects of agriculture on the breeding ecology of Lapwings *Vanellus vanellus*. *Journal of Applied Ecology* 25:487–503.

Galbraith, H., Furness, R.W. & Fuller, R.J. 1984. Habitats and distribution of waders breeding on Scottish agricultural land. *Scottish Birds* 13(4):98–107.

Gaston, K.J. & Blackburn, T.M. 2000. *Pattern and Process in Macroecology*. Blackwell Science. Oxford.

Gatke, H. 1895. *Heligoland: An Ornithological Observatory*. David Douglas. Edinburgh

Gibbons, D.W. & Dudley, S. 1993. Quail. In: *New Atlas of Breeding Birds in Britain and Ireland: 1988–1991*. T. & A.D. Poyser. London.

Gibbons, D.W., Reid, J.B. & Chapman, R.A. (eds.) 1993. *The New Atlas of Breeding Birds in Britain and Ireland: 1988–1991*. T. & A.D. Poyser. London.

Gillings, S. & Watts, P.N. 1997. Habitat selection and breeding ecology of corn buntings *Miliaria calandra* in the Lincolnshire fens. In: *The Ecology and Conservation of Corn Buntings Miliaria calandra*. UK Nature Conservation No.13. Joint Nature Conservation Committee. Peterborough.

Gladstone, H.S. 1924. The distribution of Black Grouse in Great Britain. *British Birds* 18:66–68.

Glue, D.E. 1974. Food of the Barn Owl in Britain and Ireland. *Bird Study* 21:200–210.

Glue, D.E. 1977. Feeding ecology of the Short-eared Owl in Britain and Ireland. *Bird Study* 24:70–78.

Glue, D.E. & Jordan, R. 1989. Early 20th century Barn Owl *Tyto alba* diet in Hampshire. *Hampshire Bird Report* 1988:79–81.

Glue, D.E. & Morgan, R. 1974. Breeding statistics and movements of the Stone Curlew. *Bird Study* 21: 21–28.

Glue, D.E. & Scott, D. 1980. Breeding biology of the Little Owl. *British Birds* 73:167–180.

Gooch, S.M. 1963. The occurrence of weed seeds in samples tested by the Official Seed Testing Station 1960–61. *Journal of the National Institute of Agricultural Botany* 9:353–371.

Good, J.E.G, Bryant, R. & Carlill P. 1990. Distribution longevity and survival of upland Hawthorn *Crataegus monogyna* scrub in North Wales in relation to sheep grazing. *Journal of Applied Ecology* 27:272–283.

Grant, J. 1845. A few remarks on the large hedges and small enclosures of Devonshire and adjoining counties. *J.R.A.S.E.* 5:420–429.

Grant, M.C., Orsman, C., Easton, J., Lodge, C., Smith, M., Thompson, G., Rodwell S. & Moore, N. 1999. Breeding success and causes of breeding failure of curlew *Numenius arquata* in Northern Ireland. *Journal of Applied Ecology* 36:59–74.

Grant, S.A., Torvell, L., Sim, E.M., Small, J.L. & Armstrong, J.H. 1996. Controlled grazing studies on Nardus grassland: effects of between-tussock sward height and species of grazer on *Nardus* utilisation and floristic composition in two fields in Scotland. *Journal of Applied Ecology* 33:1053–1064.

Graveland, J. 1999. Effects of reed cutting on density and breeding success of Reed Warbler *Acrocephalus scirpaceus* and Sedge Warbler *A. schoenobanus*. *Journal of Avian Biology* 30: 469–482.

Gray, D.B. 1974. Breeding behaviour of Whinchats. *Bird Study* 21:280–282.

Gray, R. & Hussey, A. 1860. The trade in Goldfinches. *Zoologist* 18:711–4.

Green, F.H.W. 1973. Aspects of the changing environment: some factors affecting the aquatic environment in recent years. *Journal of Environmental Management* 1:377-391.

Green, F.H.W. 1974. Changes in artificial drainage, fertilisers and climate in Scotland. *Journal of Environmental Management* 2:107-121.

Green, F.H.W. 1975. Ridge and furrow, mole and tile. *Geographical Journal* 141:88-93.

Green, F.H.W. 1976. Recent changes in land use and treatment. *Geographical Journal* 142: 12-26.

Green, R.E. 1978. Factors affecting the diet of farmland Skylarks *Alauda arvensis*. *Journal of Animal Ecology* 47:913-928.

Green, R.E. 1988. Stone Curlew conservation. *RSPB Conservation Review* 2: 30-33.

Green, R.E. 1993. Stone Curlew. In: *The New Atlas of Breeding Birds in Britain and Ireland: 1988-1991*. T. & A.D. Poyser. London.

Green, R.E. 1995. The decline of the Corncrake *Crex crex* continues. *Bird Study* 42: 66-75.

Green, R.E. & Robins, M. 1993. The decline of the ornithological importance of the Somerset Levels, England and changes in the management of water levels. *Biological Conservation* 66:95-106.

Green, R.E., Osborne, P.E. & Sears, E.J. 1994. The distribution of passerine birds in hedgerows during the breeding season in relation to characteristics of the hedgerow and adjacent farmland. *Journal of Applied Ecology* 31:677-692.

Green, R.E., Tyler, G.A. & Bowden, C.G.R. 2000. Habitat selection, ranging behaviour and diet of the stone curlew (*Burhinus oedicnemus*) in southern England. *Journal of Zoology, London* 250:161-183.

Greenwood, J.J.D. & Baillie, S.R. 1991. Effects of density-dependence and weather on population changes of English passerines using a non-experimental paradigm. *Ibis* 133 (supplement 1):121-133.

Gregory, R.D. 1987. Comparative winter feeding ecology of Lapwings *Vanellus vanellus* and Golden Plover *Pluvialis apricaria* in the Lower Derwent Valley, North Yorkshire. *Bird Study* 34:244-250.

Grigg, D. 1989. *English Agriculture: An Historical Perspective*. Basil Blackwell. Oxford.

Griggs, Messrs. 1794. *General View of the Agriculture of Essex*. Board of Agriculture. London.

Gurney, J.H. 1899. The Bearded Titmouse. *Transactions of the Norfolk and Norwich Naturalists' Society* 6:429-438.

Haines-Young, R.H. Barr, C.J., Black, H.I.J., Briggs, D.J., Bunce, R.G.H., Clarke, R.T., Cooper, A., Dawson, F.H., Firbank, L.G., Fuller, R.M., Furse, M.T., Gillespie, M.K., Hill., R., Hornung, M., Howard, D.C., McCann, T., Morecroft, M.D., Petit, S., Sier, A.R.J., Smart, S.M., Smith, G.M., Stott, A.P., Stuart, R.C. & Watkins, J.W. 2000. *Accounting for Nature: Assessing Habitats in the UK Countryside*. Department of the Environment, Transport and the Regions. London.

Hardy, A.R. 1992. Habitat use by farmland Tawny Owls *Strix aluco*. In: *The Ecology and Conservation of European owls*. Joint Nature Conservation Committee. Peterborough.

Harris, S., Morris, P., Wray, S. & Yalden, D. 1995. *A Review of British Mammals*. Joint Nature Conservation Committee. Peterborough.

Hart, E. 1981. *Victorian and Edwardian Farming from Old Photographs*. Batsford. London.

Harvie-Brown, J.A. 1895. The Starling in Scotland, its increase and distribution. *Annals of Scottish Natural History* 5:3–23.

Haslam, S.M. 1991. *The Historic River*. Cobden of Cambridge Press. Cambridge.

Hastings, Macdonald. 1981. *The Shotgun*. David & Charles. Newton Abbot.

Hawker, P. 1988. *The Diaries of Peter Hawker 1802–1853*. Greenhill Books. London.

Henderson, I.G., Cooper, J., Fuller, R.J. & Vickery, J. 2000. The relative abundance of birds on set-aside and neighbouring fields in summer. *Journal of Applied Ecology* 37:335–347.

Hennell, T. 1934. *Change in the Farm*. Cambridge University Press. Cambridge.

Heppleston, P.P. 1972. The comparative breeding ecology of Oystercatchers (*Haematopus ostralegus*) in inland and coastal habitats. *Journal of Animal Ecology* 41:23–51.

Hills, M.G. 1974. Fertiliser use on grassland, 1974. In: *Survey of Fertiliser Practice*. Ministry of Agriculture, Fisheries and Food. London.

Hobhouse, H. 1989. *Forces of Change: Why We Are the Way We Are Now*. Sidgwick & Jackson. London.

Hollom, P.A.D. 1940. Report on the 1938 survey of Black-headed Gull colonies. *British Birds* 33:202–221.

Holloway, S. 1996. *The Historical Atlas of Breeding Birds in Britain and Ireland 1875–1900*. T. & A.D. Poyser. London.

Hope Jones, P. 1989. The chequered history of the Black Grouse in Wales. *Welsh Bird Report 1989*: 70–78.

Hoskins, W.G. 1955. *The Making of the English Landscape*, paperback edn. 1985. Penguin Books: Harmondsworth.

Hoskins, W.G. & Stamp, L. Dudley. 1963. *The Common Lands of England and Wales*. Collins. London.

Hudson, P.J. 1988. Spatial variations, patterns and management options in upland bird communities. In: *Ecological Change in the Uplands*. Blackwell Scientific. Oxford.

Hudson, P.J. 1992. *Grouse in Space and Time*. The Game Conservancy Ltd. Fordingbridge.

Hudson, W.H. 1900. *Nature in Downland*. J.M. Dent & Sons. London.

Hudson, W.H. 1923a. *Adventures among Birds*. J.M. Dent & Sons. London.

Hudson, W.H. 1923b. *Rare, Vanishing and Lost British Birds*, edited by Linda Gardiner. J.M. Dent & Sons. London.

Hughes, R.E., Dale, J., Williams, I.E. & Rees, D.I. 1973. Studies in sheep populations and environment in the mountains of North-west Wales. *Journal of Applied Ecology* 10:113–132.

Humpheryes, J. 1981. The enclosure of Sidlesham Common. *Journal and Newsletter of the West Sussex Archives Society* 20:10–13.

Inglis, I.R., Isaacson, A.J., Thearle, R.J.P. & Westwood, N.J. 1990. The effects of changing agricultural practice upon Woodpigeon *Columba palumbus* numbers. *Ibis* 132:262–272.

James, W. & Malcolm, J. 1794. *A General View of the Agriculture of Buckinghamshire and of Surrey*. Board of Agriculture. London.

Jenkins, D. 1971. *The Agricultural Community in South-West Wales at the turn of the Twentieth Century*. University of Wales Press. Cardiff.

Jenkins, D. & Watson, A. 2001. Bird numbers in relation to grazing on a grouse moor from 1957–61 to 1988–98. *Bird Study* 48:18–22.

Jonas, S. 1846. On the farming of Cambridgeshire. *J.R.A.S.E* 7:35–72.

Jones, B. 2001. The Uplands: potential for change. *Natur Cymru* 1:4–6.

Jones, E.L. 1972. The bird pests of British agriculture in recent centuries. *Agricultural History Review* 20(2):107–125.

Jourdain, F.C.R. & Witherby, H.F. 1918. The effect of the winter of 1916–1917 on our resident birds. *British Birds* 11:266–271, 12:26–35.

Joyce, B., Williams, G. & Woods, A. 1988. Hedgerows: still a cause for concern. *RSPB Conservation Review* 2:34–37.

June Census Statistics 1866–1994. Ministry of Agriculture, Fisheries and Food. London.

Kear, J. 1963a. Wildfowl and agriculture. In: *Wildfowl in Great Britain*. HMSO. London.

Kear, J. 1963b. The history of potato eating by wildfowl in Britain. *Wildfowl* 14:54–65.

Kear, J. 1990. *Man and Wildfowl*. T. & A.D. Poyser. London.

Kerridge, E. 1954. The sheepfold in Wiltshire and the floating of the watermeadows. *Economic History Review* 6:282–289.

Kleijn, D., Berendse, F., Smit, R. & Gillssen, N. 2001. Agri-environment schemes do not effectively protect biodiversity in Dutch agricultural landscapes. *Nature* 413:723–725.

Lack, P.C. 1986. *The Atlas of Wintering Birds in Britain and Ireland*. T. & A.D. Poyser. Calton.

Lack, P.C. 1992. *Birds on Lowland Farms*. HMSO. London.

Lane, C. 1980. The development of pastures and meadows during the sixteenth and seventeenth centuries. *Agricultural History Review* 28:18–30.

Laursen, K. 1980. Bird censuses in Danish farmland, with an analysis of bird distributions in relation to some landscape elements. (In Danish with English summary). *Dansk Ornithologisk Forenings Tidsskrift* 74:11–26.

Lefranc, N. & Worfolk, T. 1997. *Shrikes: A guide to the Shrikes of the world*. Pica Press. Mountfield.

Lennon, J.J., Greenwood, J.J.D. & Turner J.R.G. 2000. Bird diversity and environmental gradients in Britain: a test of the species-energy hypothesis. *Journal of Animal Ecology* 69:581–598.

Lewis, N. 1996. A review of the Welsh sheep industry. RSPB unpublished report.

Limbert, M. 1978. The old duck decoys of SE Yorkshire. *The Naturalist* 103:95–103.

Limbert, M. 1980. Hatfield Moors: an outline History. *South Yorkshire History* 4:33–46.

Lister, M.D. 1964. The Lapwing habitat enquiry, 1960–61. *Bird Study* 11:128–147.

Lloyd, H.G. 1980. *The Red Fox*. Batsford. London.

Lock, L. 1998. Review of farmland birds and lowland pastoral systems in SW England. RSPB unpublished report.

Long, W.H. 1963. The development of mechanisation in English farming. *Agricultural History Review* 11:15–26.

Love, R.A., Webbon, C., Glue, D.E. & Harris, S. 2000. Changes in the food of British Barn Owls *Tyto alba* between 1974 and 1997. *Mammal Review* 30:107–129.

Lovegrove, R.R., Shrubb, M. & Williams, I.T. 1995. *Silent Fields: The Current Status of Farmland Birds in Wales*. RSPB. Sandy.

MacNeilage, A. 1906. Farming methods in Ayrshire. *Trans. H.A.S.* 18:1–17.

Manley, G. 1952. *Climate and the British Scene*. Collins. London.

Marchant, J.H., Hudson, R., Carter, S.P. & Whittingdon, P. 1990. *Population Trends in British Breeding Birds*. British Trust for Ornithology. Tring.

Marchant, J.H., Sanderson, F. & Glue, D.E. 1999. Changes in breeding bird populations 1997–98. *BTO News* 222:10–14.

Marchant, J.H. & Gregory, R.D. 1999. Numbers of nesting Rooks *Corvus frugilegus* in the United Kingdom in 1996. *Bird Study* 46:258–273.

Marchant, J.V.R. & Watkins, W. 1897. *Bird Protection Acts*. R.H. Porter. London.

Marquiss, M., Newton, I. & Ratcliffe, D.A. 1978. The decline of the Raven *Corvus corax* in relation to afforestation in southern Scotland and northern England. *Journal of Applied Ecology* 15:129–144.

Marshall, E.J.P., Wade, P.M. & Clare, P. 1978. Land drainage channels in England and Wales. *Geographical Journal* 144(2):254–263.

Mason, C.F. & Lyczynski, F. 1980. Breeding biology of the Pied and Yellow Wagtails. *Bird Study* 27:1–10.

Mason, C.F. & MacDonald, S.M. 1999. Habitat use by Lapwings and Golden Plovers in a largely arable landscape. *Bird Study* 46:89–99.

Matheson, C. 1953. The Partridge in Wales: a survey of gamebook records. *British Birds* 46:57–64.

Matheson, C. 1957. Further Partridge records from Wales. *British Birds* 50:534–536.

Mayled, A. 1998. Restoration of National Trust watermeadows on the Sherborne Estate Gloucestershire; In: *United Kingdom Floodplains*. Westbury Press. Otley.

McCall, I. 1988. Value-added habitat. *Country Life* 30 June:170–171.

McCracken, D.I. 1993. The potential for avermectins to affect wildlife. *Veterinary Parasitology* 48:273–280.

McCracken, D.I. & Foster, G.N. 1993. The effects of ivermectin on the invertebrate fauna associated with cow dung. *Environmental Toxicology and Chemistry* 12:73–84.

Mearns, B. & Mearns, R. 1998. *The Bird Collectors*. Academic Press. London.

Mearns, R. 1983. The status of the Raven in southern Scotland and Northumbria. *Scottish Birds* 12:211–218.

Mee, F. 1988. *A History of Selsey*. Phillimore. Chichester.

Mills, G.S. 1975. A winter population study of the American Kestrel in central Ohio. *Wilson Bulletin* 87:241-247.
Minchington, W.E. 1956. Agriculture in Dorset during the Napoleonic Wars. *Proceedings of the Dorset Natural History and Archaeological Society* 77:162-173.
Mitchell, J. 1997. Wet meadows in lowland West-central Scotland: an almost forgotten botanical habitat. *Botanical Journal of Scotland* 49:341-345.
Møller, A.P. 1983. Breeding habitat selection in the Swallow *Hirundo rustica*. *Bird Study* 30:134-142.
Møller, A.P. 2001. The effect of dairy farming on barn swallow *Hirundo rustica* abundance, distribution and reproduction. *Journal of Applied Ecology* 38:378-389.
Moorcroft, D. & Wilson, J.D. 2000. The ecology of Linnets *Carduelis cannabina* on lowland farmland. In: *Ecology and Conservation of Lowland Farmland Birds*. British Ornithologists Union. Tring.
Moore, N.W. 1962. The heaths of Dorset and their conservation. *Journal of Ecology* 50:369-391.
Moreau, R.E. 1951. The British status of the Quail and some problems of its biology. *British Birds* 44:257-276.
Moscrop, W.J. 1866. A report on the farming of Leicestershire. *J.R.A.S.E.*, series 2, 2:288-337.
Murray, G. 1868. On the farming of Huntingdon. *J.R.A.S.E.*, series 2, 4:250-277.
Murray, K.A.H. 1955. *Agriculture: History of the Second World War*. UK Civil Series. HMSO. London.
Murton, R.K. 1965. *The Woodpigeon*. Collins. London.
Murton, R.K. 1971. *Man and Birds*. Collins. London.
Murton, R.K., Westwood, N.J. & Isaacson, A.J. 1964. The feeding habits of the Woodpigeon *Columba palumbus*, Stock Dove *C. oenas* and Turtle Dove *Streptopelia turtur*. *Ibis* 106:174-188.
Murton, R.K. & Westwood, N.J. 1974. Some effects of agricultural change on the English avifauna. *British Birds* 67:41-69.
Musters, C.J.M., Kruk, M., de Graaf, H.J. & Ter Keurs, W.J. 2001. Breeding birds as a farm product. *Conservation Biology* 15:363-369.
Mutch, A. 1981. The mechanisation of the harvest in south-west Lancashire. *Agricultural History Review* 29:125-132.
Nevill, R. 1908. *Old Sporting Prints*, 1970 edition. Hamlyn. London.
Newbold, C. 1998. The nature conservation importance of floodplains in England and Wales: with particular reference to their flora. In: *United Kingdom Floodplains*. Westbury Press. Otley.
Newman, E.I. & Harvey, P.D.A. 1997. Did soil fertility decline in Medieval English farms? Evidence from Cuxham, Oxfordshire, 1320-1340. *Agricultural History Review* 45:119-136.
Newton, A. 1896. *A Dictionary of Birds*. A. & C. Black. London.
Newton, I. 1967. The adaptive radiation and feeding ecology of some British finches. *Ibis* 109:33-97.
Newton, I. 1972. *Finches*. Collins. London.
Newton, I. 1979. *Population Ecology of Raptors*. T. & A.D. Poyser. Calton.

Newton, I. 1986. *The Sparrowhawk*. T. & A.D. Poyser. Calton.
Newton, I. 1998. *Population Limitation in Birds*. Academic Press. London.
Newton, I. Bell, A.A. & Wyllie, I. 1982a. Mortality of Sparrowhawks and Kestrels. *British Birds* 75:195–204.
Newton, I., Davis, P.E. & Davis, J.E. 1982b. Ravens and Buzzards in relation to sheep farming and forestry in Wales. *Journal of Applied Ecology* 19:681–706.
Newton, I. Wyllie, I. & Mearns, R. 1986. Spacing of Sparrowhawks in relation to food supply. *Journal of Animal Ecology* 55:361–370.
Newton, I., Wyllie, I. & Asher, A. 1991. Mortality causes in British Barn Owls *Tyto alba*, with a discussion of aldrin-dieldrin poisoning. *Ibis* 133:162–169.
Nicholson, E.M. 1926. *Birds in England*. Chapman & Hall. London.
Nicholson, E.M. 1938–9. Report on the Lapwing habitat enquiry. *British Birds* 32:170–191, 207–229, 255–259.
Nicholson, E.M. 1951. *Birds and Men*. Collins. London.
Nicholson, H.H. 1943. Modern Field Drainage. *J.R.A.S.E.* 104:118–135.
Norris, C.A. 1945. Summary of a report on the status and distribution of the Corncrake (*Crex crex*). *British Birds* 38:142–148, 162–168.
Norris, C.A. 1947. Report on the status and distribution of the Corncrake. *British Birds* 40:226–244.
Norris, C.A. 1960. The breeding distribution of thirty bird species in 1952. *Bird Study* 7:129–184.
Oates, M. 1993. The management of southern limestone grasslands. *British Wildlife* 5:73–82.
O'Brien, M. 1996. The numbers of breeding waders in lowland Scotland. *Scottish Birds* 18:231–241.
O'Connor, R.J. 1985. Behavioural regulation of bird populations: a review of habitat use in relation to migrancy and residency. In: *Behavioural Ecology: Ecological Consequences of Adaptive Behaviour*. British Ecological Society Symposium No. 25. Blackwell Scientific. Oxford.
O'Connor, R.J. 1987. Environmental interests of field margins for birds. In: British Crop Protection Council *Field Margins*. Monograph No. 35. British Crop Protection Council. Thornton Heath.
O'Connor, R.J. & Mead. C.J. 1984. The Stockdove in Britain 1930–80. *British Birds* 77:181–201.
O'Connor, R.J. & Shrubb, M. 1986. *Farming and Birds*. Cambridge University Press. Cambridge.
O'Connor, R.J. & Pearman, D.N. 1987. *Long-term Trends in Breeding Success of Some British Birds*. British Trust for Ornithology Research Report No. 23. British Trust for Ornithology. Tring.
Oldham, C. 1931. Payments for 'vermin' by some Hertfordshire Churchwardens. *Transactions of the Hertfordshire Natural History Society* 19(2):1–34.
Orwin, C.S. & Whetham, E.H. 1964. *History of British Agriculture 1846–1914*. Longman. London.
Osborne, B.C. 1984. Habitat uses by red deer (*Cervus elaphus* L.) and hill sheep in the West Highlands. *Journal of Applied Ecology* 21:497–506.

Osborne, P. 1984. Bird numbers and habitat characteristics in farmland hedgerows. *Journal of Applied Ecology* 21:63–82.

Owen, M., Atkinson-Willes, G.L. & Salmon, D.G. 1986. *Wildfowl in Great Britain*, 2nd edn. Cambridge University Press. Cambridge.

Pain, D.J., Hill, D. & McCracken, D.I. 1997. Impact of agricultural intensification of pastoral systems on bird distributions in Britain 1970–1990. *Agriculture, Ecosystems and Environment* 64:19–32.

Parslow, J.L.F. 1973. *Breeding Birds of Britain and Ireland*. T. & A.D. Poyser. Berkhamsted.

Partridge, M. 1973. *Farm Tools through the Ages*. Osprey. London.

Payne-Gallwey, R. 1886. *The Book of Decoys: Their Construction, Management and History*. Van Voorst. London.

Peach, W.J., Siriwardena, G.M. & Gregory, R.D. 1999. Long-term changes in the abundance and demography of British reed buntings *Emberiza schoeniclus*. *Journal of Applied Ecology* 36:798–811.

Peakall, D.B. 1962. The past and present status of the Red-backed Shrike in Great Britain. *Bird Study* 9:198–216.

Peal, R.E.F. 1968. The distribution of the Wryneck in the British Isles. *Bird Study* 15:111–126.

Pearce, W. 1794. *A General View of the Agriculture of Berkshire*. Board of Agriculture. London.

Peers, M. 1997. The ornithological importance of common land in central Breconshire. *Welsh Birds* 1(5):56–62.

Pell, A. 1887. The making of the land in England: a retrospect. *J.R.A.S.E.*, series 2, 23:355–374.

Pell, A. 1899. The making of the land in England: a second retrospect. *J.R.A.S.E.*, series 3, 10:136–141.

Penford, N. & Francis, I. 1990. Common Land and Nature Conservation in England and Wales. *British Wildlife* 2(2):65–76.

Pesticide Usage Surveys. 1969–97. Ministry of Agriculture, Fisheries and Food. London.

Phillips, A.D.M. 1980. Mossland reclamation in nineteenth century Cheshire. *Transactions of the Historic Society of Lancashire and Cheshire* 129:93–107.

Picozzi, N. & Hewson, R. 1970. Kestrels, Short-eared Owls and Field Voles in Eskdalemuir in 1970. *Scottish Birds*. 6:185–191.

Pollard, E., Hooper, M.D. & Moore, N.W. 1974. *Hedges*. Collins. London.

Pollitt, M., Cranswick, P., Musgrove, A., Hall, C., Hearn, R., Robinson, J. & Holloway, S. 2000. The Wetland Bird Survey 1998–99: Wildfowl and wader counts. BTO, WWT, RSPB, JNCC. Slimbridge.

Pomeroy, T. 1794. *General View of the Agriculture of the County of Worcestershire*. Board of Agriculture. London.

Potts, G.R. 1970a. Recent changes in the farmland fauna with special reference to the decline of the Grey Partridge. *Bird Study* 17:145–166.

Potts, G.R. 1970b. Studies of the changing role of weeds of the genus *Polygonum* in the diet of partridges. *Journal of Applied Ecology* 7:567–576.

Potts, G.R. 1986. *The Partridge: Pesticides, Predation and Conservation.* Collins. London.
Potts, G.R. 1991. The environmental and ecological importance of cereal fields. In: *The Ecology of Temperate Cereal Fields.* Blackwell Scientific. Oxford.
Potts, G.R. 1997. Cereal farming, pesticides and Grey Partridges. In: *Farming and Birds in Europe.* Academic Press. London.
Prestt, I. 1965. An enquiry into the recent breeding status of some of the smaller birds of prey and crows in Britain. *Bird Study* 12:196–221.
Pulliainen, E. 1984. Changes in the composition of the autumn food of *Perdix perdix* in West Finland over 20 years. *Journal of Applied Ecology* 21:133–139.
Pusey, P. 1851. On Agricultural Implements; Class IX (Exhibition of 1851). *J.R.A.S.E.* 12:587–643.
Rackham, O. 1986. *The History of the Countryside.* J.M. Dent & Sons. London.
Rackham, O. 1990. *Trees and Woodland in the British Landscape,* revised edn. J.M. Dent & Sons. London.
Rademacher, B., Koch, W. & Hurle, K. 1970. Changes in the weed flora as a result of continuous cropping of cereals and the annual use of the same weed control measures since 1956. *Proceedings of the 10th British Weed Control Conference 1970* pp. 1–6.
Ramsden, D.J. 1998. Effect of barn conversions on local populations of Barn Owl *Tyto alba. Bird Study* 45:68–76.
Ratcliffe, D.A. 1976. Observations on the breeding of Golden Plover in Great Britain. *Bird Study* 23:63–116.
Ratcliffe, D.A. 1980. *The Peregrine Falcon.* T. & A.D. Poyser. Calton.
Ratcliffe, D.A. & Thompson, D.B.A. 1988. The British uplands: their ecological character and international significance. In: *Ecological Change in the Uplands.* Blackwell Scientific. Oxford.
Read, C.S. 1855a. On the farming of Oxfordshire. *J.R.A.S.E.* 15:189–276.
Read, C.S. 1855b. Report on the farming of Buckinghamshire. *J.R.A.S.E.* 16:269–322.
Read, C.S. 1858. Recent improvements in Norfolk farming. *J.R.A.S.E.* 19:265–311.
Reid, C. 1885. The geology of Holderness. *Memoirs of the Geological Survey.* London.
Relton, J. 1972. Breeding biology of Moorhens on Huntingdonshire farm ponds. *British Birds* 65:248–256.
Roberts, H.A. 1958. Studies on the weeds of vegetable crops. I. Initial effects of cropping on the weed seeds in the soil. *Journal of Ecology* 46:759–768.
Roberts, H.A. 1962. Studies on the weeds of vegetable crops. II. Effect of six years of cropping on the weed seeds in the soil. *Journal of Ecology* 50:803–813.
Roberts, H.A. 1968. The changing population of viable weed seeds in an arable soil. *Weed Research* 8:253–258.
Roberts, H.A. 1981. Seed banks in soils. *Advances in Applied Biology* 6:1–35.
Roberts, H.A. & Stokes, F.G. 1965. Studies on the weeds of vegetable crops. V. Final observations on an experiment with different primary cultivations. *Journal of Applied Ecology* 2:307–315.
Roberts, H.A. & Dawkins, P.A. 1967. Effect of cultivation on the numbers of viable seeds in soil. *Weed Research* 7:290–301.

Roberts, J.L. 2001. Breeding biology of Swallows *Hirundo rustica* on two farms straddling the North Wales border. *Welsh Birds* 3:44–56.

Roberts, R.A. 1959. Ecology of human occupation and land use in Snowdonia. *Journal of Ecology* 47:317–323.

Robinson, M. 1986. The extent of farm underdrainage in England and Wales prior to 1939. *Agricultural History Review* 34(1):79–85.

Robinson, R.A. & Sutherland, W.J. 2002. Post-war changes in arable farming and biodiversity in Great Britain. *Journal of Applied Ecology* 39:157–172.

Robinson, R.A., Wilson, J.D. & Crick, H.P.Q. 2001. The importance of arable habitat for farmland birds in grassland landscapes. *Journal of Applied Ecology* 38:1059–1069.

Rowley, J.J. 1853. The farming of Derbyshire. *J.R.A.S.E.* 14:17–66.

RSPB. 1983. *Land Drainage in England and Wales: An Interim Report.* RSPB. Sandy.

Sage, B.L. & Vernon, J.D.R. 1978. The 1975 National Survey of Rookeries. *Bird Study* 25:64–86.

Salisbury, E. 1961. *Weeds and Aliens.* Collins. London.

Schlapfer, A. 1988. Populationsökologie der Feldlerche in der intensiv gunetzen Agrarlandschaft. *Der ornithologische Beobachter* 85:309–371.

Seddon, Q. 1989. *The Silent Revolution.* BBC Books. London.

Shawyer, C.R. 1987. *The Barn Owl in the British Isles: Its Past, Present and Future.* The Hawk Trust. London.

Shawyer, C. 1994. *The Barn Owl.* Hamlyn. London.

Sharrock, J.T.R. 1976. *The Atlas of Breeding Birds in Britain and Ireland.* T. & A.D. Poyser. Berkhamsted.

Sheail, J. 1971. The formation and maintenance of watermeadows in Hampshire, England. *Biological Conservation* 3:101–106.

Sheail, J. 1976. Land improvement and reclamation: the experiences of the First World War in England and Wales. *Agricultural History Review* 24:110–125.

Sheail, J. 1985. *Pesticides and Nature Conservation: The British Experience 1950–1975.* Clarendon Press. Oxford.

Sheail, J. & Mountford, J.O. 1984. Changes in the perception and impact of agricultural land improvement: the post-war trends in Romney Marsh. *J.R.A.S.E.* 145:43–56.

Sheppard, J.A. 1958. The Hull valley: the evolution of a pattern of artificial drainage. *Geographical Studies* 5:33–44.

Shrubb, M. 1968. The status and distribution of Snipe, Redshank and Yellow Wagtail as breeding birds in Sussex. *Sussex Bird Report* 20:53–60.

Shrubb, M. 1970. Birds and farming today. *Bird Study* 17:123–144.

Shrubb, M. 1980. Farming influences on the food and hunting of Kestrels. *Bird Study* 27:109–115.

Shrubb, M. 1985. Breeding Sparrowhawks *Accipter nisus* and organo-chlorine pesticides in Sussex and Kent. *Bird Study* 32:155–163.

Shrubb, M. 1988. The influence of crop rotation and field size on a wintering lapwing *Vanellus vanellus* population in an area of mixed farmland in West Sussex. *Bird Study* 35:123–131.

Shrubb, M. 1990. Effects of agricultural change on nesting Lapwings *Vanellus vanellus* in England and Wales. *Bird Study* 37:115–127.
Shrubb, M. 1993a. Nest sites in the Kestrel *Falco tinnunculus*. *Bird Study* 40:63–73.
Shrubb, M. 1993b. *The Kestrel*. Hamlyn. London.
Shrubb, M. 1997. Historical trends in British and Irish corn bunting *Miliaria calandra* populations: evidence for the effects of agricultural change. In: *The Ecology and Conservation of Corn Buntings* Miliaria calandra. UK Nature Conservation No. 13. Joint Nature Conservation Committee. Peterborough.
Shrubb, M. & Lack, P.C. 1991. The numbers and distribution of Lapwings *V. vanellus* nesting in England and Wales in 1987. *Bird Study* 38:20–37.
Shrubb, M., Williams, I.T. & Lovegrove, R.R. 1997. The impact of changes in farming and other land uses on bird populations in Wales. *Welsh Birds* 1(5):4–26.
Siriwardena, G. 1999. Survival or breeding performance: are these causing farmland bird declines? *BTO News* 220:8–9.
Siriwardena, G.M., Baillie, S.R., Buckland, S.T., Fewster, R.M., Marchant, J.H. & Wilson, J.D. 1998. Trends in the abundance of farmland birds: a quantitative comparison of smoothed Common Birds Census indices. *Journal of Applied Ecology* 35:24–43.
Siriwardena, G., Baillie, S.R., Crick, H.P.Q., Wilson, J.D. & Gates, S. 2000. The demography of lowland farmland birds. In: *Ecology and Conservation of Lowland Farmland Birds*. British Ornithologists Union. Tring.
Smith, K.W. 1983. The status and distribution of waders breeding on lowland wet grasslands in England and Wales. *Bird Study* 30:177–192.
Smith, R.S. & Jones, L. 1991. The phenology of mesotrophic grassland in the Pennine dales, northern England: historic hay cutting dates, vegetation variation and plant species phenologies. *Journal of Applied Ecology* 28:42–59.
Smith, S. 1950. *The Yellow Wagtail*. Collins. London.
Smith, W. 1949. *An Economic Geography of Great Britain*. Methuen. London.
Smout, T.C. 2000. *Nature Contested: Environmental History in Scotland and Northern England since 1600*. Edinburgh University Press. Edinburgh.
Snow, B. & Snow, D. 1988. *Birds and Berries*. T. & A.D. Poyser. Calton.
Southwell, T. 1870–1. On the ornithological archaeology of Norfolk. *Transactions of the Norfolk and Norwich Naturalists' Society* 1:14–23.
Southwood, T.R.E. & Cross, D.J. 1969. The ecology of the partridge. III. Breeding success and the abundance of insects in natural habitats. *Journal of Animal Ecology* 38:497–509.
Spearing, J.B. 1860. On the agriculture of Berkshire. *J.R.A.S.E.* 21:1–46.
Speir, J. 1906. Changes in farm implements since 1890. *Trans. H.A.S.* 18:46–62.
Spencer, K.G. 1965. A history of the non-domestic doves and pigeons in Lancashire during the past century. *The Naturalist* 90:48–50.
St. John, C.W.G. 1848. *A Sportsman and Naturalist's Tour in Sutherlandshire*. John Murray. London.
Stamp, L. Dudley. 1955. *Man and the Land*. Collins. London.
Steele-Elliot, J. 1906. Extracts from Churchwardens' accounts of Bedfordshire. *The Zoologist*, series 4 10:161–167, 253–265.

Stoate, C. 2001. Reversing the declines of farmland birds: a practical demonstration. *British Birds* 94:302–309.
Stone, B.H. *et al.* 1997. Population estimates of birds in Britain and in the United Kingdom. *British Birds* 90:1–22.
Stone, T. 1794. *General View of the Agriculture of Lincolnshire*. Board of Agriculture. London.
Stovin, G. 1752. A brief account of the Drainage of the Levells of Hatfield Chase and parts adjacent in the Countys of York, Lincoln and Nottingham. Reprinted in *Yorkshire Archeological Journal* 1951, 37:385–391.
Straw, A. 1955. The Ancholme Levels north of Brigg: a history of drainage and its effect on land utilisation. *East Midland Geographer* 3:37–42.
Sturgess, R.W. 1966. The agricultural revolution on the English clays. *Agricultural History Review* 14:104–121.
Sturrock, F. 1982. Room for improvement; modern farming and the countryside. *Country Life 7 October*:1090–1092.
Survey of Fertiliser Practice. 1962–94. Ministry of Agriculture, Fisheries and Food. London.
Sussex Bird Reports. Annual report of the Sussex Ornithological Society.
Sydes, C. & Miller, G.R. 1988. Range management and nature conservation in the British uplands. In: *Ecological Change in the Uplands*. Blackwell Scientific. Oxford.
Symon, J.A. 1959. *Scottish Farming: Past and Present*. Oliver & Boyd. Edinburgh.
Tanner, H. 1858. The agriculture of Shropshire. *J.R.A.S.E.* 17:1–64.
Tapper, S. 1992. *Game Heritage: An Ecological Review from Shooting and Game-keeping Records*. Game Conservancy. Fordingbridge.
Taylor, C. 1975. *Fields in the English Landscape*. J.M. Dent & Sons. London.
Thirsk, J. 1953. The Isle of Axholme before Vermuyden. *Agricultural History Review* 1:16–28.
Thirsk, J. 1964. The Common Fields. *Past and Present* 29:3–25.
Thirsk, J. 1985. The agricultural landscape: fads and fashions. In:*The English Landscape: Past, Present and Future*. Oxford University Press. Oxford.
Thirsk J. (ed.) 1990. *Chapters from the Agararian History of England and Wales, 1500–1750, vol. 3, Agricultural Change: Policy and Practice*. Cambridge University Press. Cambridge.
Thirsk, J. 1997. *Alternative Agriculture: A History*. Oxford University Press. Oxford.
Thomas, G. 1976. Habitat usage of wintering ducks at the Ouse Washes, England. *Wildfowl* 27:148–152.
Thomas, J.F. 1942. Report on the Redshank Inquiry. *British Birds* 36:5–14, 22–34.
Thompson, F.M.L. 1968. The second Agricultural Revolution. *Economic History Review* 21:62–77.
Thomson, D.L. & Cotton, P.A. 2000. Understanding the decline of the British population of Song Thrushes *Turdus philomelos*. In: *Ecology and Conservation of Lowland Farmland Birds*. British Ornithologists Union. Tring.
Tiainen, J., Hanski, I.K., Pakkala, T., Piiroinen, J. & Yrjola, R. 1989. Clutch size, nestling growth and nestling mortality of the Starling *Sturnus vulgaris* in south Finnish agri-environments. *Ornis Fennica* 66:41–48.

Ticehurst, C.B. 1935. On the food of the Barn Owl and its bearing on Barn Owl population. *Ibis* 13:329–335.

Ticehurst, N.F. 1920. On the former abundance of the Kite, Buzzard and Raven in Kent. British Birds 14:34–37.

Ticehurst, N.F. 1935. Rewards for vermin-killing paid by the Churchwardens of Tenterden, 1626 to 1712. *Hastings and East Sussex Naturalist* 5:69–82.

Ticehurst, N.F. & Witherby, H.F. 1941. Report on the effect of the severe winter of 1939–40 on bird-life in the British Isles. British Birds 34:118–132, 142–155.

Ticehurst, N.F. & Hartley, P.H.T. 1948. Report on the effect of the severe winter of 1946–1947 on bird-life. *British Birds* 41:322–334.

Toms, M.P. 2001. *Project Barn Owl: Evaluation of Annual Monitoring Programme.* BTO Research Report No. 177. British Trust for Ornithology. Thetford.

Trafford, B.D. 1970. Field drainage. *J.R.A.S.E.* 131:129–152.

Trevelyan, G.M. 1944. *English Social History.* Longmans, Green & Co. London.

Trow-Smith, R. 1951. *English Husbandry.* Faber & Faber. London.

Tubbs, C.R. 1974. *The Buzzard.* David & Charles. Newton Abbot.

Tubbs, C.R. 1985. *The Decline and Present Status of the English Lowland Heaths and their Vertebrates. Focus No.11.* Nature Conservancy Council. Peterborough.

Tubbs, C.R. 1986. *The New Forest.* Collins. London.

Tubbs, C.R. 1993. An introduction to Hampshire. In: *Birds of Hampshire.* Hampshire Ornithological Society.

Tucker, G.M. 1992. Effects of agricultural practices on field use by invertebrate-feeding birds in winter. *Journal of Applied Ecology* 29:779–790.

Tucker, G.M. 1997. Priorities for bird conservation in Europe: the importance of the farmed landscape. In: *Farming and Birds in Europe.* Academic Press. London.

Tucker, G.M., Davies, S.M. & Fuller, R.J. 1994. *The Ecology and Conservation of Lapwings* Vanellus vanellus. UK Nature Conservation No. 9. Joint Nature Conservation Committee. Peterborough.

Tucker, G.M., Heath, M.F., Tomialojc, L. & Grimmett, R.F.A. 1995. *Birds in Europe: Their Conservation Status.* Birdlife International. Cambridge.

Turner, E. 1862. Ashdown Forest, or as it was sometimes called Lancaster Great Park. *Sussex Archaeological Collections* 14:35–64.

Turner, G. 1794. *A General View of the Agriculture of the County of Gloucestershire.* Board of Agriculture. London.

Turner, J.H. 1845. On the necessity for reducing the size and number of hedges. *J.R.A.S.E.*, series 2, 6:479–488.

Turner, M. 1981. Arable in England and Wales: estimates from the 1801 Crop Return. *Journal of Historical Geography* 7(3):291–302.

Tyler, G.A., Green, R.E. & Casey, C. 1998. Survival and behaviour of Corncrake *Crex crex* chicks during the mowing of agricultural grasslands. *Bird Study* 45:35–50.

Underhill-Day, J. 1993. Marsh Harrier. In: *The New Atlas of Breeding Birds in Britain and Ireland: 1988–1991.* T. & A.D. Poyser. London.

Uttley, A. 1931. *The Country Child.* Faber & Faber. London.

Vanhinsbergh, D. 2000. The butcher bird: lost but not forgotten. *BTO News* 226:14–15.

Venables, L.S.V. & Leslie, P.H. 1942. The rat and mouse populations of corn ricks. *Journal of Animal Ecology* 11:44–68.
Vera, F.W.M. 2000. *Grazing Ecology and Forest History*. CABI Publishing. Wallingford.
Verden, N. 2001. The employment of women and children in agriculture: a re-assessment of agricultural gangs in 19th century Norfolk. *Agricultural History Review* 49:41–55.
Vesey Fitzgerald, B. 1946. *British Game*. Collins. London.
Vickery, J.A., Sutherland, W.J., Watkinson, A.R., Lane, S.J. & Rowcliffe, J.M. 1995. Habitat switching by Dark-bellied Brent Geese *Branta b. bernicla* (L) in relation to food depletion. *Oecologia* 103:499–508.
Vickery, J.A. *et al.* 2001. The management of lowland neutral grasslands in Britain: effects of agricultural practices on birds and their food resources. *Journal of Applied Ecology* 38:647–664.
Vickery, J.A. & Gill, J.A. 1999. Managing grassland for wild geese in Britain: a review. *Biological Conservation* 89:93–106.
Voous, K.H. 1960. *Atlas of European Birds*. Nelson. London.
Wade Martins, S. 1991. *Historic Farm Buildings*. Batsford. London.
Waldon, J. 1982. A study of Stone Curlews *Burhinus oedicnemus* breeding on arable land in southern England. RSPB unpublished report.
Walker, D. 1795. *General View of the Agriculture of Hertfordshire*. Board of Agriculture. London.
Walpole-Bond, J. 1914. *Field Studies of Some Rarer British Birds*. Witherby. London.
Watson, D. 1977. *The Hen Harrier*. T. & A.D. Poyser. Berkhamsted.
Webb, N. R. 1986. *Heathlands*. Collins. London.
Wells, W. 1860. The drainage of Whittlesea Mere. *J.R.A.S.E.* 21:134–153.
Welsh Bird Report 1990–1999. Welsh Ornithological Society.
Whetham, E.H. 1970. The mechanisation of British farming 1910–1945. *Journal of Agricultural Economics* 21:317–331.
White, G. 1789. *The Natural History of Selborne*, Everyman's Library edition 1949. J.M. Dent & Sons. London.
Whitehead, C. 1899. A sketch of the agriculture of Kent. *J.R.A.S.E.*, series 3, 10:429–485.
Wilkinson, J. 1861. The farming of Hampshire. *J.R.A.S.E.* 22:239–346.
Williams, G., Henderson, A., Goldsmith, L. & Spreadborough, A. 1983. The effects on birds of land drainage improvements in the North Kent Marshes. *Wildfowl* 34:33–47.
Williams, G. & Bowers, J.K. 1987. *Land Drainage and Birds in England and Wales*. Conservation Review No. 1. RSPB. Sandy.
Williams, G. & Hall, M. 1987. The loss of coastal grazing marshes in South and East England, with special reference to East Essex, England. *Biological Conservation* 39:243–253.
Williams, M. 1970. *The Draining of the Somerset Levels*. Cambridge University Press. Cambridge.
Williams, P.H. 1982. The distribution and decline of British Bumble Bees (*Bombus* LATR). *Journal of Apicultural Research* 21(4):236–245.

Williamson, K. 1975. Birds and climatic change. *Bird Study* 22:143–164.
Wilson, A.M., Vickery, J.A. & Browne, S.J. 2001. Numbers and distribution of Northern Lapwings *Vanellus vanellus* breeding in England and Wales in 1998. *Bird Study* 48:2–17.
Wilson, J. 1864. Reaping machines. *Trans. H.A.S.* 1864:123–149.
Wilson. J. 1902. Half a century as a Border Farmer. *Trans. H.A.S.* 4:35–48.
Wilson, J., Evans, A., Poulsen, J.G. & Evans, J. 1995. Wasteland or oasis: the use of set-aside by breeding and wintering birds. *British Wildlife* 6(4):214–223.
Wilson, J.D., Taylor, R. & Muirhead, L.B. 1996. Field use by farmland birds in winter: an analysis of field type preferences using resampling techniques. *Bird Study* 43:320–332.
Wilson, J.D., Evans, J., Browne, S.J. & King, J.R. 1997. Territory distribution and breeding success of Skylarks *Alauda arvensis* on organic and intensive farmland in southern England. *Journal of Applied Ecology* 34:1462–1478.
Wilson, J.D., Morris, A.J., Arroyo, B.E., Clark, S.C. & Bradbury, R.B. 1999. A review of the abundance and diversity of invertebrate and plant foods of granivorous birds in northern Europe in relation to agricultural change. *Agriculture, Ecosystems and Environment* 75:13–30.
Wilson, J.J. 1992. Britain's arable weeds. *British Wildlife* 3:149–161.
Winstanley, D., Spencer, R. & Williamson, K. 1974. Where have all the Whitethroats gone? *Bird Study* 21:1–14.
Witherby, H.F. & Jourdain, F.C.R. 1930. Report on the effect of severe weather in 1929 on bird-life. *British Birds* 23:154–158.
Witherby, H.F., Jourdain, F.C.R., Ticehurst, N.F. & Tucker, B.W. 1940. *The Handbook of British Birds*. Witherby. London.
Wotton, S.R. & Gillings, S. 2000. The status of breeding Woodlarks *Lullula arborea* in Britain in 1997. *Bird Study* 47:212–224.
Wyllie, I. 1976. The bird community of an English parish. *Bird Study* 23:39–50.
Yalden, D.W. 1977. *The Identification of Remains in Owl Pellets*. The Mammal Society. London.
Yalden, D.W. 1985. Dietary separation of owls in the Peak District. *Bird Study* 32:122–131.
Yapp, B. 1981. *Birds in Medieval Manuscripts*. British Library. London.
Young, A. 1770. *A Tour of the North of England*.
Young, A. 1793. *General View of the Agriculture of Sussex*. Board of Agriculture. London.
Young, A. 1794. *General View of the Agriculture of Suffolk*. Board of Agriculture. London.
Young, A. 1805. *Annals of Agriculture*.

Index

Page numbers in bold indicate illustrations and those in italic indicate data in tables.

Aethusa Cynapium 159
afforestation, changes in breeding bird
    populations 28–30, *185*, 225, 328
Agricultural Chemicals Approval Scheme
    (ACAS) **179**, 181
agricultural revolution 15–16, 172–5
Agriculture Act (1947) 13–14
Agrostis 106
altitude 37–40
Anagallis 174
Anthemis 174
arable, extent **264**, *331*
arable farming systems
    1945–65 183–9
        three-year ley system 183–9
    1965–99 189–95
        breeding rates 195–9
    decline 62–5, 236–8
        vs. grasslands and stock 236, **247**
        recession, 1875–1939 214–17
    early systems 44–9
    lo-till, minimal cultivation 295–7, 325
    tillage
        changes in scale and timing 57–9, 190–1
    decline 216, 236
    two/three-course systems 46–7
    see also high farming
Arctium 157, 158
Arun Valley 134, 153, 155
ash 125
Ashdown Forest 45, 89
avermectins 325
Avocet 131, 307, 319
Avon, R, flood meadows **209**
Axholme, drainage, 45, *133*

binders 288–9
bird-catchers *see* trapping
Bittern 33, *34*, 94, 128, **129**, 131, *132*, 136, 149, 310, 319
Black-tailed Godwit 33, *34*, 94, 131, *132*, 136, 149, 310, 311, 319
Blackbird 21, 23, *34*, *38*, 56, 65, *96*, 119, *186*, *193*, *196*, 254, 265, 302
Blackcap 23, 73, *96*, *193*, *196*
bone dust 210
bovine spongiform encephalopathy (BSE) 16
Brambling 24, 31, 253, 254
Brassica spp. 52, *158*, 172, 174, 183, 257, *266*, *331*
breast-plough **81**
Brecks 46, 80, 205
breeding bird populations 21–6
    changes 28–31
        afforestation 28–30
        attributes, C19–1997 **29**
    geographic position and climate 31–2
    summer visitors 32
    list and numbers 22–4
    nesting habitat *48*
    population structure **20**
    reservoirs of breeding habitat 301–2
breeding success 195–7
Bromus sterilis 160
buildings
    modern 298–301
    as nest sites 302–4
    thatched **299**
    as winter roosts 304–5
Bullfinch 21, *24*, *34*–5, *38*, 65, *96*, **158**, *186*, *193*, *196*, 243, 314

buntings 55, 161, 315
Bunting 24
  Cirl Bunting 31, *34–5*, *96*, 166, *177*, *193*, 230, *254*, 265, 267
  Corn Bunting vii, *24*, 31, *34*, 36, *38*, 42, 52, 53, *96*, 120, 125, 174, *177*, 181, *186*, 187, *193*, *246*, *254*, 265, *266*, 267, **269**, 270, 285, *287*, 306, 307
  Reed Bunting 18, *24*, *38*, *96*, 142–3, 148, *177*, *186*, *193*, *246*, *254*, 267, 302
Burpleurum rotundifolium 162
Bustard, Great *48*, 51, 58, 82, **86**, 90–1, 94–5, *96*, 204
butterflies 233–4
Buzzard 22, *96*, 101, *102*, 104, *185*, *193*, 301, 315, *318*

cage-bird trade 169, 313–14
Cannock Chase 46, *78*
Capsella 174
carbamates **179–80**
carrion (and birds) 101, 106
cattle **209**
  1870s vs. 1930s **218–19**
  numbers 1930s, 1960s, 1997 **242**
  stocking rates and densities 201–3, **202**, **220–3**, **238–41**
Centaurea 157, 162
Cerastium 157, 158, 159, 173, 174
Chaffinch 24, 35, 39, 47, 55, 56, *96*, 125, *158*, 166, *167*, *186*, 187, *193*, *194*, *196*, *254*, 270
chalk grasslands, stocking rates 203–5
chamomile, red 284
champion/champaign country 44
changes
  breeding bird populations 28–31
    C19–1997 **29**
    climate-induced **34–5, 38–9**
  emerging technologies
    genetically modified crop seeds 324–5
    lo-till 295–7, 325
    veterinary medicines 325
  environmental limitations and climate change 32–7
  extent of food sources for seed-eating birds **264**
  farmland bird populations *96*, *193*, *194*, *196*, *185–6*
  fodder root crops decline 189, 257
  livestock stocking rates and densities 238–44
  most important
    chemical weed control 323
    grassland management and silage 13, 237, *287*, 323, *324*
    harvesting methods *324*

scale and timing of cultivations in high farming 57–9
sites for seed-eating birds 301, 329–32
wintering bird populations 30–1
charlock 2, *157*, 159, 163, 170, 173, 174, 189
Chenopodium 157, 158, 159, 172, 271
Chiffchaff 23, *34*, 73, *96*, *186*, *193*, *196*
Chough 24, 31, *34*, *96*, *186*, *193*, *194*, *254*
Chrysanthemum segetum 157, 162
Churchwardens' accounts 314–15, **316**
claying 81–2, 136
climate
  and bird populations 31–7, *34–5*
  Little Ice Age (1550–1700) 32
  severe winter, bird decline **38–9**
clover *64*, 172, **198**, 200, 212, 289
  leys 49–50, 52, 198–9, 213, 216, 255, 291
collecting 319–20
  by Victorians 95–6
  eggs 140, 307–8
combine harvester **280**, 293–4, **293**
Common Birds Census (CBC) 21, 124, 175, 184, 188, *196*, **258**, 267
  farmland birds, list and numbers **22–4**
  hard winters 37, *38–9*
commons
  and birds 82–97
  extent 77–9
  and livestock 99–104
  upland, long-term effects of enclosure 104–13
Compositae 165
conservation *see* environmental schemes
Convolvulus 174
Coot 302
coppicing 19, 73
corn harvesting **290**, 291–5
  mechanical reaper **292**
  switching from binder to combine **293**
Corn Laws, repeal 9, 138
Corn Production Act (1917) 13
corn-mills, grain-milling, and navigation 143–4
corncockle 160, 162
Corncrake 22, *48*, 52, 93, 94, *96*, *193*, 210, 213, 217, **224**, 232, 249, 285, 290–1
corvids 21, 55, 243, 250
couch grass 160, 163
Crane, Common 21
Crow 24, 56, *96*, *186*, *193*, *196*, *254*, 301
Crown Forests 80
Cuckoo 23, *34*, *96*, *193*, *194*, *196*, 245
cultivations
  changes in scale and timing, high farming 57–9
  minimal 295–7

Index

Curlew 22, 34, 96, 113, 185, 193, 194, 206, 217, 227, **228**, 229, 230, 231, 246, 249, 254, 310

darnel 284
depression, high farming 59–65
   see also recession
dodder 284
Dotterel 310, 311
Dove
   Collared 23, 34, 185, 193, 196, 243, 254
   Stockdove 22, 34, 48, 93, 94, 96, 158, 176, 185, 187, 193, 196, 245, 254, 271, 300, 302
   Turtle vii, 23, 34, 96, 158, 174, 176, 185, 193, 196, 245
drainage *see* land drainage
drills **4–5**, 283–5
   minimal cultivations 295–7
droving industry 104
duck 'decoys' 139, 311
Dunlin 22, 31, 96, 108, 113, 193, 194, 310
Dunnock 23, 34, 38–9, 56, 96, 193, 196, 245, 254, 265
Dutch elm disease 303

Eagle
   Golden 30, *318*
   White-tailed *318*
egg collecting 140, 307–8
elm 119, 303
enclosure 66–113
   enclosed C18 and C19 **70**
   areas **67**
   commons and birds 82–97
   commons and livestock 99–104
   effect on woodlands 72–3
   fragmentation 97–9
   hedges resulting from 71–2, 114–15
   legal process 68
   open or common fields 69–73
   Parliamentary Awards 66–8
   upland commons and long-term effects 104–13
   'waste' lands and commons 44, 46, 73–82
environmental destruction, as desirable aim 82
environmental limitations 31–42
   climate change effects 32–7
   geographic position 31–2
   soil and fertility 40–2
environmental schemes 17
   conservation of country property 327
   selectivity 326
   set-aside 326–7
Epilobium 174
Essex coastal marshes, drainage works *134*

Euphorbia 174
European Economic Community (EEC)
   Common Agricultural Policy 15
   grant aids 14
exploitation and persecution 306–21
   cage-bird trade 169, 313–14
   collecting 95–6, 319–20
   egg collecting 140, 307–8
   food use 307–13
      eggs 307–8
      waders and wildfowl 310–13
      wheatears and skylarks 309–10
   game preserving and raptors 317–19
   pest control 314–17
   trapping 169, 313–14

fallows 46, 49, **58**, 64
farm ponds 301–2
farmland, definition 1
farmland birds
   changes
      C19 **96**
      1900–40 and 1940–67 **185–6**
      1968–72 vs. 1988–91 **245–6**
      1972–98 **193, 194**
      woodland populations **196**
   early historical background 18–21
   list and numbers **22–4**
   see also breeding bird populations; wintering bird populations
farmsteads 298–302
   feeding areas 301
   reservoirs of breeding habitat 301–2
feeding sites
   preferred (Sussex) **55**
   seed-eating birds 301, 329–32
fenlands 44, 45, 100, 128
   claying 82
   drainage works 128–35, 140
   recession period 148–9
   species loss **132**, 139
   species re-establishment 149
   see also wetlands
fertilisers, modern grassland management 234–6
field boundary
   by type (1990) **122**
   hedge vs. other 114–15, 121
   removal **123**
   see also hedges
field drainage 137–45
   tile drainage 137–8, 150, **151**
   under-drainage 137
field patterns, variations, hedges 115–17
field size, and hedge removal 115–17
Field Vole 274, *275*, 276

Fieldfare 23, 254, 260, **262**
finches 55
fire, grouse moors 107–8
flax, linseed 156, 271
flood meadows 207, 208–10
   River Avon **209**
fodder root crops 50–7, **51**, 64, **264**, *331*
   modern decline 183, 189
food, for use of wild birds 307–13
foot-and-mouth disease 16, 327
foraging distances 195
forestry *see* afforestation
Fumaria 158
fungicides 177, **178**

Galeopsis 157
Galium spp. 157, 160, *174*
game preservation, and raptors 317–19
Garganey 22, 34, 96, 185, 193, 194, 210
geese, feeding behaviour **254**, 255–6
genetically modified crop seeds 324–5
geographic position, limitations to bird ranges 31–2
Geranium 174
gleaners *see* seed-eating birds
Goldcrest 23, 26, 35, 96, 186, **188**, 193, 194, 196
Goldfinch 21, 24, 35, 38, 59, 65, 93, 96, 157, 160, 167, **168**, 186, 193, 194, 245, 249, 254, 266, 269, 270, 313
Goose
   Barnacle 255
   Bean 22, 254, 255
   Brent 22, 254
   Canada 22, 96, 193, 254
   Greylag 22, 94, 96, 193, 254, 255
   Pink-footed 22, 254, 255
   White-fronted 22, 254, 255
Goshawk *318*
grain *see* winter food
grain-feeding birds 166–7, 261–6
   see also seed-eating birds
grain-milling, and navigation 143–4
grant aid, Agriculture Act (1947) 14
grasslands 200–51
   abandonment/expansion effects 216, 227–9
   after 1945 231–2
   decline of arable 236
   definitions, ley vs. permanent 200
   early grass systems, sheep/arable on chalk 203–5
   economic management 16
   extent 200
      as of improved farmland **262**
   haying 212–14

lowland heath and infield/outfield 205–6
modern grassland management 232–44
   effect on birds 244–51
   fertiliser use 234–6
   reseeding 183–7, 233–4
   stocking rates and densities 201–3, **202**, **220–3**, **238–41**
   vs. tillage (1997) **247**
pasture 210–12, 213
ploughing, 1939–45 229–31
recession 1875–1939 216–29
   cattle, 1870s and 1930s **218–19**
   grassland 1870s and 1930s **214–15**
silage 237
stocking rates, chalk 203–5
under-draining 144–5
see also flood meadows; water meadows
grazers, geese and ducks **254**, 255–6
Great Common, Radnorshire 98
Grebe
   Great Crested 302
   Little 302
Green Sandpiper 208
Greenfinch 24, 35, 65, 96, 157, 158, 160, 186, 193, 194, 195, *196*, 245, 254, 266, 269, 270
ground-nesting species **20**, 216, 246, 250
Grouse
   Black 22, 48, 62, **85**, 89, 93, 94, 96, 108, 109, 111, 184, 185, 193, 254
   Red 22, 31, 34, 40, 62, 96, 108, 109, 111–12, 185, 193, 194, 254
grouse moors, burning 107–8
Gull
   Black-headed 22, 34, 96, 193, 194, 254, 308
   Common 22, 254

habitat loss
   destruction, as desirable aim 82
   hedges 117–26
   set-aside 326–7
hand labour 81, **286**
   decline 283, 292
   weed hoeing 163–5
harriers
   Hen Harrier 22, 34, **83**, 86–9, 94, 96, 112, 126, 185, 193, 272, 273, *318*
   Marsh Harrier 22, 28, 94, 95, 96, **130**, 132, 149, 185, 193, *318*
   Montagu's Harrier 22, 28, 34, 36, 48, 58, **84**, 86–9, 93, 94, 96, 185, 193, 227, *318*
hawthorn **112**, 119
haymaking 212–14, 285–91
   labour required to harvest by hand **286**
   reaper/mower **288**
heath (*Calluna*) 90, 100, 108

Index

heaths, and infield/outfield 205–6, 229
hedges 114–26
  diversity in hedge structure **118, 126**
  as habitats 117–21
  prosperity period 120
  removal/deterioration 61, 115–17, 118
    modern stock 121–6
    rate of loss 124–6
  with standard trees 119
  total, resulting from enclosure 71–2, 114–15
  trimming 117–18
  variations in field patterns 115–17
herbicides 170–5, 190
  pre-emergent 172–3
  timing of cultivation 190
high farming 49–59
  1750–1875 4–7
    mechanisation 4
    wheat yields 6–7
  1875–1947 in decline 7–12
    collapse 145
    loss of tillage and expansion of grassland 9
    repeal of Corn Laws 9
    wheat economics 9
  arable farming
    decline 62–5
    early systems 44–9
    rotations 7–8
  arable land as percentage of improved farmland (1875) **8**
  changes in scale and timing of cultivations 57–9
  definition 6, 49
  new crops and rotations 51–7
  recession 59–65, 146–7
  weed control 59
  weeds/weed control 59, 161–9
historical background 1–21
  1750–1875 4–7
  1875–1947 7–13
    changes in farm enterprises 10–13
    high farming in decline 7–10
  1945–87 13–16
    Agriculture Act (1947) 13
    British Sugar Corporation (1936) 14
    Corn Production Act (1917) 13
    EEC membership 15–16
    estate tenancies vs. owner-occupied farms 14–15
    Milk and Potato Marketing Boards (1933) 14
    pastoral farming 16
  post-1987 16–17
Hobby 22, *34, 96, 193, 318*

hoeing, hand 163–5
hoes 163, **164**, 283–5
Honey Buzzard *318*
horses
  horse-drawn/powered combine **280**
  power source 278–80, **281, 292**

Inclosure Commissioners 138
industrial crops 3
Industrial Revolution 1
infield/outfield 47
insecticides **178**, 179–82, 187, 191
invertebrates
  as foods 53, *176*, 177, 230, 231, 257–61, 272, *273*
  loss due to grassland management 172, 244
    nitrogen fertilisers 235–6
    reseeding of old grasslands 233–4
  loss in flooding 154–5

Jack Snipe 22, *254*
Jackdaw 24, 56, *96, 186, 193, 196, 254*, 302
Jay 21, 23, 35, 73, *96, 186, 196*, 245 *254*
Juncus 106, 142

Kestrel vii, 18, 22, *96, 126, 185, 193*, 210, 245, 272, *273*, 302, 303, 304–5, *318*
Kite 101, *102*, 104, 301, 315, *318*
  Red 22, *34, 96, 185, 193*
Knot 310

labour, decline 169, 283, 292
  see also hand labour
Lamium 174
land drainage 127–55
  1750–1880 128–45
    fields 137–45
    major wetlands 128–37
  1880–1940 145–9
    limited impact? 143
  1940–present 149–55
    impact on waders 154–5
    new works 150
    post-war period 232
    under-draining 144–5
Land Drainage Act (1861) 136–7
Land Drainage Act (1918) 148–9
Lapwing vii, 22, 33, *34*, *38*, 42, 56, 58–9, 60, 72, 93, 94, *96*, 108, 113, 140, 147–8, 154, *176, 185, 193*, 195, 206, 208, 210, 227, 230, 231, 234, 236, 243, 246, 249, 250, *254*, 259, 260, **262**, 307, 308
larks 55
larkspur 284
liming 210

Linnet 24, 35, 38, 47, 48, 49, 56, 59, 93, 96, 157, 158, 160, 174, *177*, 186, *193*, 195, *196*, 245, 254, 266, 269
Little Ice Age (1550–1700) 32
liver fluke 100, 101, 107, 136
livestock, stocking rates and densities 201–3, **202, 220–3, 238–41**
changes 238–44
lo-till, minimal cultivation 295–7, 325
lowland heaths, and infield/outfield 205–6

machinery, power sources 278–83
Magpie 21, 24, 56, 96, 186, *193*, *196*, 254
Mallard 22, 38, 96, 139, 148, *193*, *194*, 254, 256, 302
manure 53, 211, 234
marling 81–2
Martin Mere 79
1750–1880, drainage works 133
meadow 64, 213
see also flood meadow; water meadow
Meadow Pipit 23, 39, 94, 96, *193*, 232, 246, 254
mechanisation 278–83
Melampyrum arvense 162
Merlin 22, 31, 94, 96, 109, 112, 185, *193*, *194*, 260, 272, *318*
Milk and Potato Marketing Boards (1933) 14
milk production 12, *16*, 234
minimal cultivations 295–7
Mink 250–1
Molinia 106
Moorhen 22, 34, 38, 96, 148, *193*, 246, 254, 302
moors
burning 107–8
gullying 108
mowing, early for silage 237
mowing/haying 212–16, 288–91

Nardus spp. 106
Neolithic land clearance 19
nest predation 250
Netherlands, waders, farmers' breeding scheme 326
New Forest 206
nidicolous and nidifugous young 195, 234
Nightingale 73, 313
Nightjar 319
nitrogen fertilisers 232, 235–6, **235**
weed control **161**
Norfolk rotation 6

oil-seed rape 15, *56*, 189, 195, 257, **258**, 271, *273*
open-field system 44, **67, 70**, 71
organochlorines/phosphates **29**, *175*, **179–80**, 181, *185*, 187, 188, 197
Osprey *318*

Owls 21
Barn Owl vii, 23, 31, 34, 96, 126, 181, *185*, *193*, 232, 233, 245, 273, 274, 275–7, 302, 304, 318
Little Owl vii, 23, 34, *185*, *193*, 245, 254, 302
Long-eared Owl 23, 34, 96, *185*, *193*, 272
Short-eared Owl 23, 96, 112, 184, *185*, *193*, 272, *273*, 274
Tawny Owl 23, 34, 96, *185*, *193*, 245, 302
owner-occupied farms, vs. estate tenancies 14–15
Oystercatcher 22, 34, 96, 113, *193*, *194*, 246, 249–50, 307

Partridge
Grey vii, 22, 38, 53, 62, **63**, 96, 157, 158, 172, 176, 182, *185*, 187–8, *193*, 195, 231, 246, 254, 265
Red-legged 22, 31, *38*, 56, 96, *193*, *194*, 195, 254
Partridge Survival Project 156
pasture, vs. meadow 213
Peregrine Falcon 181, 260, *318*
persecution *see* exploitation and persecution
pest control, persecution of birds 314–17
Pesticide Safety Precautions Scheme 181
pesticides 175–7
and bird decline 175, **176–7**
Pheasant 22, 96, 186, *193*, *194*, 254, 302
Phragmites 142
pig arks 190
pigeons 55
Pintail 22, 148, 153, 254, 256–7, 313
Plantago spp. 157, *174*
ploughing of grasslands
1939–45 229–31
post 1945 231–2
ploughing of 'waste' lands, breast plough 80, **81**
Plovers 55
Golden Plover 22, 30, 31, 34, 56, 94, *96*, 107, 108, *109*, 112, 154, *185*, *193*, *194*, 254, 259–60
Poa annua 159
Pochard 153
Polygonum spp. 157, 158, 159, 172, 173, *174*
ponds **204**, 301–2
poppies 165
potato crops 255–6
1875–1947, as percentage of tillage **10, 11**
birds feeding on 255–6
power sources 278–83
horse-drawn/powered combine **280**
steam ploughing **279**
tractors 281–3
water 144, 278
prosperity cycles 3

Ptarmigan 108
pyrethroids **179–80**

Quail 22, 34, 42, 48, 52, 71, 93, 96, 185, 193, 194, 210, 285, 287

Radnorshire, Great Common 98
Ranunculus 174
Raphanus 173, 174
raptors 272–7
  and game preservation 317–19
  pest control and persecution 314–17
  status 1900 **318**
  and wintering bird populations 272–7
rats and mice 273, 274, **275**, **276**, 277
Raven 24, 30, 96, 101, 103, 104, 186, 193, 254, 301
reapers
  mechanical **292**
  reaper-binder 288, 292
  reaper/mower **288**
reaping (painting) **290**
recession 16–17
  1875–1939, grasslands 214–29
  in high farming 59–65, 146–7
Red-backed Shrike 23, 34, 95, 96, 120, **176**, 186, 193, 230
Redshank vii, 22, 96, 140, **141**, 147, 185, 193, 206, 217, 227, 230, 231, 246, 249, 308, 310
Redwing 254, 260
relief 41–2
reseeding, loss of old grasslands 233–4
ricks *see* stackyards
rights in common 74
Ring Ouzel 23, 34, 40, 94, 96, 109, 186, 193
river navigation, and grain-milling 143–4
Robin 23, 38–9, 96, 193, 194, 254, 265, 302
Romney Marsh 152
Rook 18, 24, 33, 35, 53, 56, 96, 176, 186, 188, 193, 254, 294, **295**, **296**
roots *see* fodder root crops
Rothampsted 13, 211
rough grazing *see* Scotland
Ruff 22, 95, 131, 132, 254, 308, 310, 311, 319
Rumex spp. 157, 158, 174
Runrig Act (1695) 69
ryegrass 230, 233

sainfoin 49, 200, 204, 212, 289
saltmarshes 266
Scotland
  1750–1880, drainage works 133, 134
  1875–1947 9–10
  bird decline **109**
  cattle/sheep
    stocking rates and densities **202**

wader populations 249–50
feeding sites, seed-eating birds 301, 329–32
grouse moors, burning 107–8
raptors, pest control and persecution 317–19
rough grazing **75**
runrig vs. enclosure 69
sheep-farming 107–8
shooting 317–19
'waste' lands 74–6, **75**
zero-grazing 213
seed-drills **4–5**, 163, 283–5
seed-eating birds 261–6, 267–71
  changes in extent of food sources **264**
  effects of weed control 156–9, 165–9
  feeding sites 1870s–1994 329–32
  habitat use, winter **269**
  mean flock sizes **271**
  site changes 301
  see also winter food; wintering bird populations
seeds, cleaning 162
Senecio spp. 157, 158, 159, 165, 169
set-aside 54, 326–7, *331*
Severn, R 210
sheep
  1870s vs. 1930s **218–19**
  and birds, Wales 109–13
  chalk grasslands 203–5
  increases and effects on farmland birds 15
  losses before enclosure 101
  numbers **105**, **110**, **242**
  'rot' (liver fluke) 100, 101, 107, 136
  stocking rates and densities 201–3, **202**, 220–3, **238–41**
  upland regions **105**, 106–7
  headage payments 327–8
  watering, Wiltshire Downs **204**
  see also Scotland
sheepfolds **68**
Shelduck 302
Sherwood Forest 80, 205
Shoveler 22, 34, 96, 148, 153, 185, 193, 210, 313
shrews 273, 274, 275
silage 13, 237, 287, 324
Skylark 23, 38–9, 48, 53–4, 55, 56, 72, 93, 96, 113, 120, 157, 174, 175, **176**, 193, 197, 198–9, 205, 236, 237, 249, 254, 266, 269, 271, 291, 309–10
Snipe 22, 42, 94, 96, 108, 140, 147, 185, 193, 206, 208, 210, 217, 227, 231, 246, 249, 254, 310, 311
soil
  fertility 40–2
  grading and fertility 2
  liming and chalking effects 2

Somerset Levels 99, 101, 128, *134*, 152–3
   drainage works 128, *134*
Sonchus 174
Sparrow
   House 24, 34, 56, 96, 166, 169, 186, 193, 245, 254, 270, 302, 304, 315, *316*
   Tree 24, 34, 36, 38, 96, 174, 186, 193, 195, 245, 254, 266, 269, 270, 271
Sparrowhawk 21, 22, 96, 126, 181, 185, 186, 193, 245, 249, 260, *318*
Spergula 174
Spoonbill 21, 310
Spotted Crake 131, 149
Spotted Flycatcher 23, 73, 96, 188, 193, 196, 245, 302
stackyards **98**, 261, 263, **264**, 265
Starling 18, 24, 34–5, 55, 56, 65, 96, 125, 186, 193, 211, 245, 254, 260, **262**, 302
steam power 279–80, **279**
Stellaria 157, 158, 159, 174
Stint 310
Stone Curlew 22, 34, 42, 48, 51, 58, **87**, 91–2, 95, 96, 185, 193, 204–5, 225, 232, 308
Stonechat 23, 93, 94, 96, 186, 193, 194, 254
Stork, White 21
stubble **264**, 265, 266, 273, *331*
Suffolk Sandlings 45, 98, 205
sugar beet 12–13, 189, 274, *331*
   British Sugar Corporation (1936) 14
   geese feeding 255–6
Swallow 21, 23, 34, 96, 193, 302, 304
Swan
   Bewick's 22, 254
   Mute 22, 34, 96, 185, 193, 254
   Whooper 22, 254

Taraxacum 157, 158
Teal 22, 96, 139, 148, 153, 193, 210, 246, 254, 313
tenancies, vs. owner-occupancy 14–15
Tern
   Black 131, *132*, 308
   Sandwich 307
Thames, R 209
thistles 59, *157, 158,* 165, 167, 169, *174*
threshing 166, **263**, **276**, 278, 292
thrushes 55
   Mistle Thrush 23, 34, 38, 96, 186, 193, 196, 245
   Song Thrush 23, 34, 38–9, 55, 96, 119, 125, 175, *176,* 186, 193, 196, 197–8, 254
tile drainage 137–9, 150, **151**
tillage *see* arable
titmice 65, 188, 265
   Bearded Tit 94, 131, *132,* 149, 319
   Blue Tit 23, 35, 38–9, 96, 193, 194, 196
   Coal Tit 23, 96, 186, 193, 196
   Great Tit 23, 35, 38–9, 96, 193, 194, 196

Long-tailed Tit 23, 38, 96, 193, 196, 249
tractors 281–3
tramlines 285
transhumanance 104
trapping
   cage-bird trade 169, 313–14
   food use of wild birds 309–13
   pest control 314–17
   raptors, game preservation 317–19
   Victorian collecting 95–6, 319–20
   see also exploitation and persecution
Treecreeper 23, 38, 96, 193, 194, 196
Trent, R 209
Tufted Duck 153
turnip sawflies 51
turnips, bird food/habitat 49, 51–2, **164**, 257, 315
Twite 24, 34, 94, 96, 186, 193, 194
Tywi Valley, Wales 152, 154

undersown cereals 52, 265, *331*
upland commons, long-term effects of enclosure 104–13
upland regions, sheep **105**, 106–7

vegetable and potato crops 255–6
   1875–1947 12
   as percentage of tillage **10, 11**
Veronica 174
veterinary medicines, changes 325
Vicia 174
Viola 174
voles 248, 250, 274, 275, 276, *318*

waders
   food use and persecution 310–13
   Netherlands, farmers' breeding scheme 326
Wagtail
   Pied 23, 38, 94, 96, 186, 193, 194, 254, 302
   Yellow 23, 48, 63, 64, 64, 72, 93, 96, 139, 142, 148, 186, 193, 227, 246
Wales
   drainage works *133*
   sheep-farming and birds 109–13
   Tywi Valley 152, 154
   'waste' lands **79**
Warbler
   Dartford 206
   Garden 23, 96, 188, 193, 196, 245
   Grasshopper 23, 34, 96, 142–3, 193, 246
   Reed 23, 34, 96, 139, 142–3, 148, 186, 193, 302
   Savi's 131, *132*
   Sedge 23, 34–5, 96, 142–3, 148, 186, 193, 194, 246, 302
   Willow 23, 35, 73, 96, 193, 194, 196

warping 82
wars
　high farming in decline 9
　increase of food production 146
'waste' lands 44, 46, 73–82
　enclosure 73–82
　　England C18 **77–9**
　　reversion to 328
　　Scotland C19 74–6, **75**
　　Wales C18 **79**
water meadows 206–8
　early spring **207**
　management 213–14
water power 144, 278
Water Vole 250–1
weeds/weed control 156–61
　1870–1940 recession 169–70
　arable weeds 159–60
　characteristics 157–9
　fertilisers and **161**
　fungicides 178
　grassland weeds 160–1
　herbicides 170–5
　in high farming 59, 161–9
　　clean crops 163–5
　　clean seed/land 162–3
　　hand and tractor hoeing 163–5, **164**
　　impact on birds 165–9
　hoes 283–5
　hosts to arthropods 172
　insecticides 179–82
　pesticides 175–7
　seed numbers in arable soils 159, 165
　as source of food 156–7, **158**, **159**
wetlands
　1750–1880, drainage 128–37
　1945–present
　　drainage threats 152–3
　　stocking densities 209
　flood meadows 208–10
　see also land drainage
Wheat Act (1931) 14
Wheatear 23, 34, 39, 48, 93, 96, 186, 193, 194, 204, 226, 229, 231, 309–10
Whinchat 23, 39, 48, 93, 94, 96, 186, 193, 210, 246, 285, 287
Whitethroat 23, 34, 33, 56, 96, 119, 188, 193, 196, 245
　Lesser 23, 34, 96, 186, 193

Wigeon 22, 153, 254, 257
wilderness *see* 'waste' lands
wildfowl, food use and persecution 310–13
windmills 135
Windsor Forest 80
winter food resources 252–77
　good feeding sites **55**, **253**
　　grain store **299**
　newly turned ground **253**
　　short straw varieties 267
　stackyards **98**, 263, **264**, 265, 267–8
　stockyards 267–8
　stubble 267
　threshing **263**
winter roosts, buildings 304–5
wintering bird populations 26–7
　changes 30–1
　farmland raptors 272–7
　　feeding behaviour **254**
　gleaners **254**, 261–6
　grazers
　　geese and ducks **254**, 255–6
　　pigeons 257, 258
　immigrations 253
　invertebrate feeders **254**, 257–61
　see also seed-eating birds
Woodcock 310, 311
woodlands
　changes in farmland bird populations **196**
　coppicing 19
　effects of enclosure 72–3
　Neolithic land clearance 19
Woodlark 23, 34, **88**, 92–3, 96, 186, 193, 226–7, **226**, 231, 313
Woodpecker
　Great Spotted 23, 34, 96, 185, 193, 196, 245
　Green 23, 34, 73, 96, 185, 193, 194, 196, 243, 254, 319
Woodpigeon 23, 34, 55, 56, 96, 181, 193, 196, 254, 257, 315
Wren 23, 38, 96, 188, 193, 194, 196, 254, 265, 302
Wryneck 23, 34, 42, 96, 185, 193
Wychwood Forest 80, 81

Yellowhammer 18, 24, 48, 56, 93, 119, 177, 181, 186, 187, 193, 196, 246, 254, 265, 266, 269, 270, 271
Yorkshire moors 96, 133

Printed in the United States
By Bookmasters